A Guide to Common Freshwater Invertebrates of North America

This book is dedicated to
my daughter Stephanie,
who reminded me how much fun it is
to collect bugs.

A Guide to Common Freshwater Invertebrates of North America

J. Reese Voshell, Jr., Ph.D.

Department of Entomology
College of Agriculture and Life Sciences
Virginia Tech

Illustrated by

Amy Bartlett Wright

The McDonald & Woodward Publishing Company
Blacksburg, Virginia
2002

The McDonald & Woodward Publishing Company
Blacksburg, Virginia

A Guide to Common Freshwater Invertebrates of North America

Printed in Canada

Third Printing November 2003

09 08 07 06 05 04 03   10 9 8 7 6 5 4 3

Library of Congress Cataloging-in-Publication Data

Voshell, J. Reese
A guide to common freshwater invertebrates of North America / J. Reese Voshell, Jr.; illustrated by Amy Bartlett Wright.
p. cm.
Includes bibliographical references (p.) and index.
ISBN 0-939923-87-4 (pbk. alk. paper)
1. Freshwater invertebrates--North America--Identification. I. Title.
QL365.4.A1 V67 2002
592.176'097--dc21

2002001922

To promote a better understanding of industry's role in preserving the environment

Partial funding for the development of *A Guide to Common Freshwater Invertebrates of North America* was contributed by VMA Outreach, the non-profit affiliate of the Virginia Manufacturers Association (VMA). VMA is a trade association of Virginia's leading manufacturers, and is the leading voice of businesses in Virginia on environmental issues. VMA established VMA Outreach in October 1998 in recognition of the need to establish relationships and channels of communication with other constituencies on important environmental issues. Outreach consists of VMA members who work to promote communication, education, and outreach to the media, environmental interest groups, government, and the general public about environmental issues and the important role manufacturers play in protecting the environment.

VMA Outreach

Alcoa, Inc.
American Electric Power
APLP Richmond Electric/BFI
BP Yorktown Business Unit
BWX Technologies, Inc.
Celanese Acetate, LLC
Degussa-Goldschmidt Chemical Corporation
Dominion Resources, Inc.
Draper Aden Associates
E. I. DuPont de Nemours & Co., Inc.
Earth Tech, Inc.
Ethyl Corporation
Ford Motor Company
Georgia-Pacific Corporation
Greif Bros. Corporation
Hercules, Inc.
Honeywell
Hunton & Williams
McGuireWoods, LLC
MeadWestvaco Corporation
Merck & Co., Inc.
Northrop Grumman Newport News
Old Dominion Electric Cooperative
Olver Laboratories
Philip Morris Management Corporation
Rocco, Inc.
Siemens VDO Automotive Corporation
Smurfit-Stone Container Corporation
URS Corporation
Williams Mullen

For more information visit our web site www.vmaoutreach.com.

Table of Contents

Acknowledgments

Several units at Virginia Tech, including the Department of Entomology, the College of Agriculture and Life Sciences, and the Virginia Agricultural Experiment Station have supported the twenty-five years of research and teaching that gave me the knowledge to write this book. These units at Virginia Tech also assisted me with this book by providing the resources to prepare the manuscript and communicate with the artist and publisher.

I am grateful to Virginia Manufacturers Association Outreach for providing funding to have all of the illustrations prepared by an accomplished professional artist. In addition to the funding, the members of this group were exceptionally enthusiastic about the project as a means of promoting environmental education and stewardship, and their enthusiasm helped maintain my motivation. Jay Gilliam, Director of Virginia Save-Our-Streams, played a key role in getting this project started by endorsing the idea to VMA Outreach and helping to find a publisher.

The outstanding feature of the book is the beautiful artwork by Amy Bartlett Wright. Amy is an exceptionally talented artist, who manages to be a consummate professional about her work while balancing a busy personal life as a soccer mom and wife. She was always enjoyable to work with throughout the many hours on the telephone and the lengthy e-mail exchanges, which, invariably, were prolonged by me being unnecessarily concerned about nit-picky details.

Of course the illustrations would not have been possible without good quality specimens. Most of the specimens came from the Virginia Tech Branch of the Virginia Museum of Natural History. I deeply appreciate the work of that institution at curating the valuable material collected by Virginia Tech professors and students. The Museum of Entomology at the Florida State Collection of Arthropods loaned some of the specimens that were used for the illustrations. FSCA was especially kind to provide me with space to work during several visits that I made. The following individuals provided additional preserved specimens or color images of live organisms: Gregory W. Courtney, Iowa State University; Robert T. Dillon, Jr., College of Charleston; Richard Franz, University of Florida; James H. Kennedy, University of North Texas; Boris C. Kondratieff, Colorado State University; and Richard J. Neves, Virginia Tech. I thank Thomas F. Wieboldt at Virginia Tech for identifying the aquatic plants that were included in the habitat illustrations. I am grateful to Nathan Hiner, Stephen Hiner, and Stephanie Voshell for patiently posing for the photographs of study methods in Section 1.

Several colleagues contributed their expertise to reviewing my manuscript: Robert T. Dillon, Jr., Stephen W. Hiner, James H. Kennedy, Boris C. Kondratieff, and George M. Simmons, Jr. Their suggestions greatly improved the accuracy and clarity of the book. I take full responsibility for any remaining shortcomings.

It has been a pleasure working with Jerry McDonald and the staff of McDonald and Woodward Publishing Company. Jerry was instrumental in making the idea for this book become a reality. His personal contributions to the finished product went far beyond the traditional role of a publisher. I especially appreciate his patience. He had a much better grasp of how long it takes to prepare a book manuscript than I did. Jerry graciously tolerated almost all of the requests that I made about layout and text revision right up to the very last steps of the publishing process. Judy Moore did a fantastic job of editing the original manuscript and several successive revisions. Her work brought my writing into compliance with the accepted rules of grammar and greatly helped with making the content understandable to persons who are not scientists. Judy has an uncanny eye for finding the few typographical errors that lurk in the midst of hundreds of pages of proofs. I am very grateful for her efforts. She also inserted her keen sense of humor into marginal comments, which made my task of revising the manuscript much more enjoyable.

I would like to close with some personal expressions of gratitude to a few individuals who have made special contributions to my life. First, the wonderful teachers, without whom I would not be where I am today. There are too many to name individually, but I would be remiss not to mention two especially influential teachers. Andy Johnson was my social studies and homeroom teacher at Northside Junior High School in Norfolk, Virginia. His enthusiasm for education made a lifelong impression on me. He became a family friend and was an important source of advice when it came time to make decisions about college. George Simmons first introduced me to freshwater ecology and invertebrates in an undergraduate course during the summer of 1970. His extroverted teaching style made the subject fascinating, and I first began to think about being a professor when I noticed how much he enjoyed his livelihood. George later served as major professor for my Ph.D. studies, during which time he was a tremendous mentor, and he has remained an invaluable friend and colleague ever since.

Next, my students, because professors such as myself acquire much of our knowledge from our interactions with them. Again, there are too many to mention individually, but two have been very special to me. I first met Boris Kondratieff at a professional meeting in 1976 while he was still an undergraduate, and I was in the last few months of my Ph.D. studies. By a quirk of fate that was fortunate for me, this young man with a hyperactive obsession with insects became one of my first graduate students. I learned far more about insects from Boris than he ever learned from me. Throughout his distinguished career he has been a great friend, and he has generously helped me many times, including the preparation of this book. Stephen Hiner was fresh out of the mountains of Bath County when he took my aquatic entomology course in the spring of 1979. Evidently the subject intrigued him, because upon graduation, he decided to use his undergraduate major in fisheries science for his avocation of fly fishing, and to make his living doing research on aquatic insects. I have had the great pleasure of working with him ever since, and he now manages the aquatic entomology lab as a result of his expertise on the taxonomy and biology of aquatic insects. Stephen played a key role in almost all of the prior research and teaching experiences that I drew upon to write this book, and he provided a tremendous amount of assistance during its preparation. Most importantly, Stephen is an extraordinary friend.

And lastly my family. I thank my parents, Joe and Carolyn Voshell, for their love and support. They always put me at the forefront of their lives when I was growing up, and they have continued to be a solid foundation for me in my adult life. I am especially grateful for their tolerance and encouragement during a phase of my youth when I was infatuated with collecting insects and other invertebrates and keeping a menagerie of live creatures in jars. I eventually came back to this interest and even managed to make something productive out of it. I thank my wife Trisha and my daughter Stephanie for making life such a pleasure. Trisha was exceptionally patient and supportive during the long hours that I worked single-mindedly on this book, just as she has been about every other time-consuming aspect of my research and teaching during our marriage. I could not have done it without her. Stephanie, an aspiring scientist, gave me some good suggestions for improving the utility of the book. Our time together as a family provided the essential relief to keep me going. Maybe now I can quit taking my laptop computer on vacation.

Purpose

The purpose of this guide is to provide people who are interested in natural history, but have never studied invertebrates, with a source of information about some of the fascinating creatures that are commonly encountered in the shallows of freshwater environments. The scope of this book includes invertebrates that spend at least part of their lives in the water and are large enough to be seen with the naked eye. As a result of many years of doing outreach education and training for conservation organizations, government agencies, and schools, as well as teaching college students, I have learned what people need to identify the "bugs" that live in water and what they want to know about them. This guide makes it possible to find and identify about 100 of the most common kinds of freshwater invertebrates in North America, without using any complicated collecting equipment, expensive microscopes, or confusing taxonomic keys. I estimate that, with a little practice, users will be able to correctly name about 90% of the invertebrates they find. In addition, the guide explains in non-technical language the most important information on the biology of the organisms. Hopefully, the convenient size, simple format, and numerous realistic color illustrations will make this book a constant companion of nature enthusiasts.

This book is intended for amateur naturalists and anyone else interested in the environment, regardless of their age or education. Examples of likely users are: K-12 students and their teachers; youth organizations such as scouts, 4-H, or science clubs; citizens who are involved with volunteer water quality monitoring programs; persons who observe or photograph living things for a hobby; college students in general biology and ecology classes;

and anglers. In addition, professional aquatic biologists, environmental resource managers, and conservationists will find the quality illustrations and readily available biological information useful.

A Guide to Common

Freshwater Invertebrates

of North America

How to Use This Book

A Guide to Common Freshwater Invertebrates of North America is organized into three major sections: Introduction, Identification of Different Kinds, and Information about Different Kinds. I have assumed that most users of this book have no formal training or experience with freshwater invertebrates. Therefore, a minimum amount of essential background information is presented in Section 1 — Introduction. I suggest that beginners study this section before using the guide. The introduction explains what we mean by freshwater invertebrates, why we study them, and how they are classified within the animal kingdom. Because it is impossible to separate the study of living organisms from the study of where they live, part of Section 1 also covers some of the fundamentals of freshwater ecology. Freshwater invertebrates are a diverse group, and the different kinds lead very different lives. For brevity and utility, this book emphasizes a few of the most important aspects of their lives, which are explained in the part of Section 1 on the fundamentals of freshwater invertebrate biology. The final part of Section 1 will help novices learn how to study freshwater invertebrates by telling them how to locate and catch these organisms; how to observe these organisms in their natural environment, as well as at home or in a classroom; and where they can go to get help with their studies of freshwater invertebrates.

Section 2 — Identification of Different Kinds — tells us who the freshwater invertebrates are. This section will probably be used more than the rest of the book, so it is organized to be used independently. The taxonomic level of identification used in this guide is mostly that of order or family, or in some cases, class — levels that meet the needs of most persons interested in natural history. Identification at these taxonomic levels can be done by simple visual comparison of live, or preserved, organisms with the illustrations. For each kind of invertebrate covered, there is a detailed color illustration of the whole body. For many of the kinds, there are also black and white illustrations of the important body structures or a different view of the organism on the same page as the primary color illustration. A brief description of the features that distinguish each kind of organism accompanies each illustration, and each is cross-referenced to expanded information on identification and biology that appears in Section 3. Although many organisms described in Section 2 can be identified easily with the naked eye, a simple hand lens or magnifying glass might be necessary to identify some.

The first step in identifying an organism is to determine which of the major groups it belongs to (for example, worm, crustacean, or insect). To make this first step as easy as possible, I have developed a simple tool that I call the QuickGuide to Major Groups of Freshwater Invertebrates. This is located near the beginning of Section 2. The QuickGuide consists of a series of questions about the appearance of the organisms, charts with illustrations, and brief written descriptions of the simplest, most reliable features for distinguishing each group. By answering a few questions and matching the specified features of the organisms with the descriptions and illustrations, users will be able to place freshwater invertebrates consistently in their correct group (for example, mayflies). The QuickGuide then directs the user to the color plates where the different kinds of invertebrates in that group can be identified. I suggest that all users, regardless of experience, start out with the QuickGuide when identifying organisms. It might be handy to bookmark page 81. Section 2 also includes references to advanced books on identifying the different kinds of freshwater invertebrates, in the event that a troublesome specimen is collected.

Section 3 — Information about Different Kinds — provides expanded descriptions and details of the biology for the different kinds of invertebrates illustrated in Section 2. Information in Section 3 includes the number of genera and species and the geographic distribution within each group of organisms, which will help users when they are trying to identify what kind of invertebrate they have observed or collected. Section 3 also explains the origins of many of the common and scientific names for the major groups of invertebrates.

Most of Section 3 is devoted to describing the lives of the various kinds of invertebrates in their natural freshwater environments. The main purpose of this section is to tell us what invertebrates do, which in turn explains why they are important to us as humans and to the health of ecosystems. Most of these organisms lead complex lives, and there are many differences in the biology of the various kinds. For the sake of brevity and usefulness, this guide emphasizes six of the most important biological features of the individual kinds of freshwater invertebrates: habitat, movement, feeding, breathing, life history, and stress tolerance. Section 3 ends with references to other books that contain more details about the different kinds of freshwater invertebrates.

Section 1
Introduction

What Do We Mean By Freshwater Invertebrates?

Freshwater environments are generally considered to include all of the inland waters on the surface of the earth that are not influenced by tides from the seas. These include environments such as streams, rivers, ponds, and lakes. Excluded are marine environments, which usually go by names such as oceans, seas, and bays, and estuaries, which are the intermediate zones where freshwater rivers meet and mix with marine waters. Freshwater environments are usually characterized by low concentrations of salts, as the name implies, but there are exceptions with some inland lakes such as the Great Salt Lake.

Invertebrates are all of the animals that do not have an internal skeleton of cartilage or bone. There is a vast array of invertebrates, many of which can only be seen under a microscope with very high magnification. This guide only covers the invertebrates that are large enough to see with the unaided eye. Professional aquatic biologists use the term "macroinvertebrates" for these types of invertebrates. The title of this guide would have been more accurate as "macroinvertebrates," but I thought that using this long, tongue-twisting word would only serve to discourage persons from studying these fascinating creatures.

Why Study Freshwater Invertebrates?

One quite simple reason that we study them is because they are interesting and it is an enjoyable and useful thing to do. Anyone who has the least amount of interest in natural history will be amazed with the diversity of these organisms and their adaptations to freshwater environments. Freshwater invertebrates are very easy to find and study. Persons who enjoy learning about the different kinds of trees, wildflowers, birds, mammals, and butterflies in the terrestrial environment will also enjoy learning about some of the common invertebrate creatures in nearby freshwater environments.

Perhaps a more compelling reason for studying freshwater invertebrates is that they are important. Invertebrates account for 70% of all known species of living organisms — microbes, plants, and animals. If we just consider animals, 96% of known species are invertebrates. This percentage is as high, if not higher, in freshwater environments, which are critical to our existence. Freshwater invertebrates are different from marine and terrestrial invertebrates, and they have unique adaptations that make them successful in inland waters. Some kind of invertebrate lives in every type of freshwater environment, no matter how inhospitable. Studying these organisms helps us understand and appreciate how all living things are interrelated with the world that we live in.

Freshwater invertebrates play important roles in the communities and ecosystems of which they are part. Their best known role is serving as food for other organisms, especially fish, amphibians, and water birds. However, they are also intricately involved in subtle ecological processes such as the breakdown and cycling of organic matter and nutrients, much like the role of earthworms in soil. As a result of research in various disciplines such as environmental science, ecology, and conservation biology, scientists have come to realize that it is important to maintain the diversity of all living things, including freshwater invertebrates, in order to keep the ecosystems that we all depend upon functioning properly.

Some freshwater invertebrates, such as crayfish and river shrimp, are useful to humans because they are delicacies that we like to eat. Unfortunately, a few freshwater invertebrates are important to us for negative reasons, such as when gnats swarm in the air about us in enormous numbers or mosquitoes bite us to feed on our blood. However, the numbers of freshwater invertebrates that are pests are very small in comparison to those that are beneficial to humans and the natural world we live in.

Freshwater invertebrates are used more often than any other group of freshwater organisms to assess the health of freshwater environments. Some kinds are very sensitive to stress produced by pollution, habitat modification, or severe natural events, while others are tolerant of some types of stress. Taking samples of freshwater invertebrates and identifying the organisms present can reveal whether a body of water is healthy or ill, and the likely cause of the problem if one exists, much like an examination by a physician. This process, known as biomonitoring, has become a significant activity for professional aquatic biologists employed by government agencies, consulting companies, and universities. In addition, many grass roots biomonitoring programs have become established and involve citizen volunteers of all ages and educational backgrounds.

Freshwater fish are important for recreational angling and commercial harvesting, and invertebrates are essential food for many fish species. Freshwater angling is an enjoyable form of outdoor recreation for many people and, thus, an important economic enterprise. Many kinds of freshwater invertebrates serve as the models for artificial lures used by anglers. Anglers collect some kinds of freshwater invertebrates for use as live bait, while some kinds are even harvested and sold in bait shops. Because invertebrates are important in the diets of many fishes, fisheries biologists must be able to identify them and understand their biology in order to manage populations of sport and commercial fishes.

Because freshwater invertebrates are easy to collect and observe, they are very useful in education programs. This ranges from formal instruction in life sciences in schools and colleges to outreach activities by museums, zoos, science centers, conservation organizations, and government agencies. Freshwater invertebrates fall within subject areas such as zoology and ecology for teaching life sciences. In addition, freshwater invertebrates have been the subjects for artists, authors, poets, and historians, so they can be effectively included in interdisciplinary teaching. Because freshwater invertebrates live in many different environments and play important roles in ecosystems, they can be used easily in outreach programs concerned with conservation, biodiversity, natural resource management, and environmental stewardship.

Classification of Freshwater Invertebrates

There are more than a million species of animals in the world, and 95% of them are invertebrates. Within the insects alone, about 900,000 species have been described and scientists estimate that there is an equivalent number of species that have not been discovered. Obviously, the study of organisms requires a scheme for classifying them. Classification means organizing living things into groups of similar organisms that can be distinguished from other groups of organisms. The main categories of the groups used for classifying living things are kingdom, phylum, class, order, family, genus, and species. These categories are hierarchical, which means they have been conceived and organized in a graded or nested series. It is customary to list them in sequence, beginning with the highest category that includes the most categories below it. The species is the lowest and most fundamental category in the classification system. A species is defined as a group of organisms in nature that are similar in structure and capable of successfully producing offspring, but reproduction does not occur with other such groups. The classification system for animals is illustrated below for a terrestrial insect that is familiar to everyone — the house fly.

Kingdom: Animalia (animals)
Phylum: Arthropoda (arthropods)
Class: Hexapoda (hexapods)
Subclass: Insecta (insects)
Order: Diptera (true flies)
Family: Muscidae (muscid flies)
Genus: *Musca*
Species: *domestica*

The subject of classification (systematics) stimulates much controversy among the scientists who study it. Classification is supposed to represent the relationships of organisms according to their evolution (phylogeny), but this is always speculative. Scientists propose classification schemes chiefly by comparing the body structure of modern species to that of fossils of extinct species. It is a daunting task to solve such intriguing mysteries with very little evidence. As a result, classification schemes are constantly changing and there is never complete agreement on how organisms should be classified. This is illustrated above in the example of the classification of the house fly. At one time everyone thought that all invertebrates with six legs were a separate class known as insects (Insecta). Now, most scientists think that there are two distinct groups of invertebrates that have six legs, so the insects have been put into a subclass category, and the class of all invertebrates with six legs has been renamed as hexapods (Hexapoda). The intricate details of classification and the complex scientific names that they produce are beyond the purpose of this guide. Users just need to be aware that the many different kinds of invertebrates are organized in a system of categories within categories based on evolution, but the system will always be undergoing improvements.

The term that scientists use when referring generally to any category in the classification system is taxon (pleural = taxa); in this usage, for example, family, genus, or species are each a taxon and collectively are taxa. In this guide, I use the word "kind" as a less technical substitute for "taxon." My use of kind is based on the accepted evolutionary classification scheme presented above, but I use the word interchangeably for family, genus, and species.

The following is a list of the common names of the major groups of freshwater invertebrates that users of this guide are likely to encounter, along with the classification category and scientific name that is most widely accepted at this time.

- Flatworms (Phylum Platyhelminthes, Class Turbellaria)
- Segmented worms (Phylum Annelida)
  - Aquatic earthworms (Class Oligochaeta)
  - Leeches (Class Hirudinea)
- Mollusks (Phylum Mollusca)
  - Snails (Class Gastropoda)
  - Mussels, clams (Class Bivalvia)
- Arthropods (Phylum Arthropoda)
  - Arachnids (Subphylum Chelicerata, Class Arachnida)
    - Water mites (a group of unrelated freshwater mites, usually assigned the name Hydracarina or Hydrachnidia)
  - Crustaceans (Subphylum Crustacea, Class Malacostraca)
    - Aquatic sowbugs (Order Isopoda)
    - Scuds, sideswimmers (Order Amphipoda)
    - Crayfishes, shrimps (Order Decapoda)
  - Insects (Subphylum Atelocerata, Class Hexapoda, Subclass Insecta)
    - Mayflies (Order Ephemeroptera)
    - Dragonflies, damselflies (Order Odonata)
    - Stoneflies (Order Plecoptera)
    - True bugs (Order Hemiptera)
    - Dobsonflies, fishflies, alderflies (Order Megaloptera)
    - Water beetles (Order Coleoptera)
    - Caddisflies (Order Trichoptera)
    - True flies (Order Diptera)

The biology of organisms within a category becomes more uniform the lower you proceed in the hierarchy. For example, the organisms included within the subclass of insects eat very different foods, live in different habitats, have different life histories, etc., while organisms belonging to the house fly species are almost identical in regard to their biological characteristics. Unfortunately, accurate identification becomes more difficult the lower one goes in the hierarchy. Anyone can distinguish an insect from a spider (six legs versus eight, respectively), but species level identification of invertebrates requires a high quality microscope and formal training in the subjects of morphology and taxonomy. This guide emphasizes identification at the family level, which is a good compromise between identification difficulty and biological uniformity. Many of the common freshwater invertebrates can be identified to family by eye, or with no more magnification than a simple hand lens. In some instances, the guide distinguishes organisms only to class or order, if they are difficult to identify but there is reasonable biological uniformity at the higher levels. Users should always remember that the illustrations in this guide show common species that are representative of a family, order, or class, but there may be some slight differences in the appearance of the invertebrates that they find. There are two reasons for this. There is natural variability within species over the geographic areas where they occur, and other closely related species in the group may also be commonly collected.

Fundamentals of Freshwater Ecology

Ecology is the study of the relationship between organisms and their environment. It comes from two Greek words: *oikos*, meaning "the family household," and *logy*, meaning "study of." It is important that users of this guide know something about the household where freshwater invertebrates live in order to understand the biology of the different kinds of invertebrates. As the science of ecology teaches us, it is impossible to separate the study of organisms from the study of their environment. However, freshwater ecology is a broad and complex science, and the scope of this guide only allows for a brief overview of the most pertinent elements. References for more detailed information on freshwater ecology are provided at the end of this part. This brief overview concentrates on the ecological factors that determine where and when invertebrates occur and how abundant they are. It is useful to consider these according to three types of ecological factors: physical, chemical, and biological.

Physical Factors

The most important physical factors that affect freshwater organisms are temperature, light, water current, and composition of the bottom materials (substrate). The metabolic rate of animals varies according to temperature, with higher temperature producing higher metabolism. All enzyme activity and oxidation of carbohydrates are affected. Invertebrates are commonly referred to as being "cold blooded," which means that they cannot regulate their body temperature. Individual species are adapted to a certain range of temperatures. If temperatures are consistently outside the acceptable range for a species, reproduction and survivorship become reduced until the species can no longer exist at that location. Temperature also affects the amount of oxygen that can dissolve in water, with cold water holding more oxygen than warm water. Since most aquatic invertebrates use dissolved oxygen for breathing, water temperature determines if there is sufficient dissolved oxygen for their needs.

Light has some direct effects on the behavior of aquatic invertebrates, in that some organisms hide during the day and become active at night. However, the main effects of light on aquatic invertebrates are indirect. The amount of light reaching the water affects its temperature. More importantly, the amount of light reaching the water and how far it penetrates in the water determines how much photosynthesis takes place in aquatic plants, including flowering plants, moss, and algae. Photosynthesis by aquatic plants produces food and dissolved oxygen for invertebrates. A very important feature of aquatic environments is the relative amount of plant biomass produced within the system (autochthonous) versus the amount of plant biomass produced outside the system (allochthonous), such as tree leaves and needles that fall into the water. Light is, therefore, the primary determining factor of whether the food base for a given community is live green plants growing in the aquatic environment or decaying plant matter that originated in the terrestrial environment. The proportion of the food base that comes from these two sources is a primary determinant of what kinds of invertebrates live in an aquatic environment. The abundant plant life that develops in aquatic environments with ample light produces good habitat for invertebrates in addition to food.

Water current affects aquatic invertebrates in many direct and indirect ways, and it is interrelated with most of the other ecological factors. Current can be a direct benefit to invertebrates by bringing them suspended particles of plant matter for food and by refreshing the surfaces of their gills with water that is saturated with dissolved oxygen. However, moving water exerts a tremendous force. Small invertebrates are susceptible to being dislodged, which makes them easy prey for fish, or having their delicate body structures, such as gills, damaged by the abrasive sand that is being transported by the flowing water. Macroinvertebrates that live in swift streams or the wave-washed shores of lakes must have special adaptations to deal with the force exerted by the currents. Current also influences the composition of bottom sediment, and, in turn, the size of the mineral particles composing the bottom of aquatic environments has a profound influence on the distribution of aquatic invertebrates.

Substrate is the base on which an organism lives. In aquatic environments, substrate includes everything on the bottom or sides or projecting out into the aquatic environment. The nature of the substrate is probably the most important factor that determines the kinds of invertebrates that live in an aquatic environment and their distribution and abundance within that environment. Substrate has numerous direct and indirect effects on invertebrates, and it is interrelated with many other ecological factors. The substrate can be composed of mineral material, which ranges from large boulders to fine particles of clay, or diverse types of organic material, such as fallen trees, living rooted plants, filamentous algae, or detritus. Invertebrates depend upon it for attachment sites and hiding places. For stream-dwelling organisms, the size and arrangement of the substrate create very small refuges with a range of current velocities, so that different species all have the velocity that is appropriate for them. Substrate is also related to the food of invertebrates. Algae that grow attached to solid objects (periphyton) require firm, stable surfaces for their existence. Detritus accumulates around large pieces of substrate, such as rocks or logs, especially in flowing waters. The substrate is where nearly all invertebrates reside most of the time, so its importance cannot be overstated.

Chemical Factors

The most important chemical factors that affect freshwater invertebrates are oxygen, acidity, alkalinity, hardness, and nutrients. Like all animals, the metabolic activities of invertebrates require that they obtain oxygen and get rid of carbon dioxide. The scientific term for this physiological process is respiration. Sometimes muscular action is used to increase the rate of respiration, such as when we enlarge our chest to take in a deep breath, and that is called ventilation. In this guide, it will suffice to use the simple term breathing for all processes related to obtaining oxygen. Most aquatic invertebrates breathe oxygen that is dissolved in the water. Unfortunately, the chemical properties of water are such that it does not hold much oxygen in solution. The air that we breathe is about 21% oxygen, but less than 15 parts of oxygen will dissolve in 1 million parts of water (0.0015%) at the temperature just above freezing. Water contains only about half as much dissolved oxygen at warmer temperatures that are common in the summer under natural conditions. The oxygen requirements of invertebrates vary considerably among the different kinds, and the amount of available dissolved oxygen is a limiting factor for many of them. Species with high oxygen requirements are often restricted to places where the water remains cool, where splashing puts more oxygen in solution, or where there is sufficient current to constantly deliver oxygen and keep the organism saturated with the oxygen that is available.

The acidity of water is a reflection of the concentration of hydrogen ions in solution. When hydrogen ion concentration is high the water is acidic, whereas when hydrogen ion concentration is low, bicarbonate and carbonate ions prevail and the water is alkaline. The measure of acidity is pH, which is the hydrogen ion concentration. A pH value of 7 is neutral, below 7 is acid, and above 7 is alkaline. Most invertebrates are adapted for water that remains close to neutral. The pH scale is logarithmic to base 10, so a change of 1 pH unit represents a tenfold increase or decrease in hydrogen ion concentration. In general, pH values below 5 and above 9 are harmful to most aquatic organisms.

The alkalinity of water is the concentration of all the compounds that can shift the pH into the alkaline range. In natural inland waters, alkalinity usually comes from bicarbonate, carbonate, and hydroxide ions. Alkalinity provides buffering capacity, which is the ability of water to neutralize hydrogen ions and thereby avoid becoming acidic.

Hardness is another commonly measured chemical property of water. It is a measure of all positively charged ions (cations) in solution. The main cations found in freshwater are calcium, sodium, magnesium, and iron. Water with high concentrations of these cations is called "hard," and water with low concentrations is called "soft." As a general rule, there are more kinds of invertebrates in moderately hard waters. The specific benefits of living in water containing additional cations are not completely understood, but scientists have observed that invertebrates living in hard waters are more resistant to stress, especially from certain toxic chemicals.

The primary nutrients in surface waters are the negatively charged ions (anions) of nitrogen and phosphorus, especially nitrate and phosphate. These chemicals do not directly affect invertebrates, but they do cause important indirect effects by stimulating the growth of all types of aquatic plants, including microscopic algae and large flowering plants. Moderate increases in the concentrations of nutrients are sometimes beneficial to invertebrates because slight increases in plant growth increase the amounts and types of useable food and habitat. However, an overabundance of nutrients leads to the condition known as eutrophication, which is always detrimental to invertebrates. Eutrophication occurs when extra nutrients act like fertilizer and cause too much plant growth. This upsets the balance in the aquatic environment. Overly abundant floating algae can cause the water to be soupy, which blocks sunlight from penetrating down in the water where other plants would normally live. Solid substrates can become coated with a thick layer of slimy algae, which prevents invertebrates from attaching. In addition, the kinds of algae that come with eutrophication are undesirable for food and are avoided by invertebrates. Consequently, many invertebrates will disappear from aquatic environments affected by eutrophication because of the loss of important food sources and places to attach. Rooted flowering plants can become so thick that they clog the aquatic habitat. When these copious amounts of plants die and decompose the concentration of dissolved oxygen can become very low, or nonexistent. Thus, nitrogen and phosphorus have a direct bearing on the quality of the aquatic environment, which, in turn, determines the kinds and numbers of invertebrates that are able to live there.

Biological Factors

The biological factors that affect the occurrence and abundance of invertebrates are even more complex than the physical and chemical factors. There are so many kinds of plants and animals in freshwater environments and they interact with the environment and themselves in such diverse ways, there are countless subtle effects on invertebrates. Biological factors are especially important to invertebrates in three ways: their influence on the organic substrate, food, and relationships among species. Plants and plant materials, either living or dead, can serve as organic substrate for invertebrates. As is true of the mineral substrate, the organic substrate provides a base for organisms to live on. Typical dead plant materials that function as substrate are logs, limbs, twigs, and leaves. Live aquatic plants that are inhabited by invertebrates are usually flowering plants with complex structures, such as leaves, stems, stalks, and roots. Sometimes tangles of live filamentous algae provide substrate that can be used by invertebrates. As a general rule, the more complex the substrate the more different kinds of invertebrates will live there. Substrate complexity creates more places to hide from predators and more small spaces where different kinds of invertebrates can find the specific conditions that suit their needs. Organic materials are an excellent habitat because they provide complex substrate.

All animals are consumers, so food is always a very important biological factor that influences an organism's success in its environment. Some of the most important features of food that different kinds of freshwater invertebrates are specialized for include whether the food item is plant or animal, whether it is living or dead, where it originated, the size of the pieces, and where it is located in the water. Each one of these features, or combinations of them, make the food usable by only certain kinds of invertebrates. The only live plants that are eaten to a great extent by invertebrates are algae. Live aquatic vascular plants often have natural chemicals that make them taste bad. The microscopic cells of algae occur either floating in the water or attached to the surfaces of solid objects, especially large stones or vascular plants. Decaying plant material is called detritus, and this is a major food resource for many invertebrates. Much of the detritus in aquatic environments, especially streams, comes from trees and shrubs that grow on shore and drop their leaves, flowers, and twigs into the water. Scientists call this type of material allocthonous detritus. Detritus is also available from plants that lived in the water then died, or else their foliage died back for the winter. Detritus from plants that grew in the aquatic environment is called autochthonous detritus. Autochthonous detritus can come from flowering plants, mosses, or algae, and it is often more nutritious for invertebrates than allochthonous detritus, which tends to have a lot of unusable components such as cellulose. Size and location are other important features of detritus that affect its use as food by invertebrates. Pieces of detritus that are larger than 1 mm, such as fallen leaves, are called coarse detritus. Detritus particles that are smaller than 1 mm are referred to as fine detritus. Coarse detritus is usually located on the bottom or wedged against solid objects, whereas, fine detritus may be on the bottom or suspended in the water. Invertebrates have very different adaptations that make them specialized for acquiring and consuming the different kinds and sizes of detritus in various locations. In addition to plants, other animals are food for some invertebrates. Most invertebrates that eat animals eat them alive, but a few consume dead animal matter, which is called carrion.

When some invertebrates feed, they have a beneficial effect on the supply of food for other kinds of invertebrates, including the size and quality of the food as well as its location. Algae grow more and contain more nutrition in response to moderate levels of grazing by invertebrates. Invertebrates scrape off the top layer of cells and prevent the algae from shading itself from the sun, much like when we mow our lawn in the summer and the grass responds by becoming greener and growing more. Invertebrates that feed on large pieces of detritus, especially dead leaves from trees on shore, make small particles available to other organisms. The invertebrates that feed on large detritus have sloppy eating habits. A considerable amount of food falls out of their mouths after it has been shredded into smaller pieces, which are called orts. In addition, the feces that they eliminate become small particles of detritus that are useful food for other invertebrates. As a result of invertebrate feeding, large pieces of detritus in one location become small particles that are suspended in the water or drop to the bottom in another location in the aquatic environment.

The relationship among invertebrates in the natural environment is another important biological factor. There are many kinds of freshwater invertebrates and some kinds build up high population numbers in the aquatic environment. Living in close association makes it inevitable that invertebrates will experience some sort of relationship with each other. Competition and predation are the two types of relationships among freshwater invertebrates that are most commonly observed. Competition occurs when two or more organisms need some resource, such as food or space, and there is not enough of that resource to go around. Competition can occur between individuals of the same species (intraspecific) or between individuals belonging to different species (interspecific). Competition is always detrimental to all organisms involved and leads to increased death or decreased birth, or both. Intraspecific competition between individuals of the same species is usually more intense than interspecific competition. Predation occurs when one organism becomes food (prey) for another organism (predator). Predation is always beneficial to the predator and detrimental to the prey. Neither competition nor predation usually eliminates any kind of organism under natural conditions in relatively undisturbed environments. Long ago, the competitors, predators, and prey arrived at a balance within the environment where they lived. Usually the organisms on the receiving end of the detrimental effects are reduced in numbers and their distribution is restricted within their environment, but they are not eliminated. Competition and predation are beneficial processes that increase diversity and productivity. It is usually only when humans are involved that competition or predation eliminates species. Examples are when alien species are introduced into aquatic environments, or when pollution adds too much stress to the hardships already produced by competition or predation.

As can be seen from this brief review of freshwater ecology, there are many complex factors that affect where and when invertebrates occur and how abundant they are. Sometimes these factors are readily apparent to our eyes, but scientific measurements must be taken in some situations to explain why certain invertebrates occur in a particular place or why they are absent.

References on Fundamentals of Freshwater Ecology

Allan, J. D. 1995. *Stream Ecology.* Chapman and Hall, New York. 388 pages.

Caduto, M. J. 1985. *Pond and Brook. A Guide to Nature in Freshwater Environments.* University Press of New England, Hanover, New Hampshire. 276 pages.

Cole, G. A. 1994. *Textbook of Limnology.* 4th edition. C. V. Mosby Company, St. Louis, Missouri. 412 pages.

Cushing, C. E., and J. D. Allan. 2001. *Streams: Their Ecology and Life.* Academic Press, San Diego, California. 366 pages.

Giller, P. S., and B. Malmqvist. 1998. *The Biology of Streams and Rivers.* Oxford University Press, Oxford, United Kingdom. 296 pages.

Hynes, H. B. N. 1970. *The Ecology of Running Waters.* University of Liverpool Press, Liverpool, United Kingdom. 555 pages.

Leadley Brown, A. 1987. *Freshwater Ecology.* Heinemann Educational Books, Ltd., London, United Kingdom. 163 pages.

Reid, G. K., and R. D. Wood. 1976. *Ecology of Inland Waters and Estuaries.* 2nd edition. D. Van Nostrand Company, New York. 485 pages.

Smith, R. L. 1993. *Elements of Ecology.* 3rd edition. HarperCollins Publishers, New York. 617 pages + appendices.

Waters, T. F. 2000. *Wildstream: A natural history of the free-flowing river.* Riparian Press, St. Paul, Minnesota. 607 pages.

Wetzel, R. G. 2001. *Limnology.* 3rd edition. Academic Press, San Diego, California. 1006 pages.

Fundamentals of Freshwater Invertebrate Biology

Freshwater invertebrates lead complex lives and there are many differences in the biology of the various kinds. In an effort to simplify this complexity, this guide emphasizes six of the most important biological characteristics of these invertebrates: habitat, movement, feeding, breathing, life history, and stress tolerance. An explanation of these six major biological characteristics will be useful for persons who are just beginning to study freshwater invertebrates. This awareness will also facilitate the use of Section 3 — Information About Different Kinds.

Habitat

A simple definition of habitat is the place where an organism naturally lives and grows. Habitat is mostly determined by the physical and chemical conditions that exist at a particular place, but organic substrates, such as live plants, logs, or detritus, can also be important biological components of habitat. Potentially, many, many different habitats could be created from the various combinations of physical, chemical, and biological factors identified in the preceding subsection. Scientists have attempted to make it easier to study freshwater ecology by organizing the many different habitats into a few categories that share some of the most important ecological factors. These habitat categories are important because unique assemblages of various kinds of invertebrates live in each type of habitat.

Freshwater habitats are divided into two broad categories: running waters and standing waters. The ecological term for running waters is lotic, and for standing waters it is lentic. There are also many common names for lotic and lentic habitats.

Some examples of common names for running waters are seeps, springs, brooks, branches, creeks, streams, and rivers. There is no standard scientific definition of these terms according to the size or other features, but they are commonly distinguished by the following subjective descriptions.

Seep and spring usually refer to the smallest flowing waters that occur where groundwater first emerges to flow in a channel. If the water emerges boldly and flows with a distinct current, it is probably called a spring. If the water merely trickles out of the ground without any noticeable current, it is a seep. Springs usually flow through cracks in impervious underground rock strata, whereas seeps flow through loose soil.

Brook, branch, creek, and stream often refer to intermediate size flowing waters that can be waded from one side to the other. Of these terms, the best one to use in freshwater ecology is stream.

River is usually used for larger flowing waters that are too deep to wade across.

Some examples of common names for standing waters are bogs, marshes, swamps, ponds, and lakes. Different depths, chemistry, and plant communities distinguish these standing-water habitats, although, like the names for flowing waters, there are no standard scientific definitions. Standing water bodies are commonly distinguished by the following subjective descriptions.

Bogs are very shallow, with little open water, because they have dense growths of *Sphagnum* mosses. The water is lacking in nutrients and is acidic, with $pH < 4.5$. The bottom is covered with a thick layer of decaying organic matter, known as peat.

Marshes have a bit more water depth and open water than bogs. Emergent vegetation, such as rushes, reeds, and sedges are present, rather than mosses. The water is rich in nutrients because marshes are connected to adjacent bodies of water and receive inputs of nutrients during times of flooding. Marsh waters may be slightly acidic, and the bottom may have a thin layer of peat.

Swamps are similar to marshes, except they are somewhat deeper and have much more open water. The main difference between marshes and swamps is the dominant vegetation. Large shrubs and trees that are adapted to grow in the water, such as cypress trees in the southeastern United States, dominate swamps.

Ponds and lakes are different from the previous types of lentic habitats by being much deeper. They are confined in distinct basins, whereas bogs, marshes, and swamps are spread out. Ponds are usually thought of as being smaller than lakes, but it is difficult to differentiate between a large pond and a small lake. Some scientists distinguish ponds from lakes by the depth that light penetrates and the resulting growth of rooted plants. In this scheme, lakes are bodies of standing water in which rooted plants do not grow across the entire width of the basin, because the middle is too deep for light to penetrate to the bottom. Ponds are considered to be bodies of standing water that are shallow enough to support the growth of rooted plants across their entire bottom, because everywhere in the basin is shallow enough for light to penetrate to the bottom.

Lotic and lentic habitats are further subdivided into zones. A zone is not defined exactly in freshwater ecology, but it is generally used to describe broad areas in which the major features of a habitat are alike, but different from the major features of other broad areas in the same habitat. For our purposes, a rule-of-thumb might be that zones extend for at least several meters and perhaps as much as several hundred meters. The concept of a zone is explained better by considering the following two examples.

Lotic habitats are divided into two zones: erosional and depositional. Another name for a lotic-erosional zone is riffle. This is the zone where streams at normal flow (not just during floods) have sufficient power to pick up fine sediments, keep them in suspension, and transport them downstream. The power of the stream in a particular zone depends on the gradient (slope) and width of the channel. A lotic-depositional zone is the area where the stream does not have sufficient power at normal flow to keep fine sediments in suspension, so the small particles in transport settle out on the bottom. Another name for a lotic-depositional zone is pool.

Lentic habitats are divided into three zones: pelagic, littoral, and profundal. Pelagic refers to all of the open water, from the surface down to the bottom, but not including the bottom. The upper part of the pelagic zone is often distinguished separately as the limnetic zone. The lower boundary of the limnetic zone is the maximum depth that enough light penetrates for photosynthesis by green plants, which is usually about 1% of the available surface light. The depth of the limnetic zone varies in different lentic habitats according to the clarity of the water. The littoral and profundal zones refer to areas of the bottom. The littoral zone is the area of bottom that extends from shore down to the point of light penetration. The profundal zone is the area of bottom that begins at the depth where light no longer penetrates and extends to the maximum depth of the lake.

Zones of lotic and lentic habitats are further subdivided into microhabitats, which are the specific places where organisms reside. A rule-of-thumb about size might be that different microhabitats could be found within an area of approximately 1 square meter. Examples of microhabitats in lotic-erosional and lentic-littoral zones are illustrated in Figures 1 and 2, respectively. For invertebrates, the primary feature of microhabitats is the substrate. Substrate is either mineral or organic. The most important characteristic of mineral substrate related to freshwater invertebrates is the size of the individual pieces. Scientists have developed a standard scale (Wentworth scale) for size categories of mineral substrate, along with names for the different sizes. These are:

Boulder	> 256 mm (Figure 1F)
Cobble	64–256 mm (Figure 1G)
Pebble	16–64 mm (Figure 1H)
Gravel	2–16 mm (Figure 1H)
Sand	0.0625–2 mm (Figures 1I, 2G)
Silt	0.0039–0.0625 mm (Figure 2G)
Clay	< 0.0039 mm (Figure 2G)

In addition to the size of mineral pieces described in the Wentworth scale, the substrate can be bedrock, which is also called rock outcrop. This is the continuous solid rock bottom that occurs where flowing water has eroded the channel down to the underlying geological formations.

Invertebrate microhabitats can also be created by organic substrate, which is either live plants or detritus. Live aquatic plants that serve as microhabitats include flowering plants, moss, and filamentous algae. There are many kinds of each of these types of plants, a few of which are illustrated in Figures 1 and 2. Different species of flowering plants can protrude upright above the water surface (cattail, Figure 2A), lie mostly on the surface of the water (water lily, Figure 2B), or have the entire plant beneath the water surface (water-starwort, Figure 1B; moss, Figure 1C; pondweed, Figure 2C; elodea, Figure 2D). These are referred to as emergent, floating, and submerged aquatic plants, respectively. Another microhabitat is created by vascular plants when their roots become exposed in the water from erosion by currents in streams or by waves washing the shores of lakes. These tangles of roots are commonly called root wads. Often the root wads come from terrestrial trees and shrubs growing at the edge of the water body.

The other type of organic substrate is detritus. In addition to being food for many invertebrates, coarse detritus, such as leaves, flowers, stems, and twigs, makes very effective substrates for invertebrates to live on. Leaf packs form a preferred microhabitat for many kinds of invertebrates in lotic-erosional zones (Figure 1E). Leaf packs form when the current jams a bunch of leaves and other plant parts on a rock or dead tree limb. In lentic-littoral zones, the coarse detritus merely lies on the bottom, often in clumps (Figure 2F). Some kinds of invertebrates eat the coarse detritus while they dwell there, but others merely use it as a place to hang on and hide. Very large and stable pieces of detritus, such as logs and branches that fall into the water from trees on shore, are also used extensively as substrates by invertebrates. This material is simply called woody debris (Figures 1D, 2E).

The greatest diversity of invertebrates in lotic habitats occurs in erosional zones where the mineral substrate is predominantly loose cobbles and pebbles, with a few boulders (Figure 1). The size and irregular composition of this material provide an infinite array of hiding places, attachment sites, current velocities, surfaces for algae to grow, edges to catch coarse detritus and woody debris, and openings where fine detritus settles.

The loose composition of mineral substrate in lotic-erosional zones often extends below the stream bottom for a meter or more. This area, called the hyporheic zone, has practically no perceptible current, but there is sufficient flow to keep the water supplied with oxygen. Many invertebrates spend at least part of their lives here, where they are safe from large predators. Aquatic biologists know very little about invertebrate life in the hyporheic zone because it is so difficult to sample this area. It is speculated that some stream species are rarely collected because they spend much of their lives in this mysterious zone below the stream bottom.

The greatest diversity of invertebrates in lentic habitats occurs in littoral zones (Figure 2). Plants are the primary habitat feature responsible for this. The many species of aquatic plants, with their different shapes and structures, provide a wide variety of places to live, in the same way that mineral substrates provide assorted opportunities in lotic-erosional zones.

In this guide, the format for presenting habitat information for the various kinds of freshwater invertebrates is to first state the categories and zones where they occur (for example, lotic-erosional). Then the different sizes and types of those habitats are presented using the common names described above (for example, springs and small streams). Finally, the specific microhabitats preferred by the kinds of invertebrates are given (for example, cobble and pebble mineral substrate as well as woody debris).

Movement

Most freshwater invertebrates have specialized body shapes and behaviors that enable them to be in a place that meets their essential requirements for acquiring food and oxygen, avoiding competition with other invertebrates, and hiding from predators. Some invertebrates stay in or near one place as much as possible, while others roam around freely. This feature of their biology is referred to as locomotion, habits, or modes of existence. The categories of movement used in this guide are explained in the following paragraphs. Examples of most categories are shown in the illustrations of riffle and pond habitats (Figures 1 and 2, respectively).

Clingers are organisms with bodies that are highly modified in order to maintain a relatively fixed position on firm substrates in current. This is a challenging feat for small organisms because water exerts a strong force when it is moving. Invertebrates have several different mechanisms for being effective clingers. Some have a flat shape that offers little resistance to flowing water (flatheaded mayflies, Figure 1Q, Plate 55; water pennies, Figure 1S, Plate 78). Others have sucker-like structures that create a vacuum to hold on to the substrate (black flies, Figure 1N, Plate 90). Several kinds produce silk, which is sticky and is used to glue them to the substrate in current (common netspinner caddisflies, Figure 1P, Plates 73, 74). Almost all clingers occur in lotic-erosional habitats. Most live on coarse mineral substrate (bedrock, boulders, cobbles, pebbles), but some inhabit large pieces of woody debris, and a few attach to live plants in current. There are a few kinds that use the clinger habit to live on the wave-swept, rocky shores of lakes.

Climbers are adapted to dwell on live aquatic plants or accumulations of plant debris. They have elongate bodies with thin, spindly legs that extend down from the body (narrowwinged damselflies, Figure 2K, Plate 36). This body form allows them to maneuver up and down stalks and through dense tangles of plants and debris. They are often a greenish color, sometimes with mottled patterns, to make them camouflaged. They can usually swim, but only do so if they need to move across a short stretch of open water. Almost all climbers live in lentic-littoral and lotic-depositional habitats because that is where suitable beds of aquatic plants and accumulations of plant debris occur.

Crawlers are somewhat like clingers and climbers in that they require a firm substrate, free of fine sediment. However, they do not have special features like clingers to hold on in swift current, nor do they have elongate bodies with thin legs to climb in plants. Crawlers seek small, protected places among loose stones (boulder, cobble, pebble, gravel), woody debris, coarse detritus (especially leaves), exposed roots, and at the bases of plants that grow in tangles (especially moss). They move around slowly, using just their legs and tarsal claws. The latter provide traction. Crawlers are most common in lotic-erosional habitats, but within specific places where they are sheltered from fast currents. Examples of crawlers that live in riffles are common stoneflies and hellgrammites in loose rocks (Figure 1L, Plate 43; Figure 1O, Plate 60), giant stoneflies in leaf packs (Figure 1J, Plate 42), and spiny crawler mayflies in moss (Figure 1M, Plate 54). Some kinds also live in lotic-depositional and

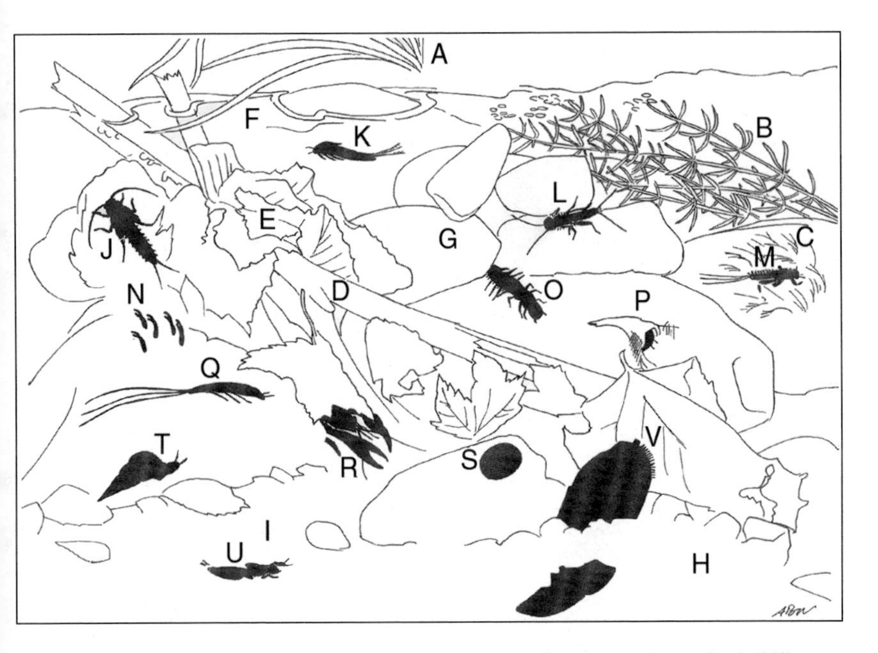

Figure 1A. Key to color figure on following two pages of a typical riffle habitat in a stream (lotic-erosional zone), illustrating the primary components of the habitat (A–I), common invertebrates that live there (J–V), and some of the biological features of those organisms, such as movement and feeding. (A) Terrestrial grass. (B) Submerged flowering aquatic plant (water-starwort, *Callitriche* sp.). (C) Submerged aquatic moss (*Fontinalis* sp.). (D) Woody debris. (E) Leaf pack. (F) Boulder. (G) Area of cobbles. (H) Area of pebbles and gravel. (I) Area of sand. (J) Giant stonefly (crawler, shredder-detritivore). (K) Ameletid minnow mayfly (swimmer, collector-gatherer). (L) Common stonefly (crawler, engulfer-predator). (M) Spiny crawler mayfly (crawler, collector-gatherer). (N) Black flies (clinger, collector-filterer). (O) Hellgrammite (crawler, engulfer-predator). (P) Common netspinner caddisfly (clinger, collector-filterer). (Q) Flatheaded mayfly (clinger, scraper). (R) Crayfish (crawler, omnivore). (S) Water penny (clinger, scraper). (T) Pleurocerid snail (clinger, scraper). (U) Clubtail dragonfly (burrower, engulfer-predator). (V) Mussel (burrower, collector-filterer).

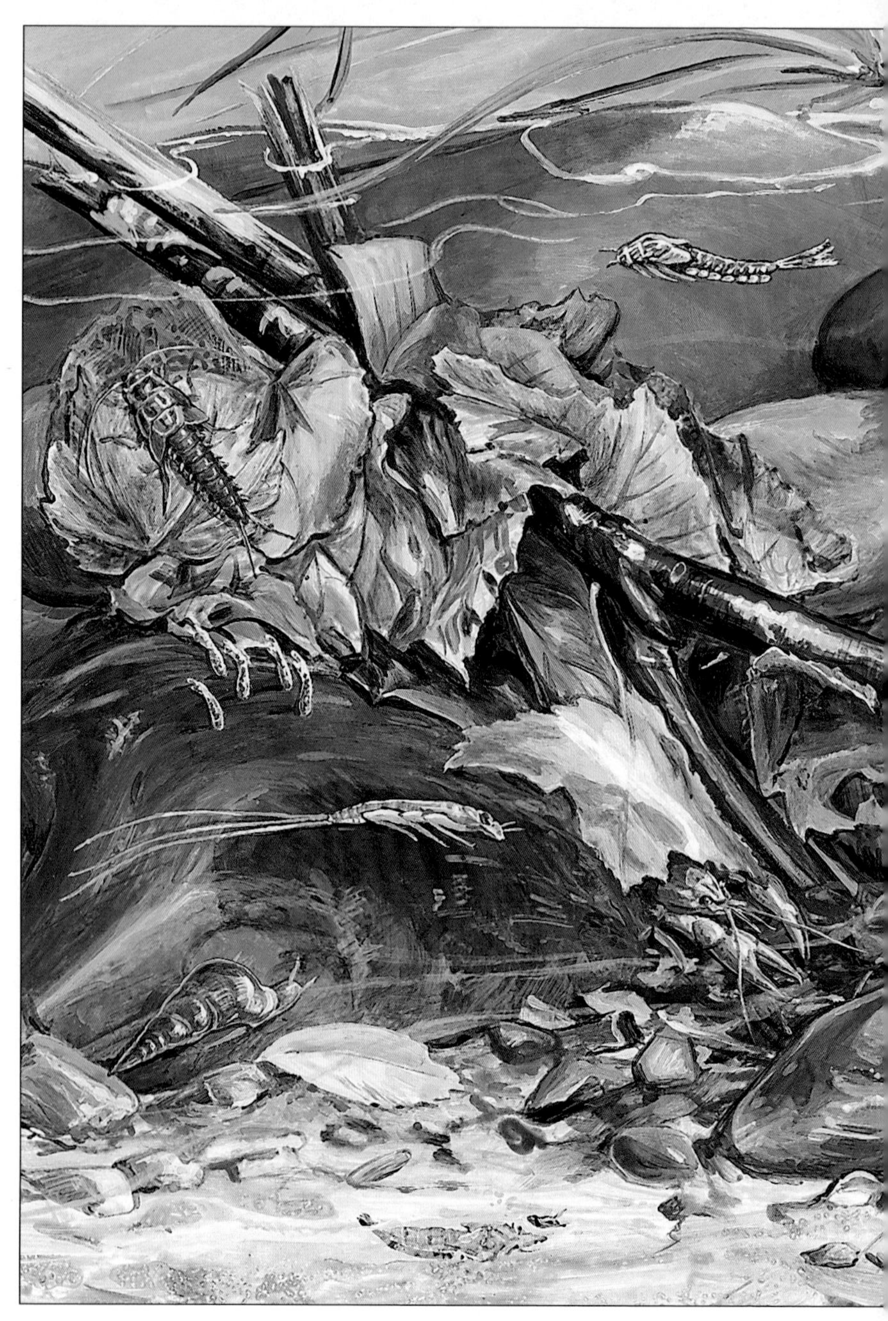

Figure 1. A typical riffle habitat in a stream.

A Wright

Figure 2. A typical habitat in the shallow water at the edge of a pond.

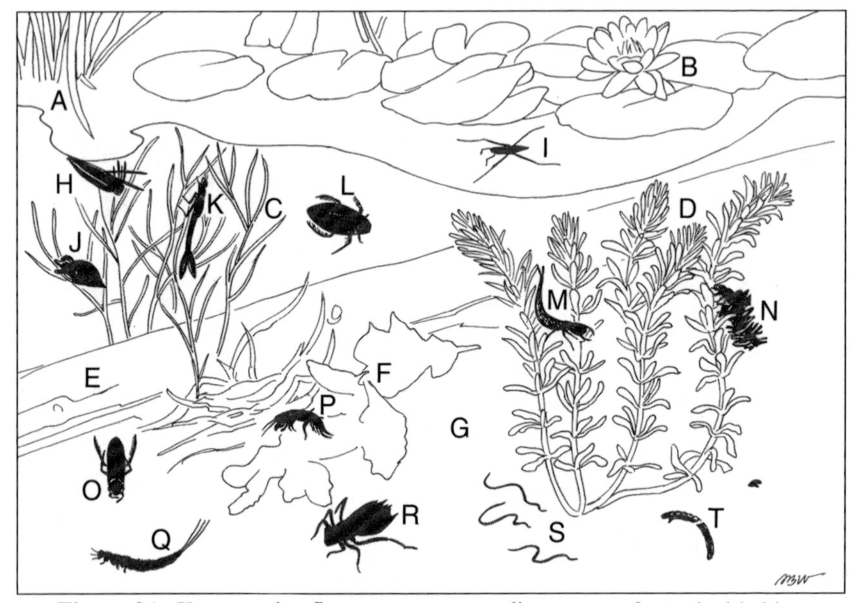

Figure 2A. Key to color figure on two preceding pages of a typical habitat in the shallow water at the edge of a pond (lentic-littoral zone), illustrating the primary components of the habitat (A–G), common invertebrates that live there (H–T), and some of the biological features of those organisms, such as movement and feeding. (A) Emergent flowering aquatic plant (cattail, *Typha* sp.). (B) Floating flowering aquatic plant (water lily, *Nymphaea* sp.). (C) Submerged flowering aquatic plant (pondweed, *Potamogeton* sp.). (D) Submerged flowering aquatic plant (elodea, *Elodea* sp.). (E) Woody debris. (F) Accumulation of dead tree leaves and aquatic plant detritus lying on bottom. (G) Bottom composed of mixture of sand, silt, clay. (H) Backswimmer (swimmer, piercer-predator). (I) Water strider (skater, piercer-predator). (J) Lymnaeid snail (clinger, scraper). (K) Narrowwinged damselfly (climber, engulfer-predator). (L) Predaceous diving beetle adult (swimmer, engulfer-predator). (M) Predaceous diving beetle larva (climber, piercer-predator). (N) Northern case maker caddisfly (climber, shredder-herbivore). (O) Water boatman (swimmer, collector-gatherer). (P) Scud (crawler, omnivore). (Q) Common burrower mayfly (burrower, collector-gatherer). (R) Skimmer dragonfly (sprawler, engulfer-predator). (S) Aquatic earthworms (burrower, collector-gatherer). (T) Non-biting midges (burrower, collector-gatherer).

lentic-littoral habitats (scuds, Figure 2P, Plate 18).

Sprawlers are adapted to live in places where the bottom consists of fine sediments (sand, silt, clay). Sprawlers have body shapes or other mechanisms to stay on top of the soft, fine sediment, rather than getting down into the sediment. In order to maintain this position, they usually have somewhat flattened bodies and legs that extend out to the sides, away from the body (skimmer dragonflies, Figure 2R, Plate 40). Sprawlers usually lie still and do not change their location very often. The body is often modified in various ways to protect the delicate gills from abrasion and clogging. Most sprawlers live in lentic-littoral and lotic-depositional habitats. Some occur in lotic-erosional habitats, but they are restricted to specific places where sand and silt accumulate, such as in eddies behind large stones and logs or near shore.

Burrowers, like sprawlers, are adapted to live in microhabitats where the bottom consists of fine sediments (sand, silt, clay). However, burrowers are adapted to dig down and reside in the soft, fine sediment, rather than lying on top. The bodies of burrowers often have special modifications for digging, such as scoop-like front legs or pointed projections on the head (common burrower mayflies, Figure 2Q, Plate 50). Others simply have elongate, pointed, flexible bodies that are effective for wiggling through the sediment (aquatic earthworms, Figure 2S, Plate 3). Some construct a distinct tunnel with walls that do not collapse (non-biting midges, Figure 2T, Plate 91), while others move through loose sediment that surrounds and touches them. Some are more stationary and simply wiggle their bodies and legs until they are almost covered by the sediment (clubtail dragonflies, Figure 1U, Plate 37). Often there are special mechanisms for obtaining dissolved oxygen from the water above them. Most burrowers live in lentic-littoral, lentic-profundal, and lotic-depositional habitats. Some occur in lotic-erosional habitats, but they are restricted to specific places where sand and silt accumulate, such as in eddies behind large stones and logs or near shore.

Swimmers are adapted for actively moving themselves through

the water. They do this in two ways. Some swimmers move their entire body to propel themselves (ameletid minnow mayflies, Figure 1K, Plate 58). This is similar to how fish swim. Most fish swim by flexing their bodies from side to side, whereas most swimming invertebrates flex their bodies up and down. Invertebrates that swim with their body are usually streamlined, without projecting hairs, gills, or other structures that would provide resistance to the water. They sometimes have long tails on the end of the body to help push them along. Other kinds of invertebrates swim by moving their legs, which is analogous to using paddles or oars to move a boat. Legs are modified for swimming (natatorial) by being flat, at least on the ends. Dense brushes of long hairs also make legs more efficient for pushing through the water. Examples of invertebrates that swim by means of modified legs are predaceous diving beetle adults (Figure 2L, Plate 32), water boatmen (Figure 2O, Plate 21), and backswimmers (Figure 2H, Plate 22). Swimmers inhabit all types of lentic and lotic habitats. Some swimmers are so efficient that they can move through very swift lotic-erosional habitats where it is difficult for us to walk. However, members of this group only swim in short bursts. Most of the time, they remain perched on solid substrates, such as rocks, pieces of wood, live plants, or coarse detritus.

Skaters are adapted to remain on the surface of water without breaking through the film created by surface tension. They usually have small bodies and very long, thin legs (water striders, Figure 2I, Plate 26). The secret to staying on the surface film usually involves special hairs on the ends of the legs and sometimes on the body. Glands at the bases of these fine hairs secrete an oily substance that keeps the hairs from becoming wet, thus they spread out on top of the surface film rather than poking through. Most skaters are capable of moving nimbly on the surface in pursuit of food. Although a few kinds of skaters are conspicuous on aquatic habitats, this habit is not common among invertebrates. Skaters are usually found in lentic and lotic-depositional habitats, but they venture out on the surface of lotic-erosional areas if the current is moderate.

Planktonic organisms lack the ability for sustained, directed locomotion. They can adjust their buoyancy to slowly rise and fall in the water, but for lateral movements they depend on the prevailing water currents produced by the wind (phantom midges, Plate 88). There is an abundant and diverse assemblage of planktonic animals in lakes and ponds, but it is composed almost entirely of microscopic organisms, especially rotifers and crustaceans such as water fleas (Cladocera). This habit is rare among the larger invertebrates that are the subject of this guide. These are usually found only in lentic habitats, but they may occur in large lotic-depositional habitats that have practically no current.

In this guide, the format for presenting information on the movements of the various kinds of freshwater invertebrates is to first state the categories that fit them (for example, mostly clingers, some crawlers). Then, further descriptions of their movements and any unique adaptations in their behavior or structure of their body are given.

Feeding

There are two ways that scientists usually consider invertebrate feeding, either the type of food that is consumed or how the food is obtained. Typical foods of freshwater invertebrates are detritus (coarse and fine particles), wood, algae, live vascular plants, and other animals. One problem with considering only the type of food consumed is that most invertebrates eat several types of food (omnivores) during their life. They tend to be opportunistic and eat whatever is most readily available. Freshwater invertebrates are much more consistent about how they obtain their food. Some scientists have found it more useful to summarize the diverse and variable information on types of food eaten by the many individual kinds of invertebrates into a handful of categories based on the body structures and behavioral mechanisms that they use to acquire their food. The term for these broad categories is functional feeding groups.

This guide emphasizes functional feeding groups, but in some cases the types of foods are also mentioned. The functional feeding groups used in this guide are explained in the following paragraphs. The illustrations of riffle and pond habitats (Figures 1 and 2, respectively) help explain some of the ways that invertebrates obtain their food.

Shredders chew on intact or large pieces of plant material. Shredders have basic mouthparts, without any special modifications. Basic mouthparts include two jaw-like structures (mandibles) for cutting and grinding and often an upper lip (labrum) and a lower lip (labium) to help keep the food inside the mouth. There are two types of shredders, based on whether the plant material is alive or dead. Shredder-detritivores feed on detritus, which is dead plant material in a state of decay (giant stoneflies, Figure 1J, Plate 42). The detritus that they consume is in large pieces (> 1 mm), often referred to as coarse detritus or coarse particulate organic matter (CPOM). Leaves, needles, flowers, and twigs that fall from trees and shrubs on shore are the most common foods of shredder-detritivores. Shredder-herbivores feed on living aquatic plants that grow submerged in the water (northern casemaker caddisflies, Figure 2N, Plate 69).

Collectors acquire and ingest very small particles (< 1 mm) of detritus, often referred to as fine detritus or fine particulate organic matter (FPOM). Collectors are divided into two types according to where the fine detritus is located in the habitat. Collector-filterers use special straining mechanisms to feed on fine detritus that is suspended in the water. Usually this involves nets they spin from silk (common netspinner caddisflies, Figure 1P, Plates 73, 74), hairs on their heads (black flies, Figure 1N, Plate 90), or hairs on their legs (brush legged mayflies, Plate 51). Some collector-filterers can use parts of their bodies to create water current for their feeding (mussels, Figure 1V, Plate 12). Collector-gatherers eat fine detritus that has fallen out of suspension and is lying on the bottom or is mixed within bottom sediments. They do not have any special adaptations to acquire their food. They either position themselves on the bottom and eat the fine detritus from the top of the sediment (non-biting midges, Figure 2T, Plate 91), or they burrow through the bottom and unselectively swallow the sediment and fine detritus as they go (aquatic earthworms, Figure 2S, Plate 3). Finger-like projections from some of the mouthparts (palps) usually help them gather the fine particles of food.

Scrapers (also called grazers) are adapted to remove and consume the thin layer of algae that grows tightly attached to solid substrates in shallow waters. These microscopic plants are very nutritious (sort of an aquatic salad bar), but invertebrates must have special adaptations to remove this material. The jaws of aquatic insect scrapers have sharp, angular edges. When these mouthparts move, the action is something like using a putty knife or paint scraper to remove old paint or varnish. After the attached algae have been loosened, the material is swept into the mouth by finger-like projections from other mouthparts. Snails scrape algae by means of a rough, tongue-like structure that works like a rasp. Examples of invertebrates that graze on attached algae are flatheaded mayflies (Figure 1Q, Plate 55), water pennies (Figure 1S, Plate 78), and snails (Figure 1T, Plate 7; Figure 2J, Plate 10).

Piercers have their mouthparts, or sometimes their entire head, elongated and protruding as modifications to puncture their food and bring out the fluids contained inside. It is informative to divide the piercers into two types, according to whether their food is plants or animals. Piercer-herbivores penetrate the tissues of vascular aquatic plants or individual cells of filamentous algae and suck the liquid contents. This is not a common functional feeding group in aquatic environments, even though it occurs extensively on land. Most aquatic piercer-herbivores simply have a small, narrow head that can be inserted into the hole that they nibble in plants (crawling water beetle adults, Plate 28; microcaddisflies, Plate 63). Piercer-predators subdue and kill other animals by removing their body fluids. These are very common in aquatic environments. Most have their mouthparts modified into one or two hard, sharp, hollow tubes that they stab into their prey, much like a medical syringe (water scorpions, Plate 23; predaceous diving beetle larvae, Figure 2M, Plate 82). They usually have other mechanisms to help with subduing their prey, such as enlarged front legs (raptorial) and poisonous saliva. Most of them pump digestive enzymes into their unfortunate prey to dissolve the internal organs into a fluid that can be sucked out through the mouthparts, similar to using a straw. When they have finished feeding, they discard the empty carcasses of their prey. Piercer-predator aquatic invertebrates usually feed on other invertebrates, but some prey on vertebrates such as fish and tadpoles. Their prey is often much larger than themselves.

Engulfer-predators feed upon living animals, either by swallowing the entire body of small prey or by tearing large prey into pieces that are small enough to consume. They typically have large jaws with pointed ends and sharp, tooth-like projections for attacking and devouring their prey. Many kinds of engulfer-predators have their mouthparts arranged to project in front of the head (prognathous) in order to maximize their effectiveness (common stoneflies, Figure 1L, Plate 43; hellgrammites, Figure 1O, Plate 60).

In this guide, the format for presenting feeding information for the various kinds of freshwater invertebrates is to first state the functional feeding groups that they belong to (for example, mostly scrapers, some collector-gatherers). Then, additional information on the specific foods that they consume and any unique adaptations in their behavior or structure of their body for feeding are given.

Breathing

Most kinds of freshwater invertebrates depend upon oxygen dissolved in the water for their breathing. Oxygen enters the organisms either through their general body surface or through gills that are specialized for this purpose, or both. This is referred to as a closed breathing system. Whether gills or body surface, dissolved oxygen passes through the skin of the organism by simple diffusion.

Some kinds of invertebrates have behavioral mechanisms, such as wiggling the body, to increase the rate of oxygen diffusion. This works by keeping the body constantly in contact with water that is completely saturated with dissolved oxygen. It is common for invertebrate gills to have muscles attached to them so they can be moved in the water, which works the same way as wiggling the body to increase the rate of oxygen diffusion.

Quite a few aquatic insects breathe oxygen from the atmosphere, reflecting their close relationship to terrestrial insects. Terrestrial insects have holes in their bodies, called spiracles, that let air in. This is called an open breathing system. All of the adult insects that live in the water, and some larvae, have open breathing systems. There are two basic ways that insects can live in the water and still breathe air by means of an open breathing system. Some kinds can attach a quantity of air to their body, commonly called an air store, and take it underwater to breathe from. The air store can be in the form of a bubble or a thin layer. The aquatic insects that carry an air store underwater hold it in a location on their body where spiracles are located, and simply breathe from the air store. These kinds of air-breathing aquatic insects usually have modifications to some of their body structures to help them hold the bubble or layer of air. The modifications include special features of their wings, legs, or hairs. The other way that some insects breathe air is to have spiracles or some type of extension on the end of their body that they push up to the surface to reach the atmosphere. The extensions for breathing this way are called breathing tubes or siphons. These kinds of insects usually rest near the surface where they can breathe continuously. When they need to go below the surface, they fill up the tubes and cavities inside their body with air and go under until they use up the oxygen. Aquatic insect larvae that obtain dissolved oxygen by means of closed breathing systems, do so in the same ways as other freshwater invertebrates.

In this guide, the first thing that is stated about how the different kinds of invertebrates breathe is whether they obtain oxygen from the water or air. If they obtain dissolved oxygen from the water, the guide explains whether they do this over their overall body surface or with gills. Then any special behaviors or structural modifications are described.

Life History

Life history is a term that scientists use to cover all of the biological events in an organism's life from birth to death. Some of these events are covered under the preceding biological aspects. In this guide, the part on life history is used to present any unique information about reproduction, growth, and development of the individual organisms. It is impossible to avoid using some specialized terms for this topic, so they are explained in this introduction.

Reproduction by invertebrates usually involves mating by a male and female of the same species. In some species, mature males and females look different (sexual dimorphism). One sex may differ from the other by having certain body structures conspicuously enlarged (usually the males) or the two sexes may be different colors or have different patterns of pigmentation. There are also numerous examples of asexual reproduction among freshwater invertebrates. The simplest type of asexual reproduction is budding or fission, in which an organism merely divides into two organisms. Fission is not common in the invertebrates that are the subject of this guide, but some flatworms do this. Parthenogenesis is a type of asexual reproduction in which egg development occurs without fertilization. This happens in some of the snails and insects. Hermaphroditic describes an individual that contains both male and female reproductive organs. This situation occurs commonly in the flatworms, aquatic earthworms, leeches, snails, and mussels. In most cases, these hermaphroditic organisms mate with another individual of their species, but self-fertilization does occur in some kinds.

Eggs result from all types of reproduction except fission. Most invertebrate females are oviparous, which means they lay their eggs somewhere outside of their body. In a few kinds of invertebrates the females retain their eggs and allow them to hatch within their body, at which time they release free-living offspring. This type of egg laying is called ovoviviparous.

After hatching from the egg, invertebrates begin life as small, immature forms that must undergo the processes of growth and development. Growth refers to increasing in size, while development refers to the changes that make them capable of reproducing. While young invertebrates are growing and developing, they are called larvae, juveniles, or just immatures. After they have reached their final stage with functional reproductive organs and external structures for mating and laying eggs, they are called adults. These topics are very important when studying freshwater invertebrates because many of the organisms that are discovered will be immature forms that are undergoing growth and development.

Growth is relatively simple in the flatworms, aquatic earthworms, and leeches because they are not impeded by any type of external protective structure. They just continuously get larger until they become adults. However, a hard, non-living, protective covering made of a complex protein (chitin) is characteristic of all arthropods. The scientific term for this outer layer is exoskeleton, but in this guide it is referred to as "skin" for simplicity. The skin of an arthropod does not grow after it has been produced. Since it is rigid, it prevents the growth of the organism inside. Thus, arthropods must periodically shed their skin (molt) in order to increase in size. When it is time to molt, much of the old skin is dissolved from the inside and the chemical components are reabsorbed by the invertebrates, sort of like recycling. At this time a new and larger skin is produced below the old one. The new skin is larger because it is soft and highly folded, thus it fits in the old, smaller skin. Arthropods break open the old skin by swallowing air or water to create internal pressure, then they climb out and swallow more air or water to enlarge their new skin that is still soft and folded. This enlargement process is kind of like ironing out the wrinkles in recently washed clothes. After the new skin has been stretched to a larger size that will accommodate the arthropod's growing body, it remains soft for a period of time. Arthropods in this soft condition after molting are referred to as being teneral. In addition to being soft, teneral arthropods also lack color. They are pale, almost white. The hardening of the new skin is caused by a chemical process called tanning, which also produces the normal color of the arthropod. Different kinds of aquatic arthropods shed their skin from as few as 3 times to as many as 45 times.

Mollusks are also enclosed in non-living protective covers produced by the organism, which are commonly called shells. The shells of mollusks are made of protein and calcium carbonate. Mollusks can enlarge their shells by producing successively larger growth rings, something like tree trunks. Thus, the bodies of mollusks can grow without shedding their shells.

The appearance of some invertebrates changes considerably between the time they hatch from the egg until they become adults. The changes that take place during development are called metamorphosis. This word comes from Greek meaning "change in form." The most dramatic examples of metamorphosis occur with the insects and mites in the arthropods and with mussels and clams in the mollusks. The metamorphosis of those groups will be described in their respective parts of Section 3.

Generation refers to a population of organisms going from the egg to the adult stage. The number of generations that different kinds of freshwater invertebrates produce in 1 year is important ecological information. Scientists can use this information to predict such things as how many fish can be supported in a body of water or how long it will take a stressed aquatic environment to recover. The overall duration of a single generation for different kinds of freshwater invertebrates ranges from as short as a few weeks to as long as several years. Thus, the number of generations produced by different kinds of aquatic insects commonly ranges from many generations each year to only one generation every several years. Specific information on the number of generations per year is presented for the individual kinds of invertebrates in Section 3.

It is also important to know if organisms remain continuously active during their life history. Some kinds of freshwater invertebrates seem to disappear from a habitat for several months or longer, then suddenly become abundant again. Usually the abrupt disappearance of organisms is a result of natural interruptions in life history. Dormancy is the broad term that is used for any period of inactivity, but there are several types of dormancy — including diapause, hibernation, and aestivation.

Diapause is a condition when invertebrates in any stage of development before the adult, including the egg stage, become dormant and cease their growth and development. Diapause is genetically determined. Some species are "programmed" to enter diapause when certain environmental conditions provide the proper cues. A combination of temperature and length of daylight in relation to darkness are common cues that trigger diapause. After diapause is initiated, it can only be broken by another set of environmental factors that have been "programmed" into the organism. The purpose of diapause is for invertebrates to avoid adverse conditions that occur regularly on a seasonal basis. Some of the naturally unsuitable conditions that invertebrates escape by means of diapause are water temperature being too hot or too cold, low flow (or no water), or not enough dissolved oxygen. Each species usually diapauses in the same stage and at the same time of year.

Hibernation and aestivation are two other types of dormancy, but they are not genetically programmed and occur irregularly, or not at all, during the larval or adult stages of invertebrates. Hibernation is a temporary response to cold, while aestivation is a temporary response to heat. These types of dormancy are much simpler and usually occur for shorter durations than diapause. If the environment where an organism lives becomes too cold for its normal activities, the organism will hibernate until the environment warms up. If the environment where an organism lives becomes too hot for its normal activities, the organism will aestivate until the environment cools off. Freshwater invertebrates must hide themselves in protected locations within their habitat before any type of dormancy is initiated. Common locations for dormancy are down in the bottom substrate, within accumulations of detritus, or in living plant tissue. If individual kinds of invertebrates commonly undergo any type of dormancy, that information is presented in the life history parts of Section 3.

Stress Tolerance

This term refers to the ability of organisms to withstand disturbances in their environment. There are many different types of disturbances that can occur in freshwater environments. Some are caused by human activities, while others are the result of the forces of nature. Pollution is the term that is most often associated with disturbances caused by human activities, but its meaning is actually restricted to substances or energy that are released into water and bring about undesirable changes. Environmental stress is a broader term that is used for any action that brings about undesirable changes. Some examples of environmental stress that do not fit the strict definition of pollution are removing water for irrigation or municipal supplies, impounding (damming) a stream, and deforestation that eliminates shade and leaf fall. However, this broad concept of environmental stress is not limited to human activities. Natural events such as floods, forest fires, and volcanoes also disturb aquatic environments.

Different kinds of freshwater invertebrates vary widely in their ability to cope with environmental stress. Professional aquatic biologists take advantage of this situation when they conduct biomonitoring to assess environmental health. If samples of the living assemblages in a body of water contain many kinds of organisms that are known to be sensitive to stress, then that indicates a healthy environment. If samples reveal just a few kinds of organisms, all of which are known to be tolerant to stress, that indicates an unhealthy environment. There are many intermediate conditions between these two extremes in which samples contain a mix of tolerant and sensitive organisms. Professional aquatic biologists derive mathematical indices from equations that incorporate numerical tolerance scores for each kind of invertebrate. However, this numerical approach is usually only applicable for a particular geographic region and requires many years of research by professional aquatic biologists to develop. This guide provides subjective information based on many years of field research on pollution and environmental stress that I have conducted, as well as that reported by other scientists. The various kinds of organisms are placed into the following categories:

Very sensitive — usually found only in nearly pristine environments; quickly eliminated if any disturbance occurs; do not occur in high numbers anywhere.

Somewhat sensitive — similar to facultative; will be in pristine environments but can also withstand a limited amount of disturbance; usually do not occur in high numbers.

Facultative — occur in environments with conditions ranging from pristine to moderate levels of disturbance; often occur in high numbers under conditions of moderate disturbance.

Somewhat tolerant — similar to very tolerant but they cannot survive in severely disturbed environments; occur in high numbers but they do not dominate the community as completely as the very tolerant kinds.

Very tolerant — seldom found in pristine environments; occasionally found in moderately disturbed environments; exceptionally high numbers in environments with severe disturbance; can withstand almost anything; flourish where conditions are so bad that they probably have only one or two competitors, or none.

If lower taxonomic units (genus or species) within the higher group that is presented in the guide (usually family) are known to vary in their tolerance to stress, then the range of stress tolerance categories within the family is also given. Many kinds of freshwater invertebrates have different levels of tolerance for particular types of disturbance. For example, a kind of invertebrate might be tolerant of the poisonous effects of a particular chemical but very sensitive to fine sediment that clogs their gills. Such distinctions between the effects of different types of pollution or environmental stress are beyond the scope of this guide. The categories of stress tolerance that are used in this guide are generalizations about a group's overall tolerance to stress from all types of disturbance. In some instances, descriptive statements are added about some kinds of organisms that are widely known to be exceptionally sensitive or exceptionally tolerant to particular types of disturbance.

References on Fundamentals of Freshwater Invertebrate Biology

Borrer, D. J., C. A. Triplehorn, and N. F. Johnson. 1989. *An Introduction to the Study of Insects.* 6th edition. Saunders College Publishing, Philadelphia, Pennsylvania. 875 pages.

Doris, E. 1993. *Entomology.* Thames and Hudson, Inc., New York. 64 pages.

Johnson, S. 1989. *Water Insects.* Lerner Publications, Inc., Minneapolis, Minnesota. 48 pages.

McCafferty, W. P., 1981. *Aquatic Entomology. The Fisherman's and Ecologists' Illustrated Guide to Insects and Their Relatives.* Science Books International, Boston, Massachusetts. 448 pages.

Merritt, R. W., and K. W. Cummins, editors. 1996. *An Introduction to the Aquatic Insects of North America.* 3rd edition. Kendall/Hunt Publishing Company, Dubuque, Iowa. 862 pages.

Mound, L. 1990. *Insect.* Eyewitness Books, Alfred A. Knopf, Inc., New York. 63 pages.

Smith, D. G. 2001. *Pennak's Freshwater Invertebrates of the United States: Porifera to Crustacea.* 4th edition. John Wiley & Sons, Inc., New York. 648 pages.

Taylor, B. 1992. *Pond Life.* Dorling Kindersley, Inc., New York. 29 pages.

Thorp, J. H., and A. P. Covich, editors. 2001. *Ecology and Classification of North American Freshwater Invertebrates.* 2nd edition. Academic Press, San Diego, California. 1056 pages.

Ward, J. V. 1992. *Aquatic Insect Ecology. 1. Biology and Habitat.* J. Wiley and Sons, New York. 439 pages.

Watts, B. 1988. *Dragonfly.* Stopwatch Books, Silver Burdett Press, Englewood Cliffs, New Jersey. 25 pages.

Williams, D. D., and B. W. Feltmate. 1992. *Aquatic Insects.* C·A·B International, Wallingford, United Kingdom. 358 pages.

Wyler, R. 1990. *Puddles and Ponds. An Outdoor Science Book.* Julian Messner, Englewood Cliffs, New Jersey. 32 pages.

How To Study Freshwater Invertebrates

With a little background information and some practice, freshwater invertebrates are easy to study with simple equipment and supplies. Keep in mind that some lands under the jurisdiction of state or federal agencies have regulations about collecting or disturbing plants and animals, including invertebrates. Some kinds of freshwater invertebrates have been given special protected status because they are rare and their continued existence is threatened. This mostly involves freshwater mussels. It is usually permissible to study freshwater invertebrates and collect a few specimens, but it is always wise to check with the appropriate authorities before beginning any type of activity. On private lands, the rights of property owners should always be respected. Rarely does a landowner object to anyone collecting invertebrates, if permission is sought beforehand.

Much of the equipment that is needed to study freshwater invertebrates can be constructed inexpensively from common household items. This guide gives brief instructions on how to construct some types of equipment, and there are references at the end of this part that provide more detailed instructions. There are also many stores and mail order companies from which you can purchase equipment. The equipment that is available ranges from rather expensive, but very sturdy, equipment that aquatic biologists would use in a professional capacity to less expensive, but effective, equipment that is intended for science education and nature interpretation. The intention of this guide is not to steer users to individual suppliers of equipment. The types of companies that sell professional grade equipment usually advertise themselves as scientific supply houses. Some specialize in equipment and supplies for studies of natural history, biology, environmental science, forestry, or wildlife. Some of the scientific supply houses also sell equipment for science education and nature interpretation, but there are many other companies that specialize in practical equipment that will get the job done for teachers, students, nature interpreters, and natural history hobbyists. This category of suppliers of this type usually has words such as nature, environment, conservation, education, K-12, outreach, interpretation, learning, activities, field studies, interactive, or hands-on associated with the name of their company or their advertisements. The major part of their inventory is often books and other educational materials. Shops associated with nature museums and zoos often sell this kind of equipment and supplies. Because of the popularity of fly fishing and the importance of invertebrates in that form of recreation, some outfitters for anglers sell equipment to collect and study invertebrates. The equipment sold by outfitters is often designed to be small and portable, so anglers can carry it along with their fishing gear.

Observing Live Organisms in the Natural Environment

The simplest way to study freshwater invertebrates is to observe them where they live, without disturbing anything. This works best in still waters, either around the edges of ponds or shallow pools and backwaters of streams. Just stand still, or perhaps kneel, on shore at the edge of the habitat. If you look closely and concentrate a few moments, you will usually begin to see invertebrates moving around. With some practice and patience, you will even begin to see the organisms that lie still. This is the most effective way to observe how organisms move or maintain their position. It is often possible to see them engaging in some of their essential activities, such as eating or breathing. Careful scrutiny may reveal them moving their gills, positioning their breathing tube at the surface, or capturing a bubble of air to take underwater with them. This sometimes works in running water too, if there is a shallow riffle with very clear water, but you may need to wade out from the shore to see the organisms.

An underwater viewing device makes it easier to see invertebrates in their natural setting (Figure 3). The viewer allows you to see underwater much more clearly because it eliminates the glare from the water surface, and it decreases the amount of water you have to look through. An underwater viewer consists of a tube 4–6 in (10.2–15.2 cm) in diameter and 7–24 in long (17.8–61.0 cm), with glass or clear plastic on one end and the other end open. To construct an underwater viewer (Figure 3B), take a large metal can (32 oz or larger), one such as juice comes in, and remove both ends with a can opener. Cover one end of the can with clear plastic wrap that is used to cover leftover foods. Stretch the plastic wrap tightly and make sure there are no folds or wrinkles. Then fasten the plastic wrap tightly to the sides of the can with some type of strong tape that is waterproof. Duct tape and clear vinyl tape are two kinds that work well for this purpose. A waxed milk carton (quart or half gallon) is an alternative to a metal can. To use the viewer, just push the end covered with glass or plastic underwater and look through the open end. Wearing polarizing sunglasses will also help you to see below the surface of the water.

Collecting Organisms

Many freshwater invertebrates lead secretive lives and remain too well hidden to be observed in their natural surroundings. Therefore, it is usually necessary to collect them and remove them from their habitat. This is an enjoyable way to study these organisms because it adds the challenges of finding and capturing creatures that are often very well camouflaged or that swim exceptionally fast. It is also the most informative way to learn about their biology

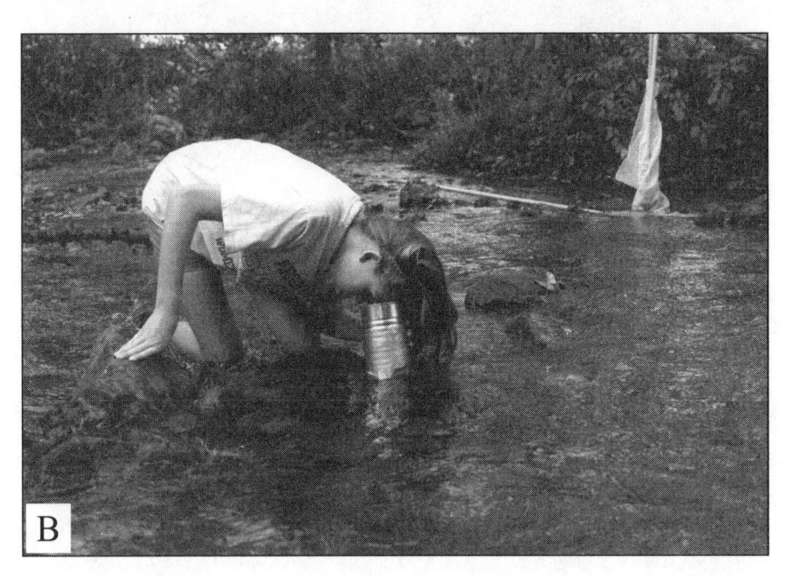

Figure 3. Two devices for viewing underwater. **(A)** Commercial product. **(B)** Homemade device made from metal can, clear plastic wrap, and tape.

because you can isolate them and examine them up close.

The simplest way to collect freshwater invertebrates is to reach into the water and pick up pieces of substrate. Rocks, sticks, leaves, exposed roots, drooping terrestrial vegetation, and aquatic plants are some of the likely invertebrate residences that can be easily removed and examined (Figure 4). Many of the freshwater invertebrates will drop off or crawl out of the substrate after it is brought out of the water. For this reason, it is best to place the solid object or loose material in some type of shallow, light-colored pan, so the invertebrates do not get lost on the ground. I have found that 6 x 9-in (15.2 x 22.9-cm) trays, such as those shown in Figure 4, work very well. These are sold as drawer organizers.

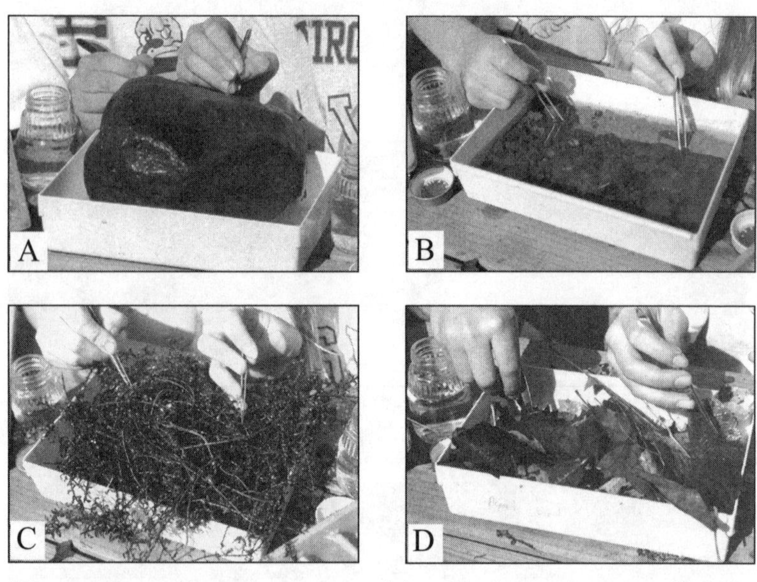

Figure 4. Examples of bottom materials where invertebrates are likely to reside. The aquatic substrates can be taken out of the water by hand or sampled in the water with a net. **(A)** Cobble-sized stone. **(B)** Gravel. **(C)** Living aquatic plants. **(D)** Dead leaves and twigs that fell into the water from trees on shore.

At first, do not put any water in the pan. If the substrate is a solid object, examine all sides carefully. If it is loose material, poke through it. Pick up the object or loose material to see if anything migrated and is sitting on the bottom of the pan. Then add a few centimeters of water to the pan and look for swimming or crawling invertebrates after you move the substrate in the pan or splash water on the substrate. The organisms should be removed and placed in another container for identification and observation (Figure 5). Most invertebrates can be picked up safely by hand, without harming them or you, but many are small and hard to catch and pick up. Some tools that are handy for picking up invertebrates are forceps (tweezers), turkey basters, kitchen sieves, or aquarium nets. Any type of container, such as glass jars or plastic food containers, will work for observing freshwater invertebrates that you have captured. The individual compartments of plastic ice cube trays are

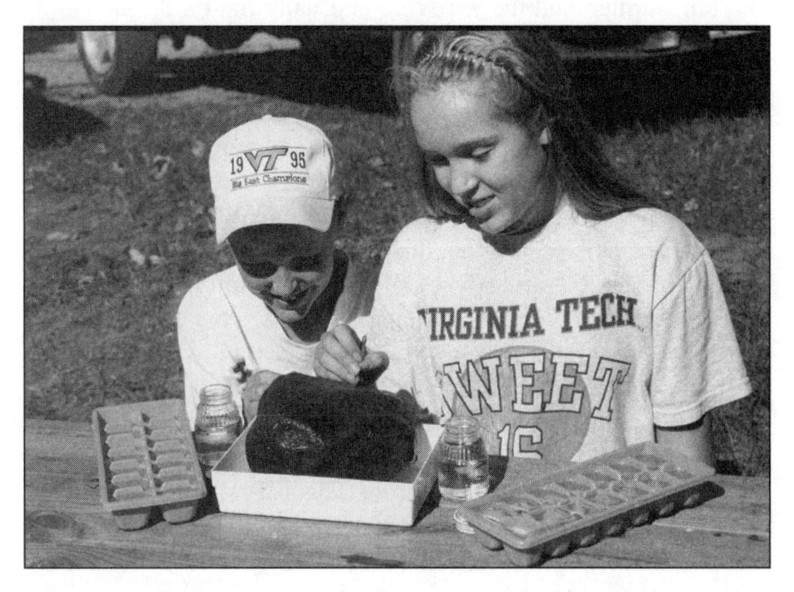

Figure 5. Technique and tools for finding and observing invertebrates on bottom materials that have been removed from the water.

efficient for comparing different kinds of invertebrates. An inexpensive hand lens (magnifying glass) will help you see small organisms or special structures. Many different types of magnifying devices are made specifically for examining small invertebrates in the field. These range from small plastic boxes with a magnifying lens in the top to miniature plastic microscopes with moving parts to focus the view.

It will not be possible to collect some invertebrates by picking up pieces of the substrate, so it will eventually become necessary to use a mesh net to find different kinds. Nets are the most effective way to collect a lot of different kinds of freshwater invertebrates. There are two basic types of nets for catching freshwater invertebrates. One has a single handle, a wire ring about 12 in (30.5 cm) in diameter attached to the end of the handle, and a bag-like net attached to the wire ring (Figure 6). It is something like a butterfly net, but sturdier, and the wire ring is usually flat on the side away from the handle so the net will fit snugly against the bottom of the aquatic habitat. This type is commonly called a D-frame dip net because of the shape of the wire ring. When collecting in flowing water, simply place the net on the bottom and move the substrate in front of the net with your hands (Figure 6A). You can also stand in front of the net and wiggle your feet to move the substrate. Because this resembles dancing, this technique goes by names such as "mayfly mash," "stonefly stomp," "hellgrammite hustle," or "crawdad shuffle." The disadvantage to using your feet is that the grinding action often damages the organisms. If a rock or log is too large to move, rub the surface with your hand.

When collecting in the still waters of slow-moving streams or the shallows of ponds and lakes, the net must be moved across the bottom or through plants and debris. This is best done my making quick, short jabs with the net (Figure 6B). Most of the invertebrates that live in habitats with soft, fine substrate stay in the upper 1–2 cm of the mud. It is not productive to dig deep in soft sediment with the net. Doing so makes it nearly impossible to find the invertebrates

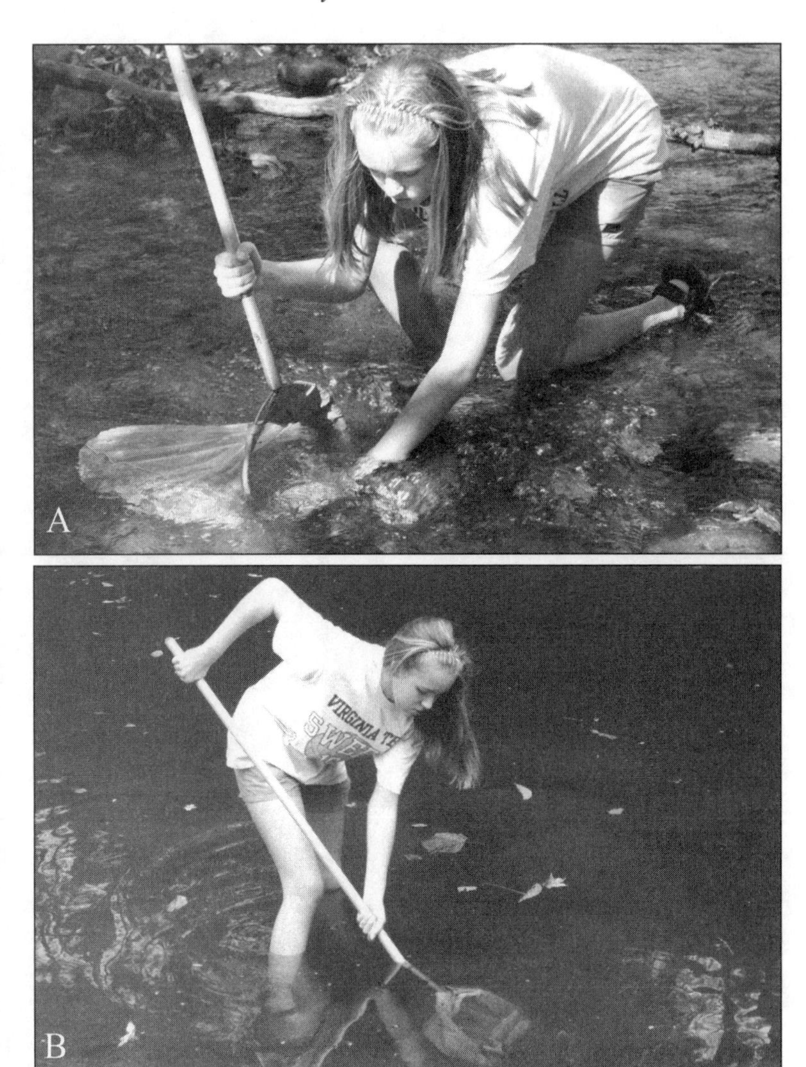

Figure 6. Techniques for using a dip net to collect freshwater invertebrates. **(A)** In flowing water, place the dip net on the bottom, move substrate in front, let current wash dislodged organisms into net. **(B)** In standing water, make quick jabs with the dip net into the sediment or other material lying on the bottom.

after they become mixed in with the fine sediment, and the water in the sorting pan becomes muddy. If any mud is scooped up in the net, make the material to be examined as clean as possible by swishing the net back and forth in the water until the water that drips out of the net is no longer muddy. The D-frame dip net is particularly effective for sampling individual microhabitats (Figure 7).

The second basic type of net for collecting freshwater inverte-

Figure 7. Using a dip net to sample some of the microhabitats where many kinds of invertebrates are often found. **(A)** Leaf pack. **(B)** Rock outcrop. **(C)** Tree roots and undercut bank. **(D)** Terrestrial vegetation hanging in water.

brates consists of mesh stretched between two poles. This is commonly called a kick net or kick screen (Figure 8). These are usually about 3 ft (0.9 m) wide and 2–3 ft (0.6–0.9 m) tall. Kick nets require two persons for collecting and are only effective in flowing water. One person stands behind the net, facing upstream, and holds it in the current, with the bottom edge against the substrate, by means of the two poles. The other person moves the substrate in front of the net with the hands or feet (Figure 8A). After some material has accumulated on the net, the two people pick it up together. The person behind the kick screen holds on to the tops of the two poles and lowers them, while the person in front grabs the bottoms of the two poles and raises them. Carry the kick net to shore and either lay it on the ground for direct examination (Figure 8B) or wash the contents into a pan. If you lay the kick net on the ground, it is helpful to put a white plastic sheet under it to catch some of the worm-like invertebrates that will wiggle through the mesh. A shower curtain liner works well for this. The kick net is particularly useful for collecting the larger invertebrates that live under large stones and logs, such as hellgrammites and crayfish. It also collects high numbers of invertebrates, which increases your chances of finding less common kinds.

Regardless of the type of net you are using or the type of habitat you are collecting in, do not let too much material accumulate in the net before examining it for invertebrates. It is much easier to find invertebrates if you repeatedly examine small amounts of material. The single best piece of advice for collecting freshwater invertebrates is to try as many different microhabitats as possible. Refer to the information on habitat given above to get ideas about where to collect with a net.

Both types of nets can be constructed at home. For the D-frame dip net, use a broom handle or any kind of round pole. Shape the net frame from strong wire that can be bent, such as that sold for clotheslines (#9 galvanized works well). Attach the wire frame to the handle with tie wire or threaded hose clamps and then wrap

Figure 8. Techniques for using a kick net to collect freshwater invertebrates. **(A)** Collecting — 1 person holds the net by its 2 poles while the other moves the substrate in front of the net with their hands or feet. **(B)** Sorting — the net is laid on the ground, preferably with a white plastic sheet under it, and the contents examined carefully.

the junction with duct tape, or some other kind of strong tape that will hold up in the water. The mesh catch net can be made from material that is available at fabric stores. One such material is tulle, which is a stiff rayon or nylon net that is used to make veils or ballet costumes. The size of the mesh should be no larger than 1/8 in (3.2 mm) to capture most of the common kinds of freshwater invertebrates. Some of the small kinds of invertebrates will not be retained in the net unless a finer mesh is used, 1/32 in (0.8 mm) or less. Sew the net into a bag-like shape, then sew it to the wire frame.

The kick screen is easier to make than the D-frame dip net because it is flat. Simply attach the same type of mesh to two poles with staples or thread. Nylon window screen also works very well for a kick screen. It can be cut with scissors, and the edges can be sealed with a flame. Kick screens work best if the bottom edge is weighted to hold it against the substrate. Sewing a chain along the bottom edge or attaching fishing weights will accomplish this. Minnow seines work fairly well as kick screens for collecting larger invertebrates.

After you have collected organisms and transferred them to containers for observation, you should identify them by using Section 2 of this guide. Live invertebrates are easier to identify than dead ones, because their colors change or fade soon after they die. The color illustrations in this guide represent live specimens. After identifying the specimens, observe them. Notice how they move or don't move. Then look for their structures, such as antennae, mouthparts, legs, suckers, claws, gills, breathing tubes, or tails, and see how they use them. Read the information in Section 3 about the particular kinds of organisms you have found, and try to observe the biological features that are mentioned. When you have finished observing and studying the invertebrates, try to return them alive to their original habitat. You may want to preserve a few specimens of each kind and keep them in a collection, as described below.

Maintaining Live Organisms Indoors

It is very educational and enjoyable to keep live freshwater invertebrates in the home or classroom, where they can be observed consistently over longer periods of time than is possible in the field. The organisms that live in ponds and lakes are very easy to maintain because any aquarium or large glass vessel can simulate their natural habitat. It is best to bring back some of the water from the habitat where you found them because the chemistry of that water is what they are used to. In addition, there are fine particles of detritus and microbes suspended in natural water that will serve as food for some of the invertebrates. Bottled spring water is a good alternative, if you cannot transport water from the place where you collect the invertebrates. Tap water from municipal supplies contains chlorine that is toxic to most invertebrates. If you must use treated water, you should bubble air from an aquarium pump through it for 24 hours to eliminate the chlorine.

Aquarium gravel makes a good substrate for invertebrates. Add a few sticks, stones, dead leaves, or live plants, preferably from the original habitat, for places the organisms can hide and hold on to. A little bit of the sediment from the original habitat should be added to serve as food for detritus feeders. Also, this material will probably contain some small invertebrates that can serve as food for predators. It is usually not necessary to aerate the water for pond dwellers, but an air stone with a small amount of air coming out will keep the water saturated with dissolved oxygen and will keep anaerobic decay processes, and their accompanying odors, to a minimum. Be gentle with the organisms when you collect and transfer them to the aquarium. Never squeeze them with forceps. Put the tips of the forceps under the organism and lift rather than squeeze. Better yet, use a small aquarium net to lift them or carefully suck them up in a turkey baster to transfer them. A screen cover on the aquarium is a good idea to keep anything from escaping, while keeping the water in contact with the air to maintain saturated levels of dissolved oxygen. Do not worry about moderate amounts of algae or scum forming in the aquarium. This is natural, and some of the organisms will probably use it as food. Excessive amounts of this material can be avoided by changing about one-third of the water volume every 2 weeks and by removing any excess food that is not eaten within 24 hours. A few snails will also help keep the algae and scum under control.

Some of the pond organisms included in this guide are easy to keep in an aquarium. Examples include:

- Dragonflies and damselflies
  - Skimmer dragonflies, darner dragonflies, narrowwinged damselflies
- True bugs
  - Water boatmen, backswimmers, water striders, water scorpions, giant water bugs
- Water beetles
  - Predaceous diving beetle adults and larvae, water scavenger beetle adults and larvae, whirligig beetle adults
- Caddisflies
  - Longhorned case makers, northern case makers, giant case makers
- True flies
  - Non-biting midges
- Crustaceans
  - Scuds, aquatic sow bugs
- Flatworms
- Segmented worms
  - Leeches, aquatic earthworms
- Mollusks
  - Snails

Read about the biology of these organisms in Section 3 of the guide so that you can provide the appropriate microhabitats and foods in your aquarium. Notice that many of these pond organisms are predators, so you will need to provide a lot of prey organisms. If you have trouble finding aquatic invertebrates for prey, most pet stores carry other live organisms that can be substituted. True bug and water beetle predators will eat live crickets. Just drop them in the water. Some of the aquatic invertebrate predators will even feed on dead crickets, so you can freeze some fresh ones and drop a few in the water as needed. Carrion scavengers, such as whirligig beetles and water scavenger beetles, will feed on small pieces of raw meat. Dragonfly and damselfly larvae do very well on small earthworms that you can find in the soil or mulch piles. If you cannot find any earthworms, pet stores usually sell several other kinds of small live worms for fish food. Many predators commonly eat water boatmen and midge larvae, and you can collect these easily in the shallow water of almost any pond. If you use dechlorinated tap water or bottled spring water to start your aquarium, there might not be enough fine detritus to feed any organisms that collect this material for their food. Good sources of this type of food are dry dog food that has been ground up, powdered milk, or fish flakes. Only add a few pinches of this material because it stimulates microbial growth that can accumulate in a thick layer. No matter how much food you provide, it is inevitable that some of the predators will attack and consume their own kind as a result of being kept in a relatively small, artificial environment. Never put any of these invertebrate predators in with tropical fish that you care about, because the fish will probably become prey within 24 hours.

It is enlightening to raise some of the immature aquatic insects to the adult stage and hopefully be able to watch them make the transformation. Dragonflies and damselflies will probably provide the highest rate of success for larvae transforming to adults in an aquarium. Try to collect larvae that are as mature as possible. These have the largest wing pads on the thorax. Collecting mature larvae cuts down on the length of time that you have to provide them with prey organisms for food. When it is time to transform to the adult stage, all dragonflies and damselflies have to climb out of the water before they shed their skin. Therefore, you must put a stick or rock that extends from the bottom of the aquarium up above the water surface. The substrate must be fairly rough so the emerging organisms can hold on with the claws at the end of their legs. A strip of wire screen hung on the inside of the aquarium also works well for an emergence site. It is best to leave the aquarium uncovered if you are trying to raise dragonflies and damselflies to the adult stage. If the aquarium is covered, they often fall in the water and drown when they first try to fly. Just let them fly away in the room. They will always be found on the inside of a window, because they fly towards the light. You will know that they have emerged because they leave their empty larval skin at their emergence site. After observing the adult for a while, capture it carefully with your fingers or a net and release it outdoors.

Making a Collection of Preserved Specimens

Although there is no point in killing invertebrates unnecessarily, it is useful and educational to build and maintain a preserved collection that contains representatives of all the different kinds you encounter in your studies. A good quality collection serves several purposes. If the specimens are properly labeled, you can use them to help identify freshwater invertebrates that you find in the future. In addition to the illustrations in this guide, you will have actual organisms for reference material. You can use your preserved specimens to practice your identification skills during periods when you are not able to get out in the field. Because they do not move, preserved specimens are especially useful for studying the special structures that invertebrates have for feeding, breathing, and other activities. Lastly, if you put appropriate information on the labels about where and when you collected the organisms, your collection will contain important ecological data. If you use common sense about not collecting too many organisms and not collecting in protected environments, preparing a preserved collection of common freshwater invertebrates is a good scientific practice that has no adverse effects on conservation of natural resources, environmental protection, or maintenance of biodiversity.

Most of the freshwater invertebrates have soft bodies that are best preserved in liquid. Most do not keep very well if you try to dry them on a pin, like is commonly done for terrestrial invertebrates such as butterflies. Although some of the organisms, such as water beetle adults, can be pinned, most students of freshwater invertebrates just put them all in liquid for convenience. Rubbing alcohol is the best liquid to use because it preserves the specimens well, and it is relatively safe, easy to purchase, and inexpensive. Rubbing alcohol, which has the chemical name of isopropanol or isopropyl alcohol, can be purchased any place where pharmaceutical supplies are sold. The rubbing alcohol that you buy at a pharmacy should not be diluted any further. It is usually sold at a concentration of 60–70% isopropanol, which is an effective concentration. Any lower concentration will not keep the organisms from decomposing, and any higher concentration will make the specimens so brittle that they easily break. Of course, follow any safety precautions on the label of the rubbing alcohol, such as keeping it away from small children and remembering that it is a flammable substance. The best containers for your preserved specimens are small jars or vials made of clear glass and having tight fitting, screw-on lids. These will probably have to be purchased from the types of suppliers mentioned at the beginning of this part, but you may be able to start with something from around the house, such as pill bottles or baby food jars.

The organisms that you want to preserve for your collection should be put directly into the small containers of alcohol as soon as you collect them in the field. It is all right to put several specimens of the same kind in one container, but make sure that the specimens do not occupy more than half of the volume of the container. The water contained in the bodies of the specimens dilutes the strength of the alcohol, so you should pour off the original alcohol and add more within a day or two. You should put a paper label with the appropriate information in each container before you leave the site where you collect the specimens. At a minimum, you should include:

Name of the organism
Name of the body of water (if there is no name, state the type of habitat)
Type of zone and microhabitat
Name or number of the nearest road and distance from that road
County or city and state or province
Date the specimens were collected
Name of the collector

In the field, you will probably want to quickly write labels with a pencil and drop them in the containers, then later prepare neater labels with permanent ink that will be easier to read than pencil labels. However, you must be careful about the type of ink that you use. Most inks, including that used in all ball point pens, will dissolve in alcohol, leaving you with tinted alcohol, stained specimens, and blank pieces of paper in the containers. Felt tip pens with fine points (size 00 is best) and permanent ink are available where art and drafting supplies are sold. You should always test the ink in alcohol for at least 24 hours to see if it fades. You can also prepare very neat and permanent labels with a computer and a laser printer. Again, it is always best to test any label for permanency in alcohol. Alcohol will eventually evaporate from any type of container, so you should check your collection several times a year to see if the alcohol in the containers needs to be replenished. If specimens preserved in alcohol are allowed to dry out, they become so shriveled and brittle that they are of no use.

Where To Get Assistance

If you are new to the study of freshwater invertebrates, you will probably have questions that are not addressed in this guide. The first place to turn is to more advanced and specialized books on the subject. References for information on various topics can be found in the respective sections of this guide. These books should be available in libraries, especially if you have a college or university nearby. When you begin to identify freshwater invertebrates it is very helpful if a knowledgeable person checks your identifications. More than likely, there are people with this expertise in the general vicinity of where you live. Colleges and universities are always a good place to start. However, the department in which freshwater invertebrates might be studied is not consistent among institutions of higher education. Some of the departments that you might want to try are biology, ecology, entomology, fisheries, systematics, and zoology. Nature or natural history museums, science centers, and zoos are other good places to ask for help. Sometimes state and federal agencies employ scientists who know about freshwater invertebrates and are willing to provide assistance. The names of the agencies vary in different states, provinces, and countries but they usually have something to do with conservation of natural resources, environmental protection, or fisheries. Please keep in mind that all of the persons mentioned above have many responsibilities in their jobs. While they enjoy helping individuals with their questions about freshwater invertebrates, they must fit this in around other assignments. Please be courteous enough to contact them ahead of time to schedule an appointment, and be patient if they ask you to leave the specimens for later identification.

References on How To Study Freshwater Invertebrates

Dunn, G. A. 1993. *Caring for Insect Livestock: An Insect Rearing Manual.* Young Entomologists Society, Inc., Lansing, Michigan. 96 pages.

Firehock, K. 1994. *Hands On Save Our Streams. The Save Our Streams Teacher's Manual for Grades 1 through 12.* 2nd edition. Izaak Walton League of America, Gaithersburg, Maryland. 215 pages.

Headstrom, R. 1964. *Adventures with Freshwater Animals.* Dover Publications, Inc., New York. 217 pages.

Hickman, P. M. 1990. *Bugwise. Thirty Incredible Insect Investigations and Arachnid Activities.* Addison-Wesley Publishing Co., Reading, Massachusetts. 96 pages.

McCafferty, W. P. 1981. *Aquatic Entomology. The Fisherman's and Ecologists' Illustrated Guide to Insects and Their Relatives.* Science Books International, Boston, Massachusetts. 448 pages.

Merritt, R. W., and K. W. Cummins, eds. 1996. *An Introduction to the Aquatic Insects of North America.* 3rd edition. Kendall/Hunt Publishing Company, Dubuque, Iowa. 862 pages.

Simon, S. 1975. *Pets in a Jar. Collecting and Caring for Small Wild Animals.* Puffin Books, New York. 95 pages.

Stokes, D. W. 1983. *A Guide to Observing Insect Lives.* Little, Brown and Company, Boston, Massachusetts. 371 pages.

United States Environmental Protection Agency. 1997. *Volunteer Stream Monitoring: A Methods Manual.* EPA 841—B—97—003. United States Environmental Protection Agency, Office of Water, Washington, D. C. 210 pages.

Section 2
Identification of Different Kinds

Identification of Different Kinds

The most reliable way to identify a freshwater invertebrate you have observed or collected is to first determine which of the major groups it belongs to. The QuickGuide below is designed for this purpose. Begin by answering the questions in Block 1. Depending on your answers, proceed to Block 2 or Chart A. If you go to Block 2, answer that question, then proceed to Chart B or C. When you arrive at Chart A, B, or C, determine which of the generalized drawings most closely matches the organism you have observed or collected. Then, carefully read all of the written descriptions of the various features in that row and determine if the organism in question has those features. If the organism fits the descriptions in that row, you have placed the organism into a major group of invertebrates. The cell at the end of the row, with the small drawings, directs you to the plates with color whole body illustrations and black and white illustrations of important body structures.

The plates are used to confirm the major group and to identify the different kinds (usually families) in that group. Some of the major groups are not identified to lower levels because of their small size and complexity of their features. The plates also have brief written descriptions of the organisms. The statements written in red are the most important diagnostic features that distinguish the particular kind of invertebrate. The two vertical bars show the actual size range of the organism. The range in size is usually the result of there being different species within the group. The exceptionally long tails on some organisms made it necessary to not show most of the tail length on the color whole-body illustrations of those kinds. In these instances, the organism is also shown in a small color illustration that has the entire length of the tails. The plates are arranged in the same sequence as the major groups in the QuickGuide. For the families within the major groups, the sequence of the plates is arranged so that the most distinctive families come first. Experienced invertebrate biologists may find this sequence unusual, because they are used to organisms being arranged in evolutionary (phylogenetic) sequence. However, I am certain that the sequence of the different kinds used in Section 2 of this guide will be easier for beginners because it is based on the appearance of the organisms rather than their evolution.

QuickGuide
to Major Groups of Freshwater Invertebrates

Block 1: All Freshwater Invertebrates

Does it have 3 or more pairs of hard jointed legs?

Does it have a recognizable head, or at least some visible jaws or hooks for feeding?

If the answer to either of these questions is yes, go to Block 2: Arthropods.

If the answer to both of these questions is no, go to Chart A: Invertebrates That Are Not Arthropods.

Block 2: Arthropods

How many pairs of hard jointed legs does it have?

If 4 or more, go to Chart B: Arthropods That Are Not Insects

If 3 or none, go to Chart C: Insects

Chart A: Invertebrates That Are Not Arthropods

Body Texture; Shape	Body Arrangement	Suckers	Hard Shells Enclosing Bodies
Soft; flat from top to bottom and elongate	All areas almost alike, no individual segments or specialized regions	None	None
Muscular; flat from top to bottom and elongate	Many similar segments arranged in a row, no specialized regions	2 on bottom, 1 at front and 1 at rear	None
Soft; cylindrical and elongate	Many similar segments arranged in a row, no specialized regions	None	None
Soft; irregular but usually not visible	Several irregular regions but usually not visible	None	1, usually coiled but sometimes a short broad cone
Soft; irregular but usually not visible	Several irregular regions but usually not visible	None	2, shaped like shallow bowls, opposite one another and connected by a hinge so that they seal tightly

Chart B: Arthropods That Are Not Insects

Hard Segmented Legs	Antennae	Body Size; Shape
4 pairs	None	Very small, usually 3 mm or less, round, spherical, look like spiders
7 pairs, first pair of legs with small claws	2 pairs, 1 pair much longer	Small, 5-20 mm; flattened from top to bottom
7 pairs, first 2 pairs of legs with small claws	2 pairs, about equal length	Small, 5-20 mm; flattened from side to side
5 pairs, first 2 or 3 pairs of legs with claws, first pair of claws sometimes very large	2 pairs, 1 pair much longer	Large, usually 25-150 mm; mostly cylindrical

	Group
	Flatworms (Phylum Platyhelminthes, Class Turbellaria) Plate 1
	Leeches (Phylum Annelida, Class Hirudinea) Plate 2
	Aquatic Earthworms (Phylum Annelida, Class Oligochaeta) Plate 3
	Snails (Phylum Mollusca, Class Gastropoda) Plates 4-11
	Mussels, Clams (Phylum Mollusca, Class Bivalvia) Plates 12-15

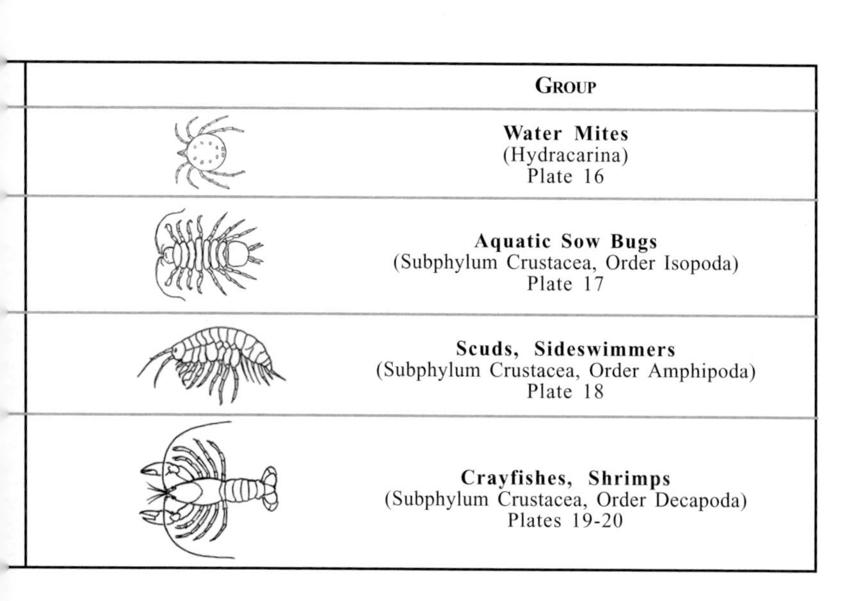

	Group
	Water Mites (Hydracarina) Plate 16
	Aquatic Sow Bugs (Subphylum Crustacea, Order Isopoda) Plate 17
	Scuds, Sideswimmers (Subphylum Crustacea, Order Amphipoda) Plate 18
	Crayfishes, Shrimps (Subphylum Crustacea, Order Decapoda) Plates 19-20

Chart C: Insects

Mouthparts	Wings on Thorax	Hard Segmented Legs; Claws	Gills	Structures on End of Abdomen; Other Features
1 sharp pointed beak or 1 blunt cone	Fully developed (capable of flying), reduced, developing in wing pads, or none	3 pairs; 2 claws	None	None or pair of elongate breathing straps or tubes
2 opposing jaws	Fully developed (capable of flying), front wings greatly modified into thick hard protective covers	3 pairs; 2 claws	None	None
2 opposing jaws, also a large, elbowed lower lip with 2 hooks on end, lower lip folded under head at rest	Wing pads (developing wings) present but hard to distinguish on young larvae	3 pairs; 2 claws	Damselflies have 3 leaf-like gills on end of abdomen, dragonflies have none	Damselflies have 3 leaf-like gills, dragonflies have none
2 opposing jaws	Wing pads (developing wings) present but hard to distinguish on young larvae	3 pairs; 2 claws	Single or branched filaments on bottom of thorax or none	All have 2 tails
2 opposing jaws	Wing pads (developing wings) present but hard to distinguish on young larvae	3 pairs; 1 claw	Flat plates or filaments on at least some of abdomen segments	Most have 3 tails but a few have 2 tails
2 opposing jaws, protrude conspicuously in front of head	No wings or wing pads (developing wings)	3 pairs; 2 claws	Slender pointed gills on sides of abdomen, some also have tufts of filaments on bottom of abdomen	2 short fleshy projections with 2 claws on each or 1 long tapering tail

	Group; Stage
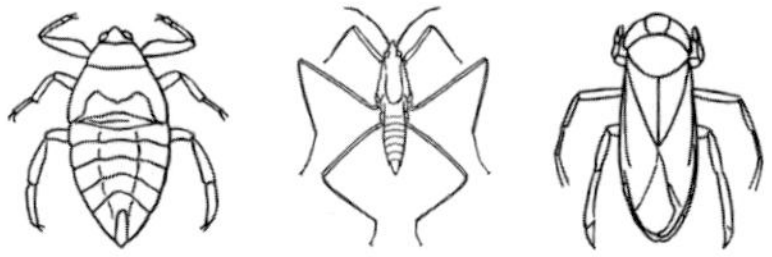	**True Bugs; Adults and Larvae** (Order Hemiptera) Plates 21-27
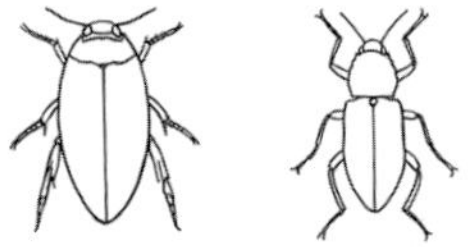	**Water Beetles; Adults** (Order Coleoptera) Plates 28-33
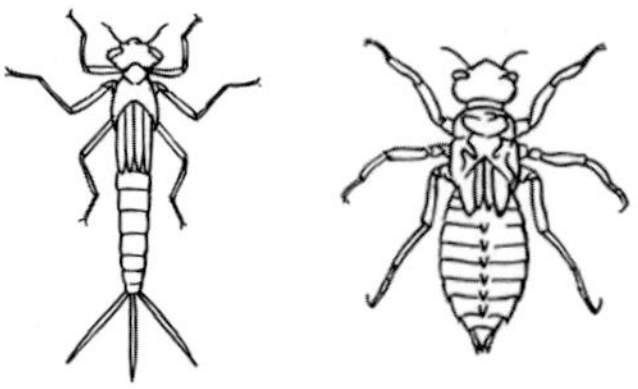	**Dragonflies, Damselflies; Larvae** (Order Odonata) Plates 34-40
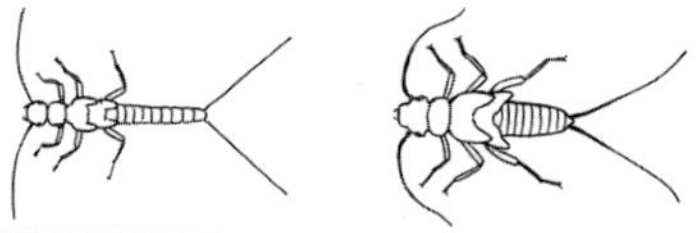	**Stoneflies; Larvae** (Order Plecoptera) Plates 41-49
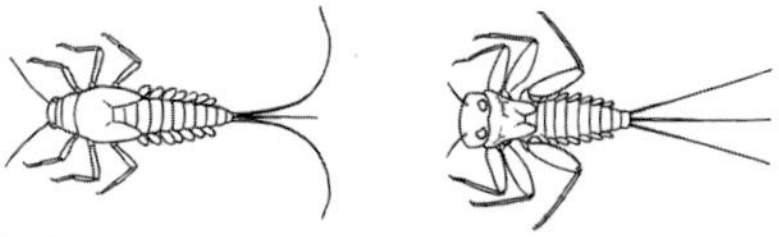	**Mayflies; Larvae** (Order Ephemeroptera) Plates 50-59
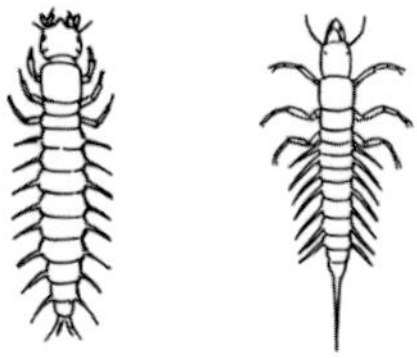	**Dobsonflies, Fishflies, Alderflies; Larvae** (Order Megaloptera) Plates 60-61

chart continues on next page

Chart C: Insects *(continued from previous page)*

Mouthparts	Wings on Thorax	Hard Segmented Legs; Claws	Gills	Structures on End of Abdomen; Other Features
2 opposing jaws	No wings or wing pads (developing wings)	3 pairs; 1 claw	Finger-like on abdomen or none	2 short fleshy projections with 1 claw on each or just 2 claws; most kinds live in portable case or attached retreat (collecting destroys retreats)
2 opposing jaws	No wings or wing pads (developing wings)	3 pairs; 1 or 2 claws	Most with none, some with small tufts of filaments on bottom or end of abdomen	Most with none, some with 2 tails
2 opposing jaws or 2 vertical hooks (like snake fangs)	No wings or wing pads (developing wings)	None	A few finger-like gills on various parts of body, or none	Various lobes, finger-like projections, pointed filaments, or 1 very long breathing tube
None or not visible if present	Wing pads (developing wings) present	3 pairs held very close to body; claws not visible	Finger-like on abdomen, filaments on thorax, paddles on end of abdomen, or none	Some with paddle-like gills or 2 short tails; some common kinds in 2 orders have aquatic pupae, difficult to distinguish, both resemble mummies

	GROUP; STAGE
 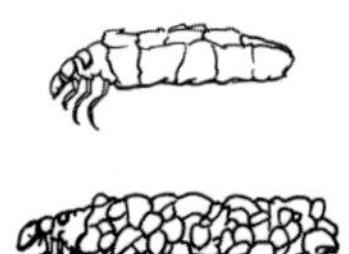	**Caddisflies; Larvae** (Order Trichoptera) Plates 62-77
 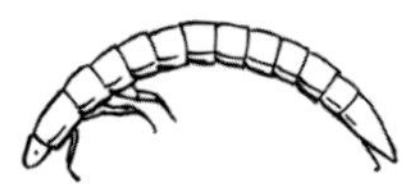	**Water Beetles; Larvae** (Order Coleoptera) Plates 78-83
 	**True Flies; Larvae** (Order Diptera) Plates 84-100
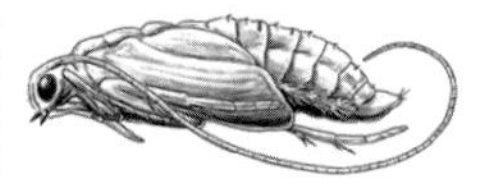 	**Caddisflies, True Flies; Pupae** (Orders Trichoptera, Diptera) Plates 73, 84, 85, 87, 90, 91, 94, 98

Plate 1

Flatworms

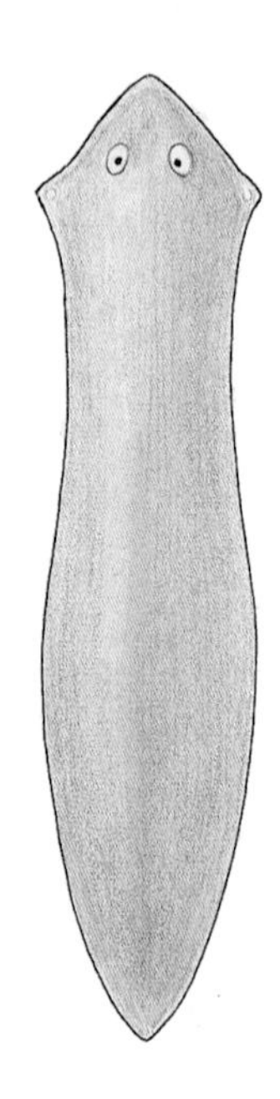

Whole body, top view. (Platyhelminthes: Turbellaria)

Distinguishing Features — Body length usually 5–20 mm, range <1–30 mm. The body is soft, elongate, flattened from top to bottom, and has no individual segments. The sides of the body are slightly constricted near the front, forming a head that is somewhat triangular, resembling an arrowhead. There are usually two eyespots on top of the head, and some common kinds look "cross-eyed." There is only one opening to the digestive tract, which is called the mouth. The mouth is usually on the bottom side, positioned about one-fifth to three-fifths the length of the body. Often the mouth is on an extendible tube (pharynx) that can be pushed out almost one-half the length of the body. Most are dark shades of gray, brown, or black on the top side, often with stripes, spots, or mottling. The bottom side is usually light and without patterns. — Page 193

Leeches

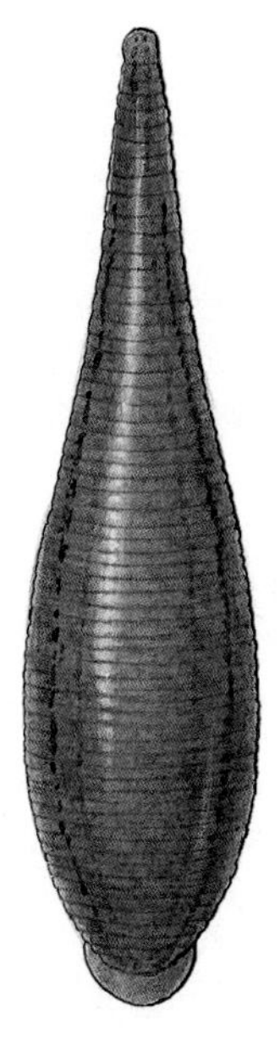

Whole body, top view. (Annelida: Hirudinea)

Distinguishing Features — Body length 4–450 mm, fully extended. Leeches have bodies that are somewhat soft, flattened from top to bottom, very muscular, and composed of 34 actual segments arranged in a row. There appear to be many more segments because a lot of the segments have fine lines running across them (annuli), creating secondary subdivisions that look like actual segments. Almost all kinds have some pattern on the top of their body, and many are brightly colored. Color patterns include red, yellow, orange, and green arranged in stripes, zigzags, spots, or splotches. Background colors are usually various shades of tan, brown, gray, or black. There are no bundles of tiny hairs on the body. A number of small eyespots are present on top of the first few segments at the front of the body. There are two distinct suckers on the bottom of the body, one on the front and one on the rear. The mouth is within the front sucker. The anus is on the top of the body, just in front of the rear sucker. — Page 202

Plate 3

Aquatic Earthworms

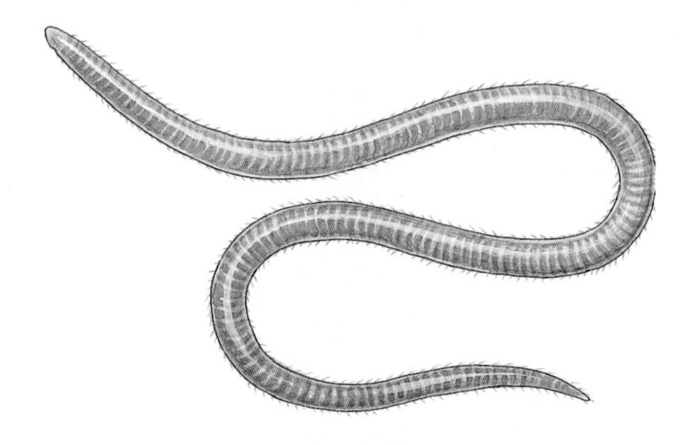

Whole body, top/side view. (Annelida: Oligochaeta)

Distinguishing Features — Body length usually 1–30 mm, up to 150 mm. The body is soft, muscular, elongate, and cylindrical and consists of many similar, round, ring-like segments arranged in a row. There are bundles of tiny hairs (chaetae) on each segment after the first. There are no suckers or eyespots on the body. — Page 198

Plate 4

Bithyniid Snails

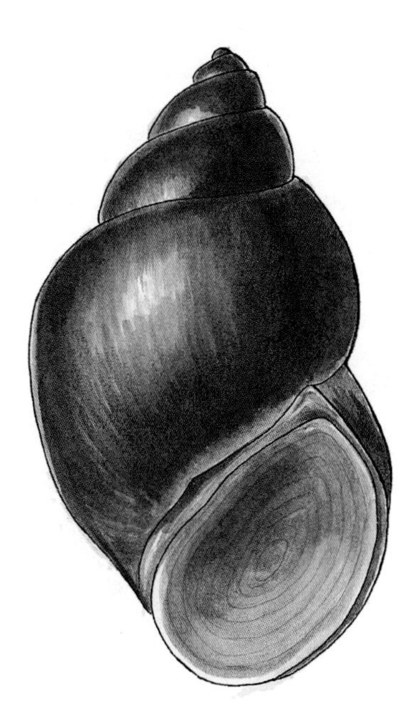

Whole body, bottom view. (Gastropoda: Prosobranchia: Bithyniidae)

Distinguishing Features — Shell 6–13 mm high (adults). There is a flat, lid-like structure, called an operculum, that can seal the body of the snail inside the shell. The operculum is not flexible because it has a high concentration of calcium carbonate (calcareous). The operculum is broadly oval to nearly round in shape. The lines on the operculum are arranged in a series of separate, successively smaller, elongate circles that are inside one another (concentric). The concentric lines are almost centered, like a target with a bull's eye. The shell has its opening on the right when the narrow end is up (dextral). The whorls of the shell bulge out distinctly on the sides (inflated). This makes the sides of the shell a series of broad curves with deep incisions between the coils. — Page 215

Plate 5

Viviparid Snails

Whole body, bottom view. (Gastropoda: Prosobranchia: Viviparidae)

Distinguishing Features — Shell 20–60 mm high (adults). There is an operculum, which is slightly hardened, but still flexible, because it is composed of more protein than calcium carbonate (corneous). The shape of the operculum is elongate and somewhat oval. In some kinds, one edge of the operculum is almost a straight line. The operculum has a series of oval lines that are successively smaller inside one another (concentric). The concentric lines are off center. The shell has its opening on the right when the narrow end is up. The whorls of the shell bulge out distinctly on the sides (inflated). This makes the sides of the shell a series of broad curves with deep incisions between the coils. — Page 216

Plate 6

Hydrobiid Snails

Whole body, bottom view. (Gastropoda: Prosobranchia: Hydrobiidae)

Distinguishing Features — Shell 3–7 mm high (adults). There is an operculum, which is slightly hardened, but still flexible, because it is composed of more protein than calcium carbonate (corneous). The mark on the operculum is a single continuous line that loops several times in a complete spiral before ending at the edge (multispiral). The shell has its opening on the right when the narrow end is up. The whorls of the shell bulge out distinctly on the sides (inflated). This makes the sides of the shell a series of broad curves with deep incisions between the coils. — Page 216

Plate 7

Pleurocerid Snails

Whole body, bottom view. (Gastropoda: Prosobranchia: Pleuroceridae)

Distinguishing Features — Shell 10–40 mm high (adults). The shell is very thick and solid. There is an operculum, but it is often not visible because it is small and the snail pulls it back deep in the shell. The operculum is slightly hardened, but still flexible, because it is composed of more protein than calcium carbonate (corneous). The operculum is marked with a continuous curved line that loops only a couple of times before ending at the edge (paucispiral). However, the paucispiral line may not be as conspicuous as the many other slightly curved lines that radiate outward from the center of the paucispiral line and end at the edge. The shell has its opening on the right when the narrow end is up. The whorls of the shell do not bulge out distinctly on the sides (flattened). This makes the sides of the shell almost straight lines with only shallow incisions between the coils. — Page 216

Plate 8

Ancylid Snails

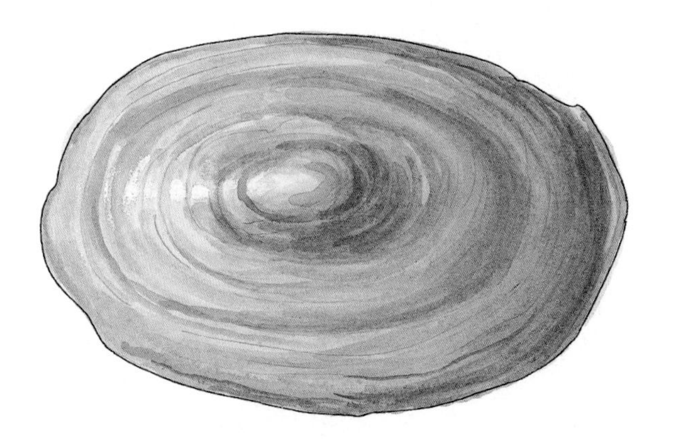

Whole body, top view. (Gastropoda: Pulmonata: Ancylidae)

Distinguishing Features — Shell 3–7 mm long (adults). There is no operculum. Instead of being coiled, the shell is shaped like a low, flat cone. — Page 217

Plate 9

Planorbid Snails

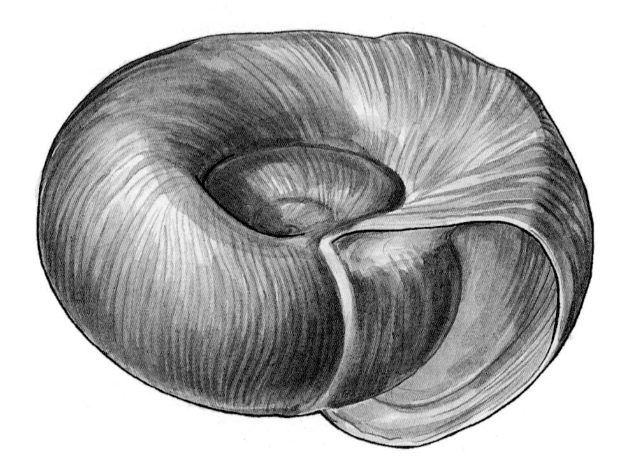

Whole body, bottom/side view. (Gastropoda: Pulmonata: Planorbidae)

Distinguishing Features — Shell 3–30 mm across coil (adults). There is no operculum. The shell is coiled flat instead of being extended into a spiral. When alive, the body of many kinds is a reddish color, which shows through the shell. — Page 218

Plate 10

Lymnaeid Snails

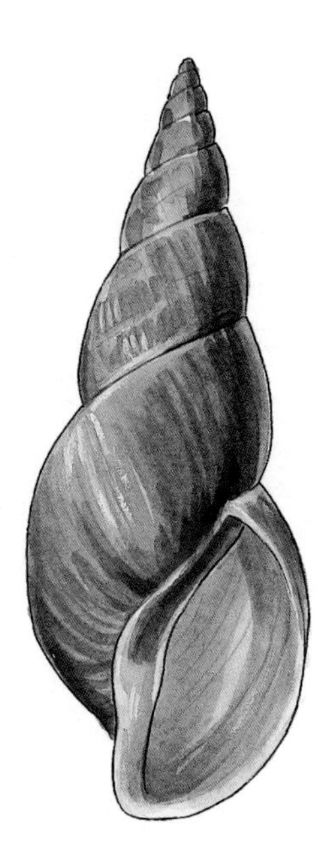

Whole body, bottom view. (Gastropoda: Pulmonata: Lymnaeidae)

Distinguishing Features — Shell 5–50 mm high (adults). There is no operculum. The shell has its opening on the right when the narrow end is up.
— Page 217

Plate 11

Physid Snails

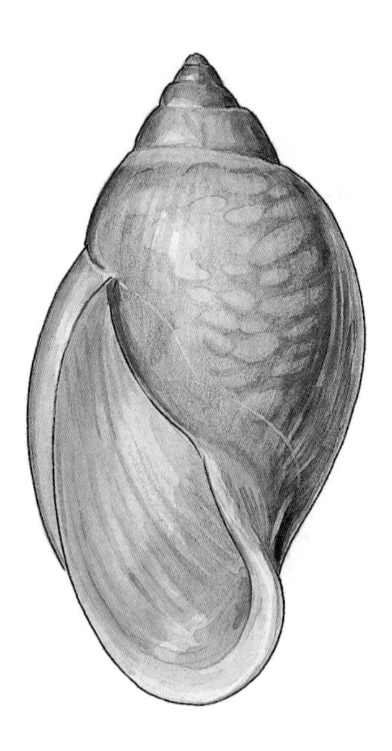

Whole body, bottom view. (Gastropoda: Pulmonata: Physidae)

Distinguishing Features — Shell 5–20 mm high (adults). There is no operculum. The shell has its opening on the left when the narrow end is up (sinistral). — Page 218

Plate 12

Mussels

Whole body, side view. (Bivalvia: Unionacea: Unionidae and Margaritiferidae)

Distinguishing Features — Shell 30–250 mm long (adults). The shell is usually thick and strong. In side view, different kinds have various shapes, including elongate, oval, round, triangular, trapezoidal, and rectangular. The color of the shell can be light yellow-green, dark green, brown, or blackish. Some kinds have bumps, wrinkles, or ridges on their shells, as well as pigmented rays or blotches. The shells of large, old mussels usually have some of the dark outer protective layer worn away on the raised area at the top of the shell (umbo), which exposes the white calcium carbonate that lies underneath. — Page 229

Plate 13

Fingernail or Pea Clams

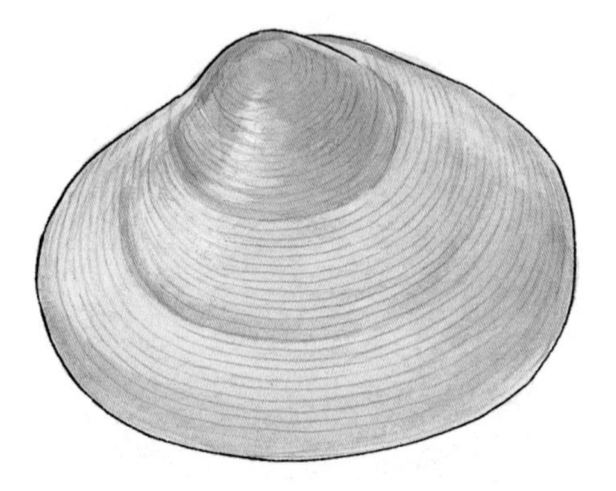

Whole body, side view. (Bivalvia: Sphaeriidae)

Distinguishing Features — Shell usually 2–10 mm long, rarely up to 20 mm (adults). The shell is usually thin and fragile. The color of the shell ranges from whitish or cream to light tan or light gray. The growth rings are very close together and hardly raised, thus, the shell feels smooth when you run your fingernail across the lines. — Page 228

Plate 14

Asian Clam

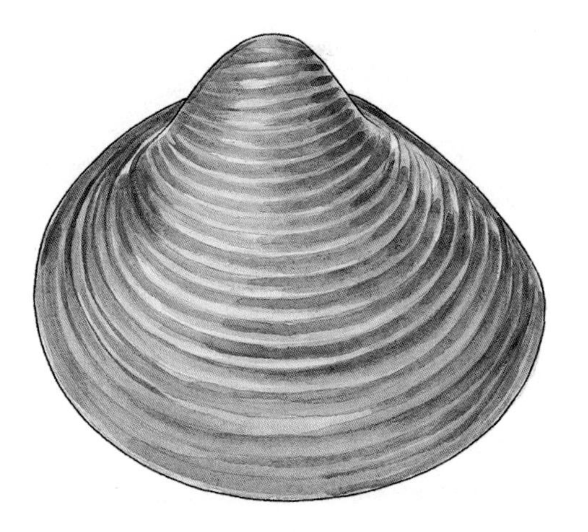

Whole body, side view. (Bivalvia: Corbiculidae: *Corbicula fluminea*)

Distinguishing Features — Shell usually 35–50 mm long, range 10–63 mm (adults). The shell is thick and strong. The color of the shell is most commonly blackish, but may also be light yellow-green, dark green, or brown. The growth rings are spaced far apart and are distinctly raised in successive ridges, thus, the shell feels rough when you run your fingernail across the lines. — Page 227

Plate 15

Zebra Mussel

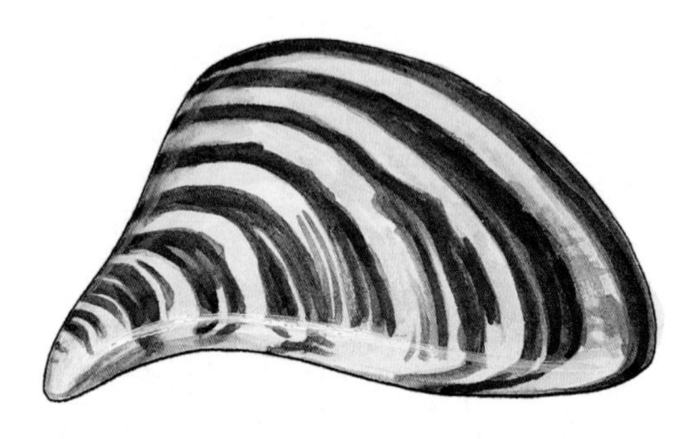

Whole body, side view. (Bivalvia: Dreisseniidae: *Dreissena polymorpha*)

Distinguishing Features — Shell usually 6–25 mm long, up to 51 mm (adults). The shell has a distinctive shape, resembling a letter D lying on its flat side. There are usually alternating light and dark bands of color, hence the common name. The light bands are pale gray or yellowish, and the dark bands are purple or brownish. However, there is considerable color variation within this species and some shells are almost uniformly dark. Live specimens have a bundle of fine threads (byssal threads) hanging out of the flat, bottom side. — Page 230

Plate 16

Water Mites

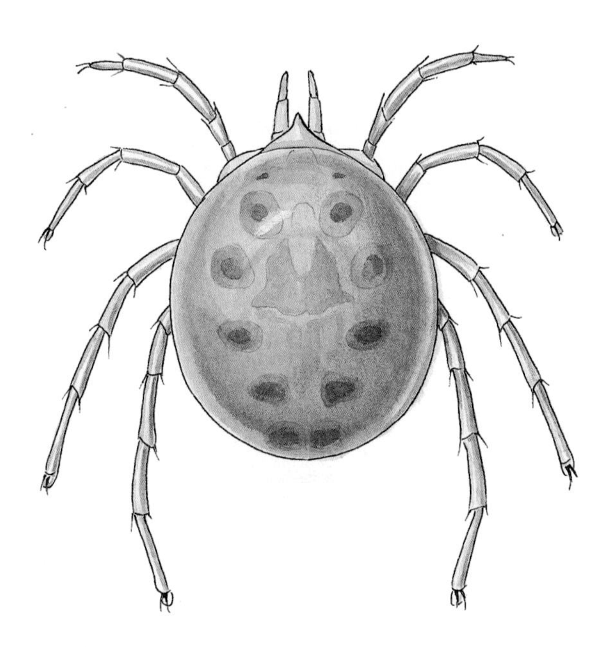

Whole body, top view. (Arachnida: Acariformes: Hydracarina or Hydrachnida)

Distinguishing Features — Body usually 2–3 mm, range <1–7 mm (adults). The body is composed almost entirely of the abdomen, which is usually a sphere without any visible segments. Water mites frequently have brightly colored patterns of green, blue, orange, yellow, or red, but some are only dull shades of brown or black. On the front edge of the body, a pair of finger-like structures (pedipalps) projects forward in the middle. Two pairs of widely separated, simple eyes may be visible on the front margin. There are no antennae. Four pairs of segmented legs project out to the sides. The legs come out from a point of attachment that is underneath the body. Note that water mite larvae only have three pairs of legs, but usually the four-legged adults are collected. — Page 234

Plate 17

Aquatic Sow Bugs

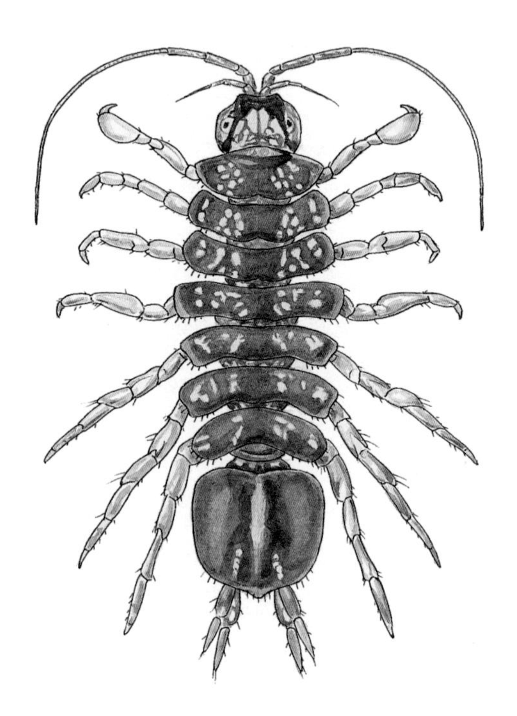

Whole body, top view. (Crustacea: Isopoda: Asellidae)

Distinguishing Features — Body length 5–20 mm (adults, not including antennae and tails). The body is strongly flattened from top to bottom. Medium to dark gray is the most common color, but they can also be blackish or brownish. Some kinds have mottled markings on the body. One pair of antennae is much longer than the other. There are seven pairs of long walking legs, one on each segment of the thorax. The first pair of walking legs has a hinged claw. The remaining six pairs of walking legs have a simple, pointed claw on the end. The abdomen segments are fused into a relatively short, broad, shield-like region to the rear of the thorax. Two flat, forked, tail-like structures stick out behind the abdomen. — Page 244

Scuds, Sideswimmers

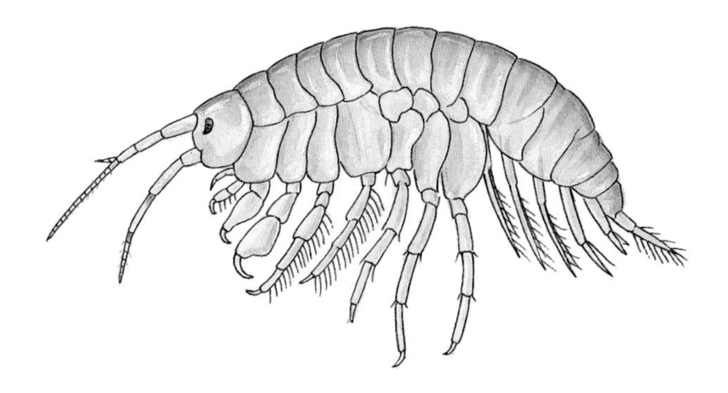

Whole body, side view. (Crustacea: Amphipoda)

Distinguishing Features — Body length 5–20 mm (adults, not including antennae and tails). The body is strongly flattened from side to side. When alive, the color is usually a creamy, translucent, light gray or brown. The two pairs of antennae are about the same length. There are seven pairs of walking legs, one on each segment of the thorax. The first two pairs of walking legs have a hinged claw on the end. The remaining five pairs of walking legs have a simple, pointed claw on the end. The abdomen consists of six individual segments to the rear of the thorax. There are short appendages on the bottom of all abdomen segments. — Page 247

Plate 19

Crayfishes

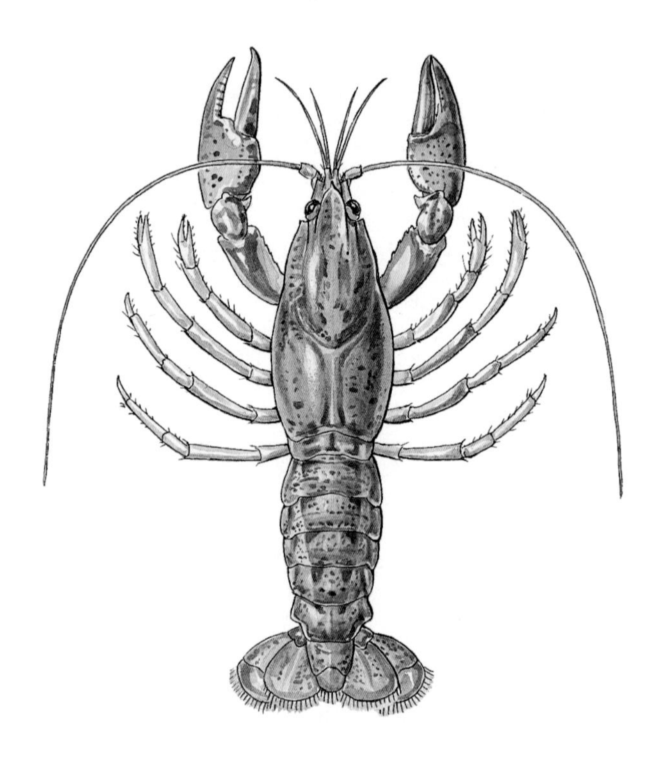

Whole body, top view. (Crustacea: Decapoda: Astacidae and Cambaridae)

Distinguishing Features — Body length 10–150 mm (adults, not including antennae). The front half of body is more or less cylindrical, and the rear half is moderately flattened from top to bottom. Crayfish are usually brownish green, but this ranges from blackish to red or orange. They are often speckled or mottled. The skin of the body and appendages is noticeably thick and brittle. There are five pairs of walking legs, with hinged claws on the ends of the first three pairs of walking legs. The first pair of claws is greatly enlarged and very hard, like those of crabs and lobsters. There is a broad flipper on the end of the body. — Page 254

Shrimps

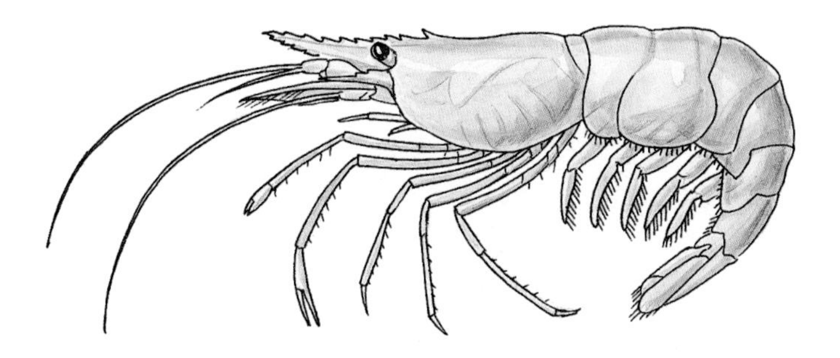

Whole body, side view. (Crustacea: Decapoda: Palaemonidae)

Distinguishing Features — Body length usually 35–165 mm, range 25–240 mm (adults, not including antennae). The entire body is more or less cylindrical, but slightly flattened from side to side. Shrimp are usually light gray or cream colored and translucent. Some kinds are almost transparent, with the internal organs visible. The skin of the body and appendages is moderately hardened and remains somewhat flexible. There are five pairs of walking legs, with hinged claws on the ends of the first two pairs of walking legs. Both pairs of claws are only slightly enlarged, and both pairs are about the same size. There is a broad flipper on the end of the body. — Page 258

Plate 21

Water Boatmen

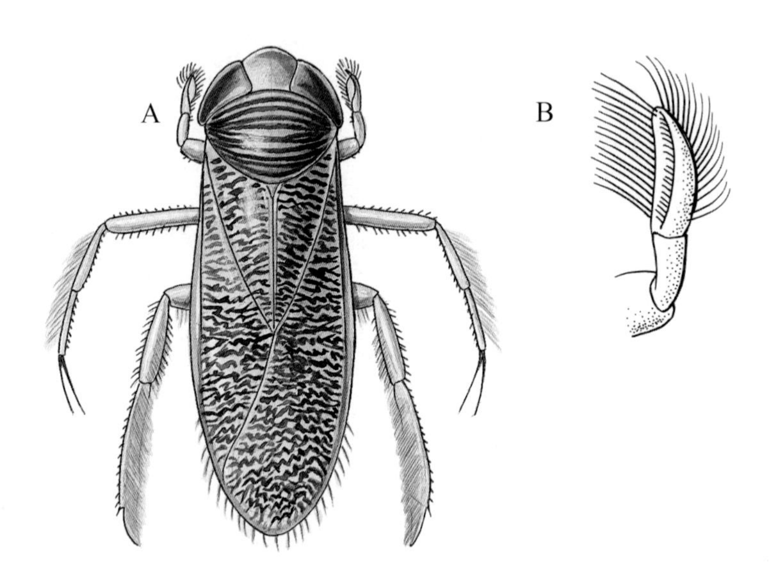

(A) Whole body, top view. **(B)** Front leg, enlarged. (Insecta: Hemiptera: Heteroptera: Corixidae)

Distinguishing Features of Adults and Larvae — Body length 3–11 mm (adults). The body is elongate, oval, and flat on top. They are usually dark gray or brown with fine, wavy, yellowish lines across the top of the body. The eyes are large and often a reddish color in live organisms. The front legs end in a broad, scoop-like segment that has a fringe of long hairs (B). They have a short, stout, cone-shaped beak that is not composed of separate segments but may have fine transverse lines (best seen from a bottom view). The beak is not movable and looks like a continuation of the head. Water boatmen somewhat resemble backswimmers (Notonectidae). — Page 337

Backswimmers

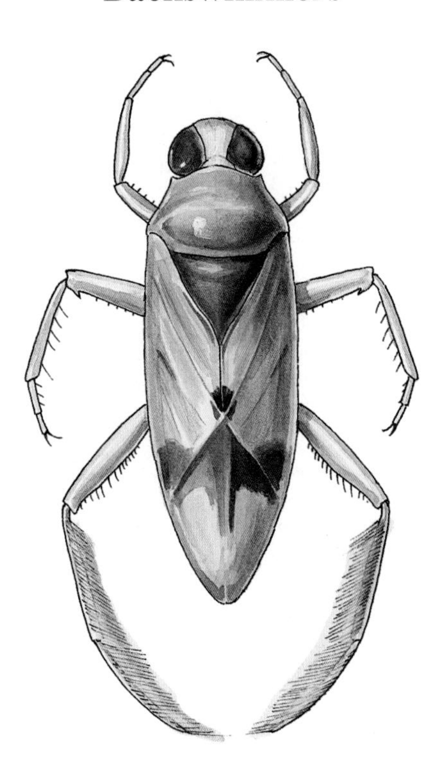

Whole body, top view. (Insecta: Hemiptera: Heteroptera: Notonectidae)

Distinguishing Features of Adults and Larvae — Body length usually 5–15 mm, range 4–17 mm (adults). The body is elongate, thick from top to bottom, somewhat flat on the bottom, and very rounded from side to side on the top. Their shading is reversed, dark on the bottom and light on top, because they swim upside down.. Most kinds have rather colorful patterns of blue-gray with white, orange, red, yellow, or black markings on top. The front and middle legs have no special modifications and are much shorter than the long, oar-like hind legs. Backswimmers somewhat resemble water boatmen (Corixidae). — Page 331

Plate 23

Water Scorpions

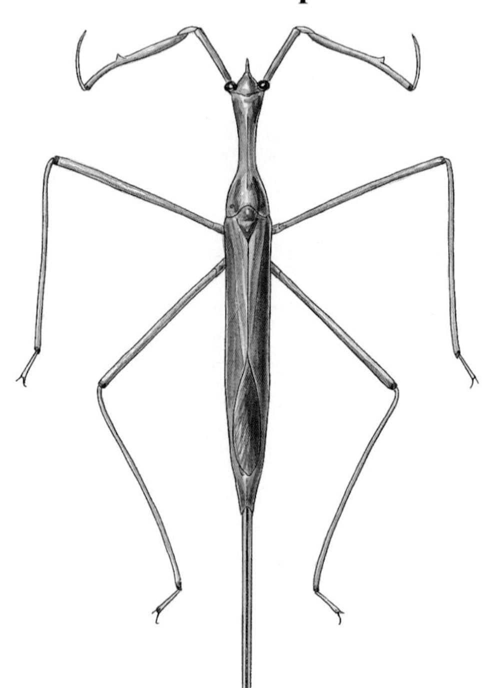

Whole body, top view. (Insecta: Hemiptera: Heteroptera: Nepidae)

Distinguishing Features of Adults and Larvae — Body length usually 15–35 mm, range 14–45 mm (adults, not including breathing tubes). The most common kinds have long, slender, almost cylindrical bodies. The color ranges from light to dark brown or gray, with traces of green. A thin, gray crust often occurs on the body, partially obscuring the color. The front legs are widened and elbowed for grabbing and holding prey. The middle and hind legs are very long and thin and are adapted for climbing on plants and debris. There is a long, cylindrical breathing tube composed of two slender filaments at the end of the abdomen. There are other less common kinds of water scorpions that have wider, somewhat oval, bodies, but they can still be distinguished by the same features of the legs and breathing tube. — Page 339

Plate 24

Giant Water Bugs

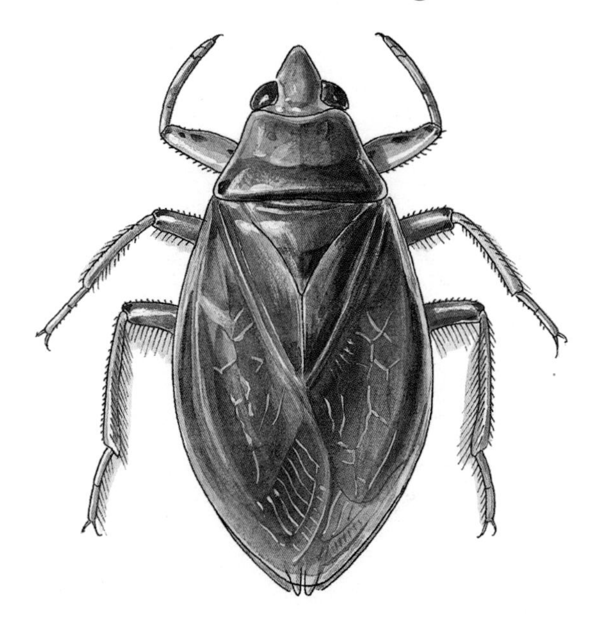

Whole body, top view. (Insecta: Hemiptera: Heteroptera: Belostomatidae)

Distinguishing Features of Adults and Larvae — Body length usually 25–45 mm, range 14–65 mm (adults). The body is broad and oval (especially toward the rear), very much flattened from top to bottom, and brownish in color. The narrow head makes the overall body shape somewhat pointed on the front. The eyes protrude outward on the sides of the head. The front legs are greatly widened and elbowed for grabbing and holding prey. The middle and hind legs are fringed with long hairs to help propel them through the water. The membranous portion of the front wings (toward the tip) has veins. At the end of the abdomen there is a pair of flat, strap-like appendages that are used for breathing air. These air straps can be withdrawn into the body when they are not being used to breathe. Some of the small species could possibly be confused with creeping water bugs (Naucoridae). — Page 335

Plate 25

Creeping Water Bugs

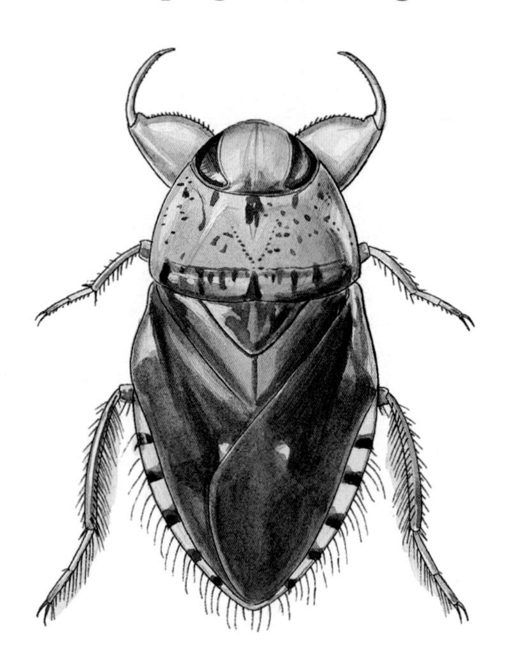

Whole body, top view. (Insecta: Hemiptera: Heteroptera: Naucoridae)

Distinguishing Features of Adults and Larvae — Body length usually about 8 mm, range 5–15 mm (adults). The body is broadly oval in top view and somewhat flattened from top to bottom, but the top is moderately convex. The color ranges from light to dark brown or gray, with traces of green. The broad, rounded head gives the body a smoothly oval shape on the front. The eyes do not protrude outward on the sides of the head. The front legs are greatly widened and elbowed for grabbing and holding prey. The middle and hind legs are fringed with long hairs to help propel them through the water. The membranous portion of the front wings (toward the tip) has no veins. There are no breathing structures on the rear of the body. Creeping water bugs could possibly be confused with some of the small species of giant water bugs (Belostomatidae). — Page 334

Plate 26

Water Striders

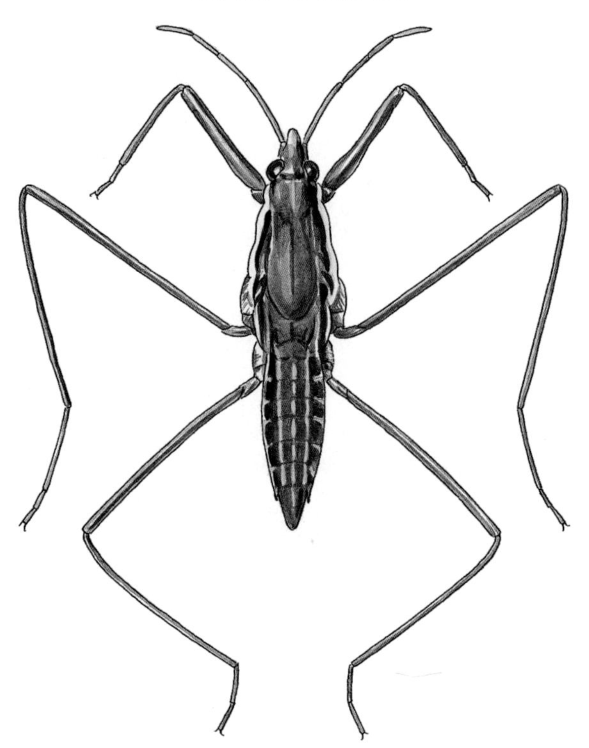

Whole body, top view. (Insecta: Hemiptera: Heteroptera: Gerridae)

Distinguishing Features of Adults and Larvae — Body length usually 3–18 mm, range 2–20 mm (adults). The body of common kinds is elongate and slender. They are usually dark gray to black with brownish areas and silvery markings. Other less common kinds have a short and broad shape or light yellow-brown coloration. Conspicuous antennae stick out in front of the head. The long, thin legs that project away from the body give them a spider-like appearance. On the hind legs, the first noticeable section (femur) is long and reaches far beyond the end of the abdomen when the leg is extended straight back. Some of the small species of water striders might resemble broad-shouldered water striders (Veliidae). — Page 340

Plate 27

Broad-Shouldered Water Striders

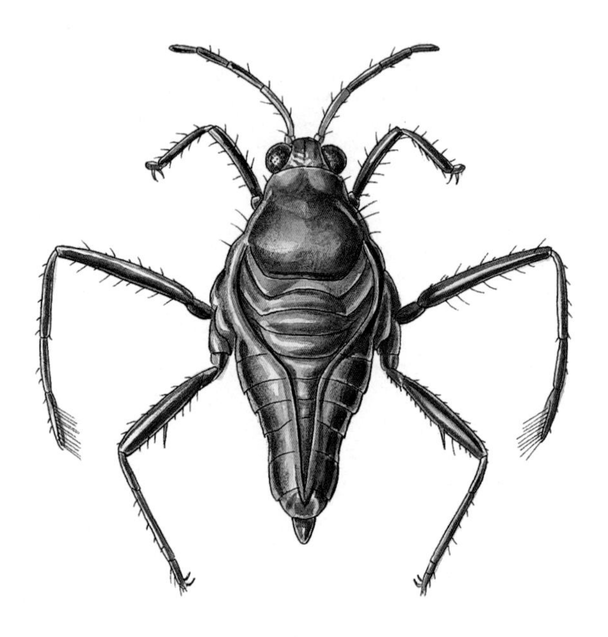

Whole body, top view. (Insecta: Hemiptera: Heteroptera: Veliidae)

Distinguishing Features of Adults and Larvae — Body length usually 2–6 mm, range 1–12 mm (adults). The kinds that are found most often have a wide thorax and a strongly tapered abdomen, but some kinds have stockier bodies that are not as tapered toward the rear. They are usually dark brown to black with silvery markings. Conspicuous antennae stick out in front of the head. The long, thin legs that project away from the body give them a spider-like appearance. On the hind legs, the first noticeable section (femur) is not long enough to reach beyond the end of the abdomen when the leg is extended straight back. Unless the hind legs are examined, some kinds of broad-shouldered water striders could be confused with small water striders (Gerridae). — Page 332

Plate 28

Crawling Water Beetles

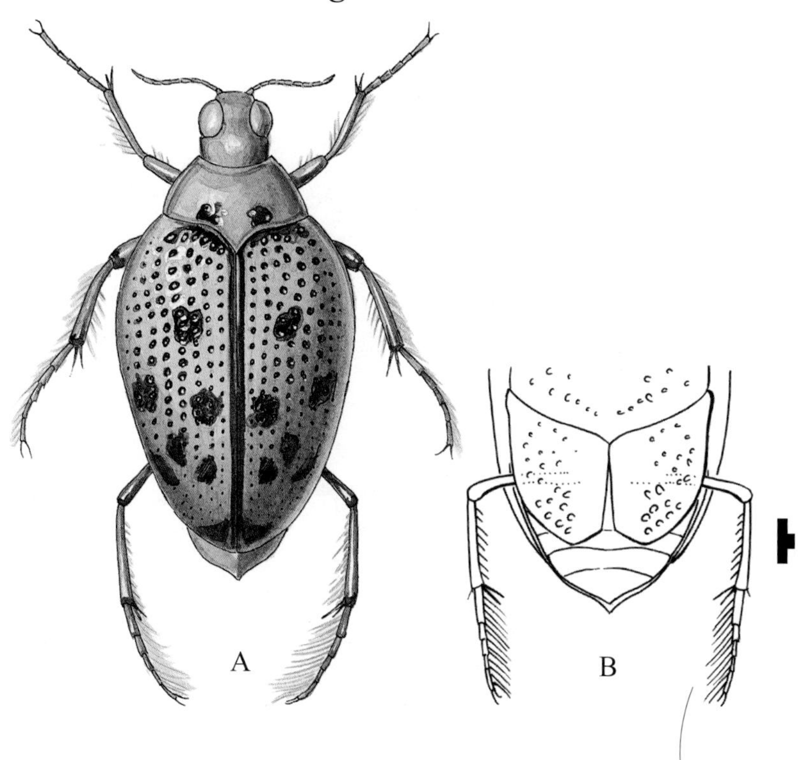

(A) Whole body, top view. **(B)** Rear half of body, bottom view. (Insecta: Coleoptera: Haliplidae)

Distinguishing Features of Adults — Body length 2–6 mm. The body is small, oval, and convex. They have a yellowish or brownish background color with black spots. The antennae are uniformly slender, not clubbed. There are numerous longitudinal rows of very small indentations, such as would be made by the point of a needle, on the hardened front wings. The hind legs have a fringe of long hairs for swimming. On the bottom side, there are broad, flat plates on the rear legs near the body (B). These plates cover about half of the hind legs and the first two or three of the abdomen segments behind the legs. — Page 359

Plate 29

Whirligig Beetles

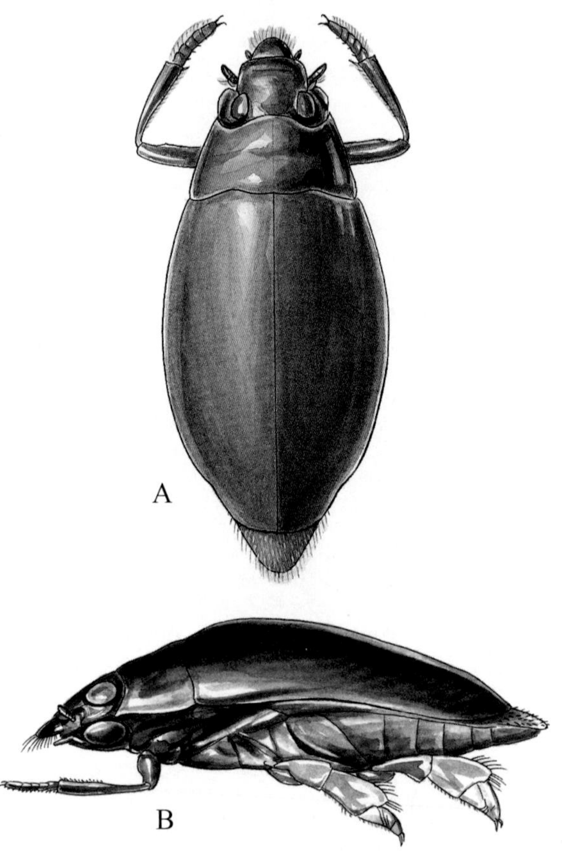

(A) Whole body, top view. **(B)** Whole body, side view. (Insecta: Coleoptera: Gyrinidae)

Distinguishing Features of Adults — Body length 3–16 mm. The body is oval, flattened, and black. The eyes are divided into separate top and bottom sections, so that they appear to have two pairs of eyes (B). The antennae are short, and the segments are larger on the end away from the head (clubbed). The front legs are a normal size and shape, but the middle and hind legs are very short and flat, like paddles (B). — Page 368

Plate 30

Riffle Beetles

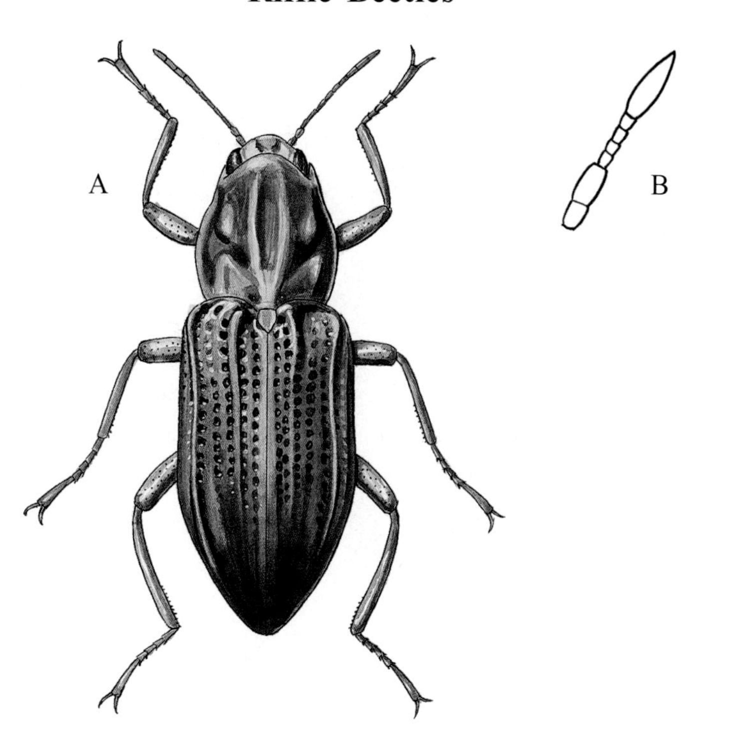

(A) Whole body, top view. **(B)** Antenna found on other kinds, enlarged. (Insecta: Coleoptera: Elmidae)

Distinguishing Features of Adults — Body length 1–8 mm. Most kinds are small, hard-bodied, and somewhat cylindrical in shape. The body is usually dark brown or red-brown, with color patterns or various metallic tints. There are numerous longitudinal rows of very small indentations, such as would be made by the point of a needle, on the hardened front wings. They have long legs, in relation to their small bodies, with two prominent claws on the ends of the legs. The head may be partially withdrawn in the thorax. The antennae are usually uniformly slender (A), but in some kinds the segments get larger toward the end away from the head, like a club (B). — Page 364

Plate 31

Long-Toed Water Beetles

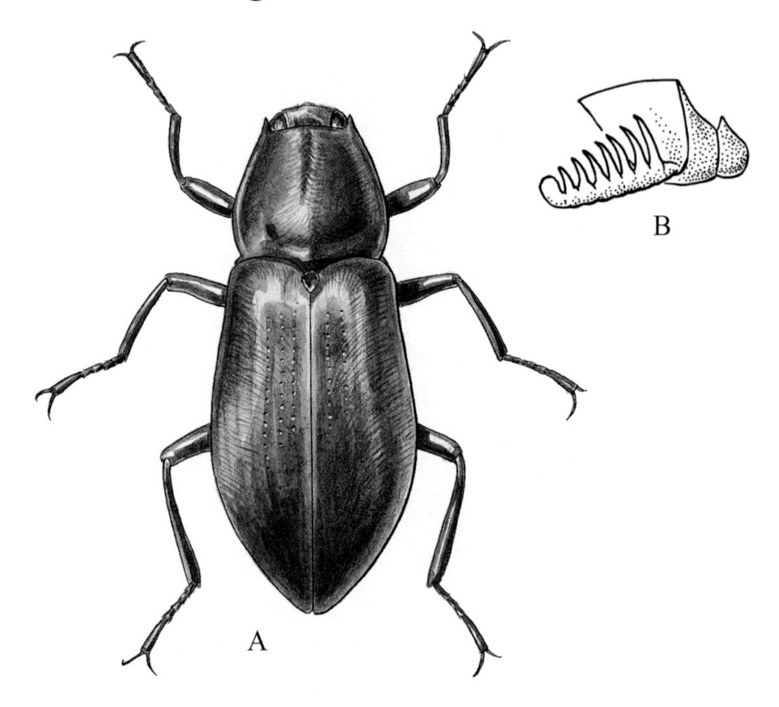

(A) Whole body, top view. **(B)** Right antenna, bottom view, enlarged. (Insecta: Coleoptera: Dryopidae)

Distinguishing Features of Adults — Body length usually 4–8 mm, range 1–10 mm. For the common kinds, the body is dull gray or brown and often covered with very fine hairs. The head is mostly withdrawn in the thorax. The bottom of thorax segment one is expanded forward beneath the head. The antennae are thick and about as long as the head. The antennae are distinctive because the first and second segments (nearest the body) are enlarged and hardened, forming a shield that covers the remaining segments, which look like the teeth on a comb (B). The hind legs do not have fringes of long hairs for swimming. There are also other kinds with different features within this family. — Page 361

Plate 32

Predaceous Diving Beetles

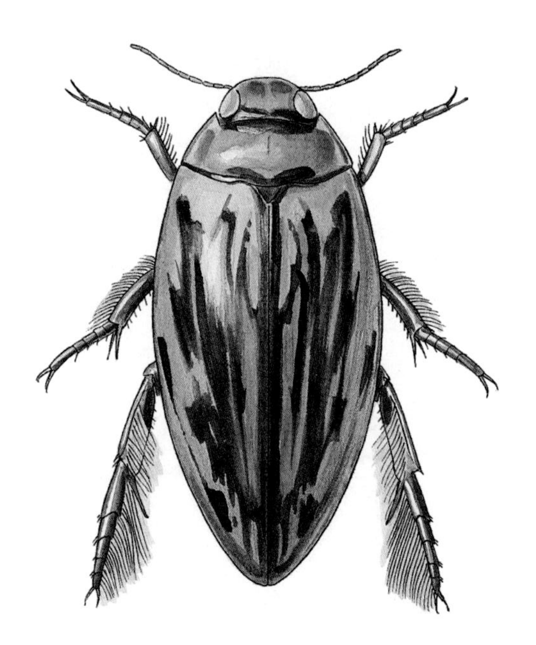

Whole body, top view. (Insecta: Coleoptera: Dytiscidae)

Distinguishing Features of Adults — Body length usually 3–25 mm, range 1–40 mm. In side view, the body curves outward (convex) on the top and bottom. There are conspicuous long, slender antennae and very short finger-like structures projecting forward from the mouthparts. These finger-like structures are usually not visible in top view. The hind legs, and usually the middle legs, are flattened and have a dense fringe of long hairs to assist with swimming. Predaceous diving beetle adults and water scavenger beetle (Hydrophilidae) adults are similar in appearance and can be confused. — Page 362

Plate 33

Water Scavenger Beetles

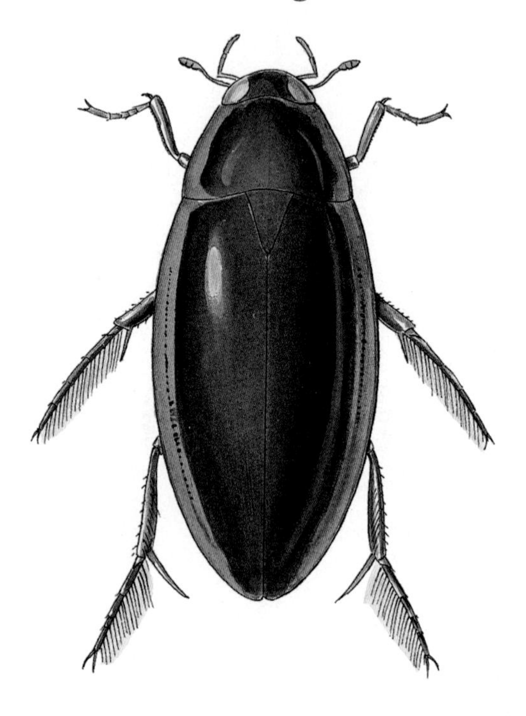

Whole body, top view. (Insecta: Coleoptera: Hydrophilidae)

Distinguishing Features of Adults — Body length 1–40 mm. In side view, the body curves outward (convex) on the top but is flat or curves in (concave) on the bottom. These adults have short, clubbed antennae on the sides of the head and very long, conspicuous, finger-like structures projecting in front of the head. The club is held in a cup-like segment of the antenna that precedes the club (toward the head). The finger-like structures are about as long, or longer, than the antennae and are easy to see in top view. The hind legs are usually flattened with a fringe of hairs for swimming. Water scavenger beetle adults and predaceous diving beetle adults (Dytiscidae) are similar in appearance and can be confused. — Page 367

Plate 34

Broadwinged Damselflies

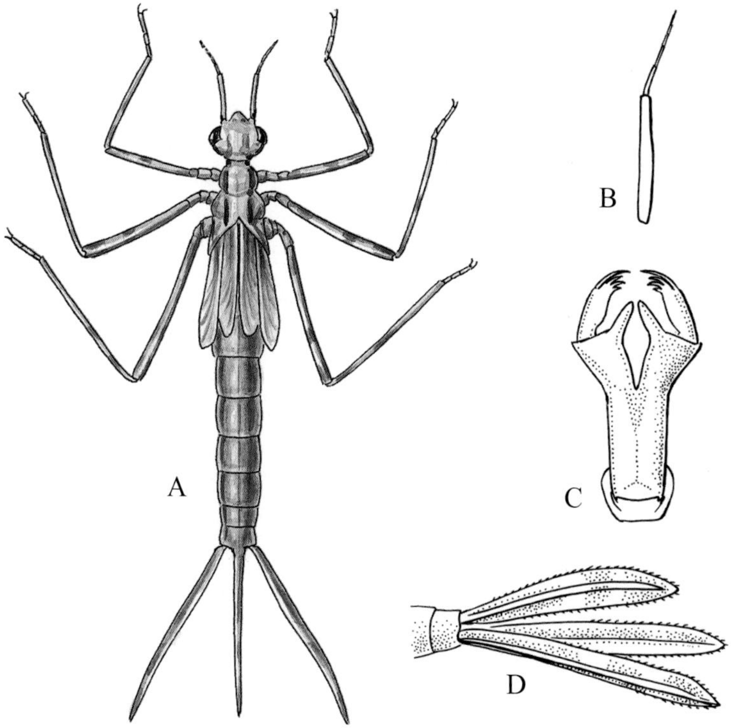

A B C D

(A) Whole body, top view. **(B)** Antenna, enlarged. **(C)** Lower lip, bottom view, enlarged. **(D)** Gills on end of abdomen, side view, enlarged. (Insecta: Odonata: Zygoptera: Calopterygidae)

Distinguishing Features of Larvae — Body length 25–50 mm (mature larvae, not including gills). The antennae stick out conspicuously, resembling antlers. The first segment of the antennae (closest to head) is very long, as long as the combined length of the other segments (B). The lower lip has a deep notch on the front edge in the middle, which is best seen in bottom view (C). The two side gills are longer than the middle gill (D). Veins are not visible within the gills. — Page 303

Plate 35

Spreadwinged Damselflies

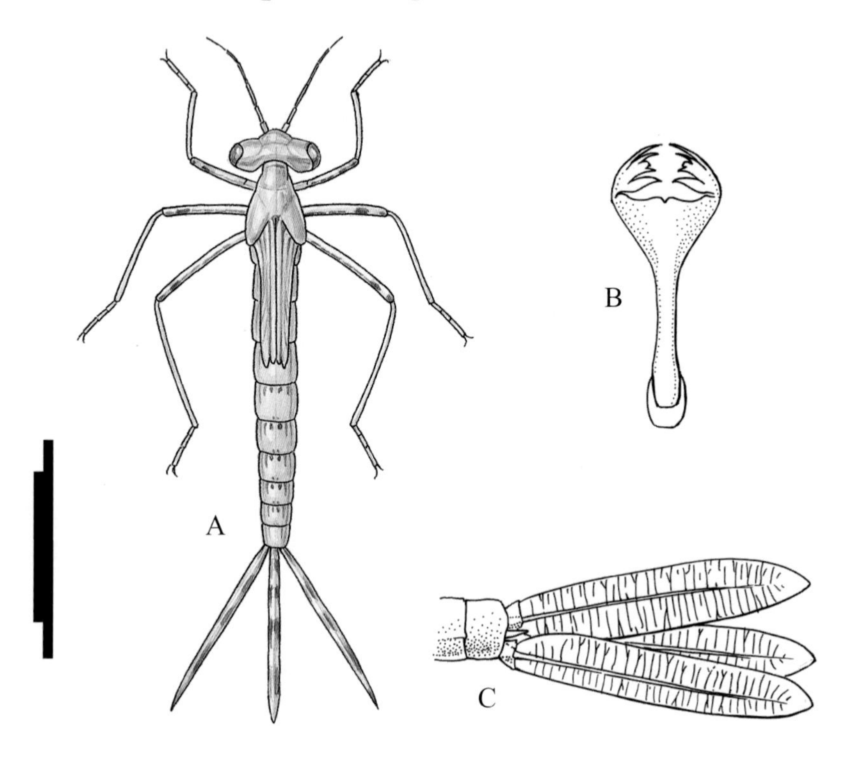

(A) Whole body, top view. **(B)** Lower lip, bottom view, enlarged. **(C)** Gills on end of abdomen, side view, enlarged. (Insecta: Odonata: Zygoptera: Lestidae)

Distinguishing Features of Larvae — Body length 20–29 mm (mature larvae, not including gills). All segments of the antennae are about the same length. The lower lip does not have a notch on the front edge in the middle (best seen in bottom view). The lower lip is elongated and expanded on the end away from the head, resembling a spoon (B). All of the gills are about the same length. The small veins within the gills are parallel and join the main longitudinal stem at a right angle (C). — Page 305

Plate 36

Narrowwinged Damselflies

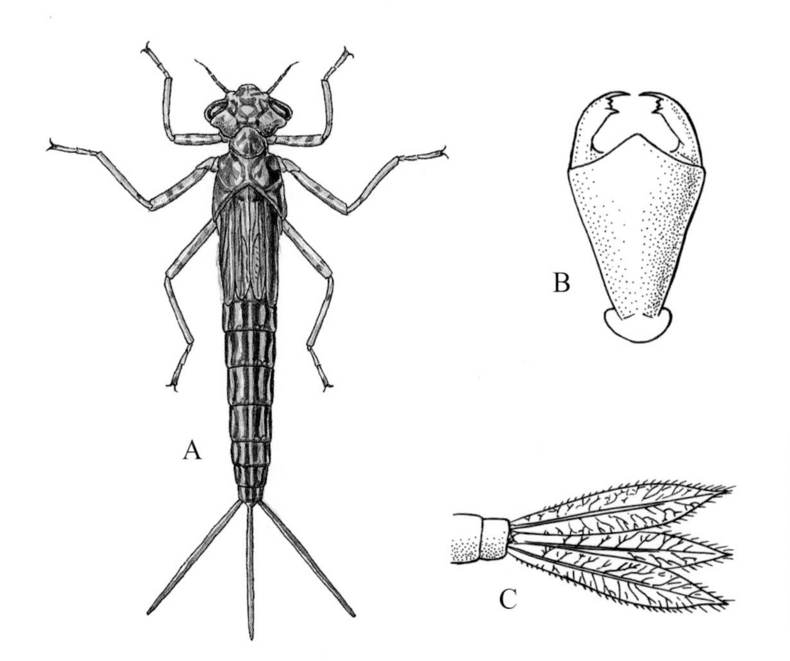

(A) Whole body, top view. **(B)** Lower lip, bottom view, enlarged. **(C)** Gills on end of abdomen, side view, enlarged. (Insecta: Odonata: Zygoptera: Coenagrionidae)

Distinguishing Features of Larvae — Body length 13–25 mm (mature larvae, not including gills). All segments of the antennae are about the same length. The lower lip does not have a notch on the front edge in the middle (best seen in bottom view). The lower lip is thick and somewhat triangular (B). All of the gills are about the same length. The small veins within the gills are highly branched and join the main longitudinal stem at a slant (C). — Page 304

Plate 37

Clubtail Dragonflies

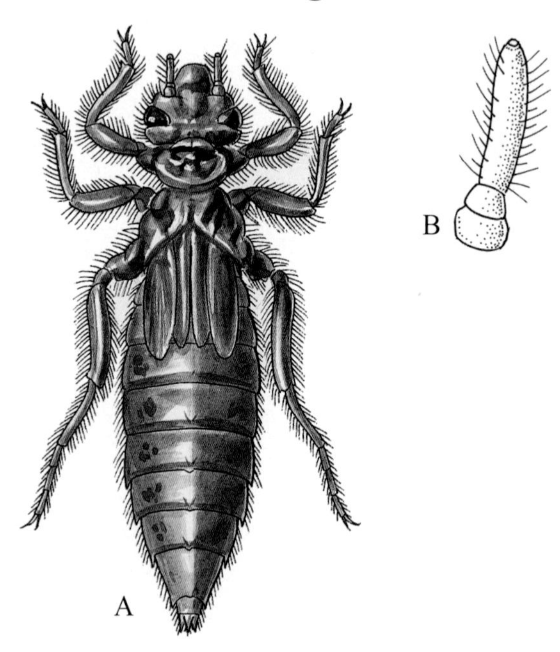

(A) Whole body, top view. **(B)** Antenna, enlarged. (Insecta: Odonata: Anisoptera: Gomphidae)

Distinguishing Features of Larvae — Body length usually 23–42 mm, range 20–68 mm (mature larvae). The body shape is variable, but common kinds are often somewhat flattened, especially their abdomen. The rear end of the abdomen is often rounded, or tapers to a blunt point. The lower lip is flat and not shaped like a spoon or scoop, which is best seen in side view (Plate 38B). The antennae have four segments, with the third segment (away from head) conspicuously larger and possibly a different shape, while the fourth segment is very small and barely visible (B). The larger third segment can be cylindrical or nearly oval. The very small fourth segment looks like a button or cap on the end of the third segment, if it is visible at all. — Page 300

Plate 38

Darner Dragonflies

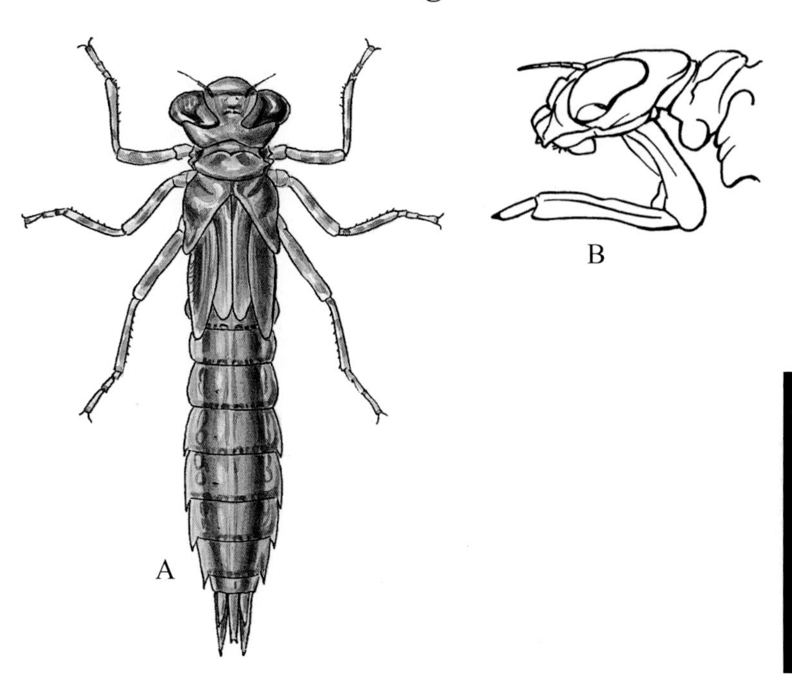

(A) Whole body, top view. **(B)** Head with lower lip pulled down, side view, enlarged. (Insecta: Odonata: Anisoptera: Aeshnidae)

Distinguishing Features of Larvae — Body length usually 30–40 mm, range up to 62 mm (mature larvae). The body is rather elongate and mostly cylindrical. The end of the abdomen tapers to a sharp point. The lower lip is flat and not shaped like a spoon or scoop, which is best seen in side view (B). The antennae have six or seven segments, with all segments slender and similar in size and shape. — Page 301

Plate 39

Skimmer Dragonflies

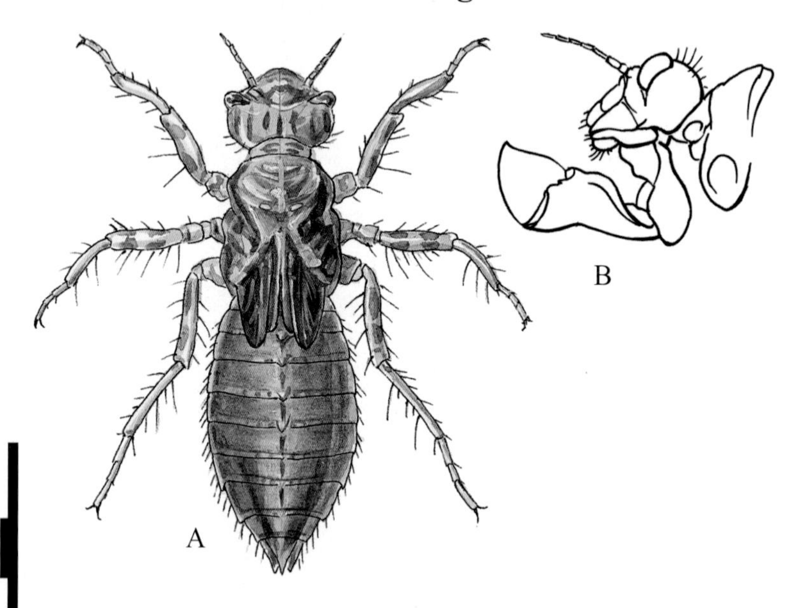

(A) Whole body, top view. **(B)** Head with lower lip pulled down, side view, enlarged. (Insecta: Odonata: Anisoptera: Libellulidae)

Distinguishing Features of Larvae — Body length 8–29 mm (mature larvae). The lower lip is curved on the sides and shaped like a spoon or scoop (B), which is best seen in side view. The top of the abdomen may be bare, have stubby spines that are mostly straight and project back (B), or have curved hooks that stick up from the middle of most segments (Plate 40B). The spines and hooks are best seen in side view. It may be necessary to push the wing pads out of the way to see the first four or five spines or hooks. On the sides of the next to last abdomen segment (number nine), there are either short points (A) or long, flat spines (Plate 40A), which are best seen in top view. Note that abdomen segment ten is sometimes small and hard to see. It is often mostly withdrawn into the middle of abdomen segment nine. The three stout pointed structures at the middle of the end of the abdomen are attached to segment ten. — Page 302

Skimmer Dragonflies

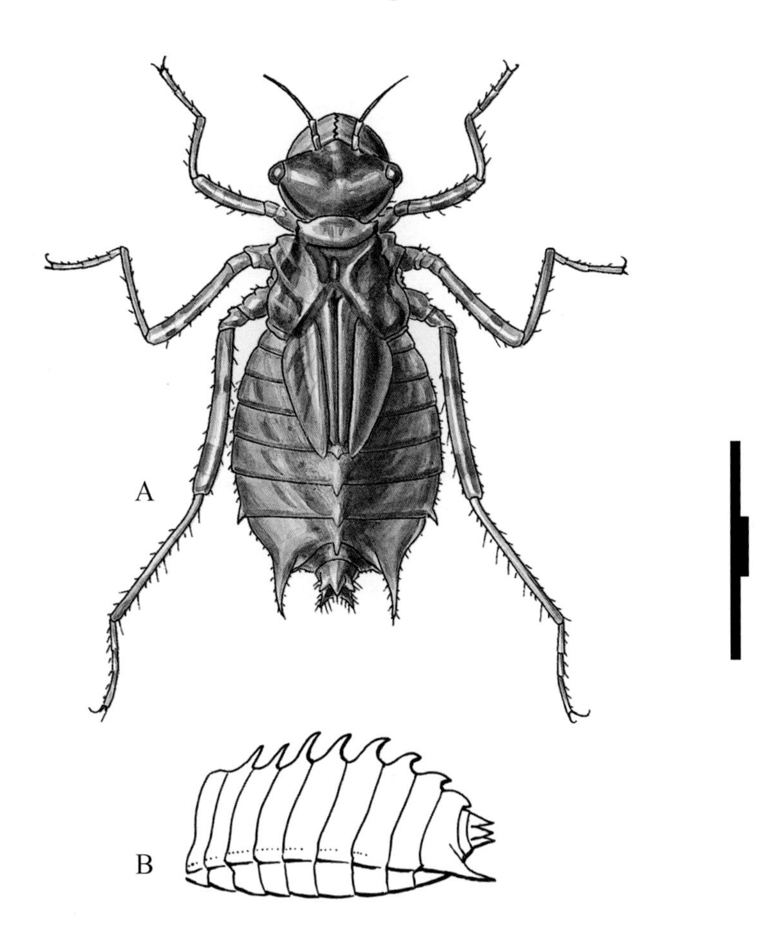

(A) Different kind, whole body, top view. **(B)** Abdomen, side view. (Insecta: Odonata: Anisoptera: Libellulidae)

Distinguishing Features of Larvae — Same as described for Plate 39. — Page 302

Plate 41

Roachlike Stoneflies

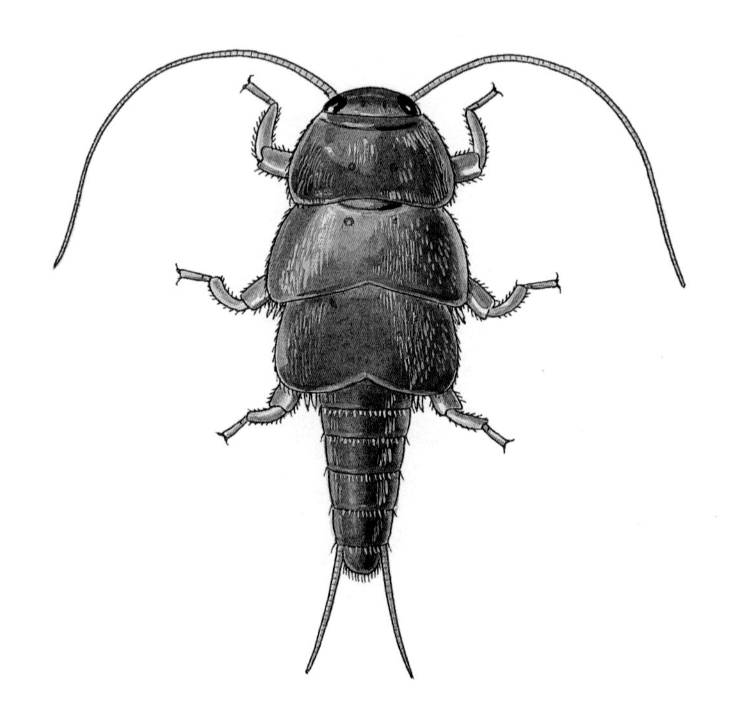

Whole body, top view. (Insecta: Plecoptera: Peltoperlidae)

Distinguishing Features of Larvae — Body length usually 6–11 mm, range up to 21 mm (mature larvae, not including antennae or tails). The body is flattened and brown in color. Larvae have a roach-like appearance because there are large, shield-like plates on the top of the thorax that cover the bases of the head, legs, and abdomen. There are a few tapering finger-like gills on the bottom of the thorax at the sides, where the legs attach to the body. There are no dense tufts of branching filamentous gills on the thorax or abdomen. — Page 321

Plate 42

Giant Stoneflies

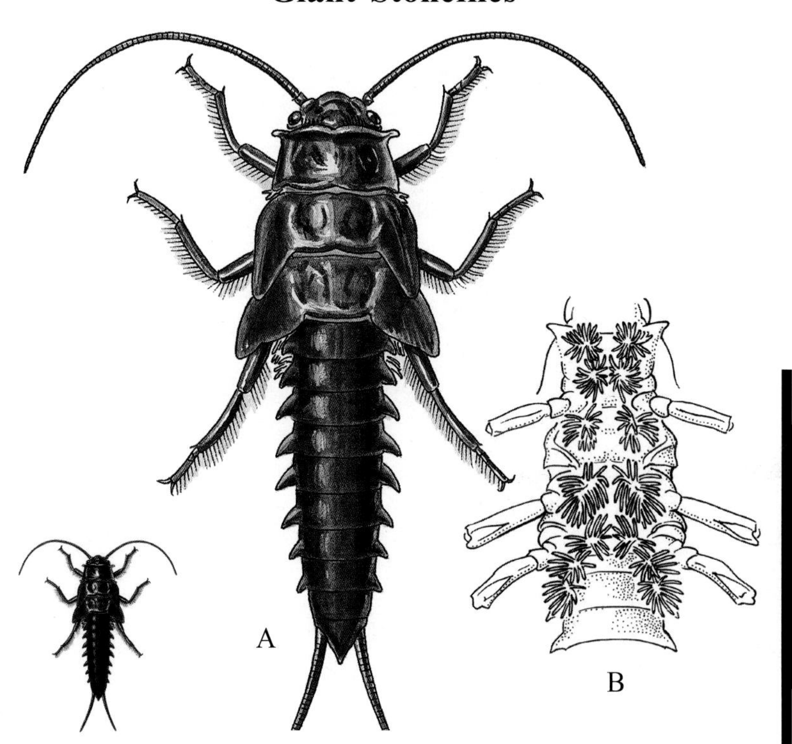

(A) Whole body, top view. **(B)** Thorax and first three abdomen segments, bottom view, enlarged. (Insecta: Plecoptera: Pteronarcyidae)

Distinguishing Features of Larvae — Body length of widespread kinds 35–50 mm (mature larvae, not including antennae or tails); some western kinds smaller, 6–22 mm. The brown or black-brown body is stocky and somewhat cylindrical. The plate on the top of thorax segment one usually projects outward at the corners. There are dense tufts of branching filamentous gills on the bottom of the thorax at the sides, where the legs attach to the body (B). In addition, there are branching filamentous gills on the bottom at the sides of the first two or three abdomen segments. — Page 316

Plate 43

Common Stoneflies

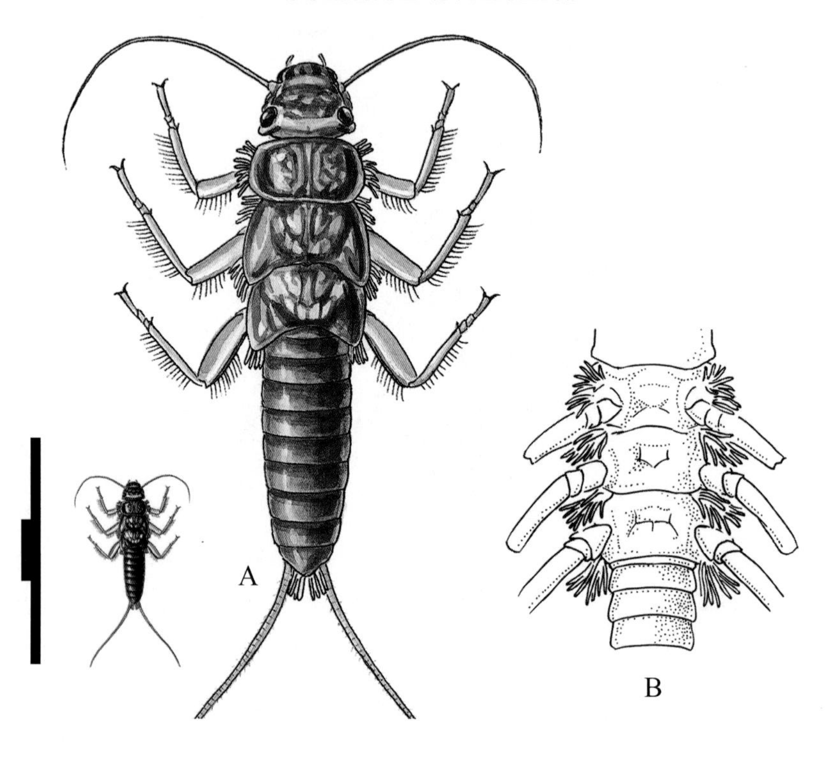

(A) Whole body, top view. **(B)** Thorax and first three abdomen segments, bottom view, enlarged. (Insecta: Plecoptera: Perlidae)

Distinguishing Features of Larvae — Body length 8–30 mm (mature larvae, not including antennae or tails). The body is somewhat flattened and is usually brownish with yellowish markings. The middle portion of the end of the lower lip, between the two finger-like projections, has one deep notch that splits the end, which is best seen in bottom view (Plate 44B). There are dense tufts of branching filamentous gills on the bottom of the thorax at the sides, where the legs attach to the body (B). There are no branching filamentous gills on the abdomen. Some common stoneflies look like perlodid stoneflies (Perlodidae), in top view. — Page 315

Perlodid Stoneflies

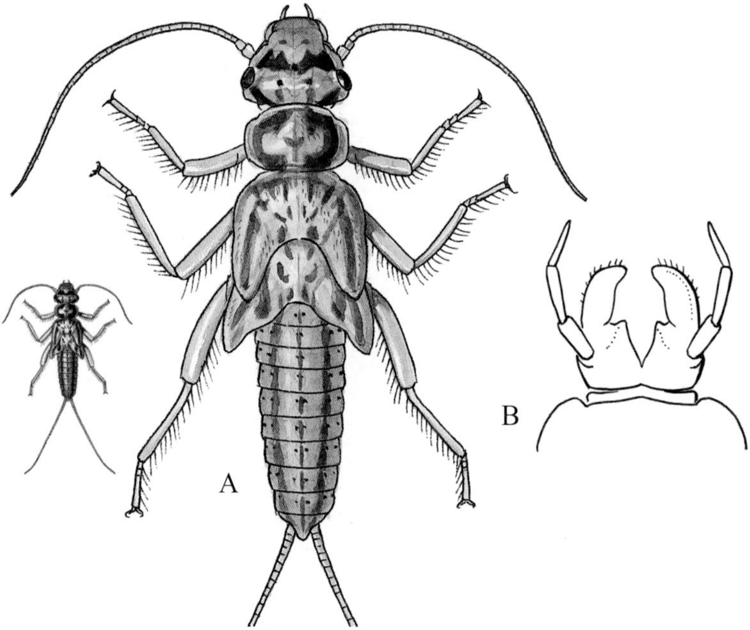

(A) Whole body, top view. **(B)** Lower lip, bottom view, enlarged. (Insecta: Plecoptera: Perlodidae)

Distinguishing Features of Larvae — Body length 6–26 mm (mature larvae, not including antennae or tails). The body is somewhat flattened, and often there are distinct contrasting light and dark patterns on the head and thorax. The middle portion of the end of the lower lip, between the two finger-like projections, has one deep notch that splits the end, which is best seen in bottom view (B). There are no dense tufts of branching filamentous gills on the thorax or abdomen. The hind wing pads diverge at an angle from an imaginary straight line running from the front to the back of the body. The front wing pads do not diverge. The tails are as long, or longer, than the length of the abdomen. Some kinds of perlodid stoneflies look like common stoneflies (Perlidae), in top view. — Page 319

Plate 45

Green Stoneflies

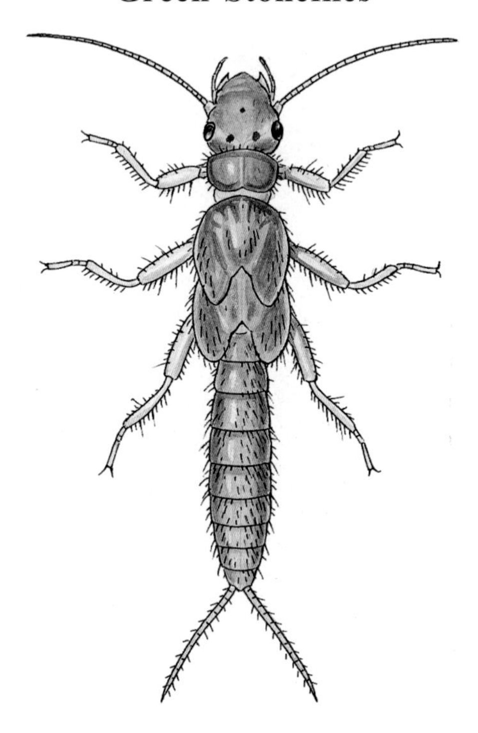

Whole body, top view. (Insecta: Plecoptera: Chloroperlidae)

Distinguishing Features of Larvae — Body length usually 6–18 mm, range up to 25 mm (mature larvae, not including antennae or tails). The body is mostly cylindrical, and usually there are no distinctive color patterns on the head or thorax. The middle portion of the end of the lower lip, between the two finger-like projections, has one deep notch that splits the end, which is best seen in bottom view (Plate 44B). There are no dense tufts of branching filamentous gills on the thorax or abdomen. Both pairs of wing pads do not diverge at an angle from an imaginary straight line running from the front to the back of the body. The tails are three-fourths as long, or less, than the length of the abdomen. — Page 317

Plate 46

Nemourid Stoneflies

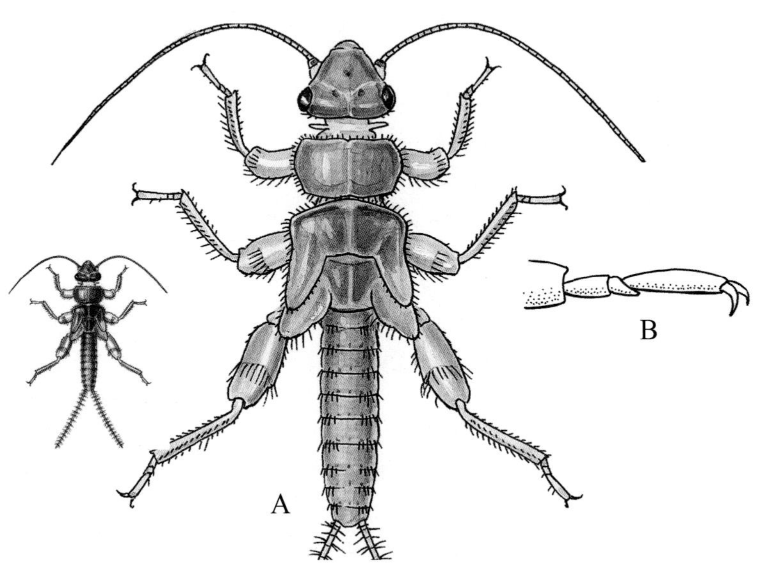

(A) Whole body, top view. **(B)** End of leg, side view, enlarged. (Insecta: Plecoptera: Nemouridae)

Distinguishing Features of Larvae — Body length 4–9 mm (mature larvae, not including antennae or tails). The body is short and stocky, especially the thorax, and often noticeably hairy. The middle portion of the end of the lower lip, between the two finger-like projections, is mostly blunt on the end with three small notches, which is best seen in bottom view (Plate 47B). Gills may be present in the neck region between the head and thorax. Both pairs of wing pads, especially the hind wing pads, diverge at an angle from an imaginary straight line running from the front to the back of the body. If the hind legs are pulled straight back, the ends of the legs reach about to the tip of the abdomen or a bit beyond. Close examination of the three small segments on the end of the legs (tarsus) reveals that the first segment (proceeding away from the body) is much longer than the second segment (B). The claws on the end of the legs do not count as a segment. — Page 318

Plate 47

Winter Stoneflies

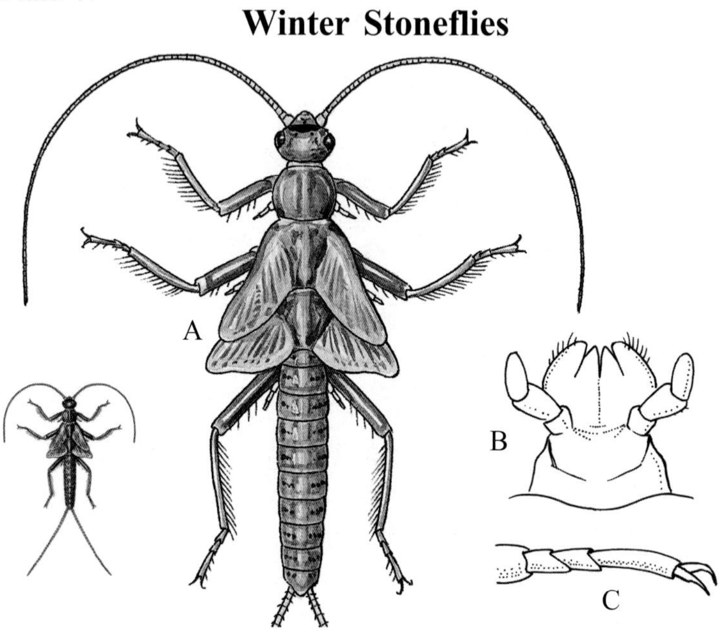

(A) Whole body, top view. **(B)** Lower lip, bottom view, enlarged. **(C)** End of leg, side view, enlarged. (Insecta: Plecoptera: Taeniopterygidae)

Distinguishing Features of Larvae — Body length 7–15 mm (mature larvae, not including antennae or tails). The body is short and stocky, especially the thorax. The middle portion of the end of the lower lip, between the two finger-like projections, is mostly blunt on the end with three small notches (best seen in bottom view). There are no gills in the neck region between the head and thorax. Both pairs of wing pads, especially the hind wing pads, diverge at an angle from an imaginary straight line running from the front to the back of the body. If the hind legs are pulled straight back, the ends of the legs reach about to the tip of the abdomen or a bit beyond. Close examination of the three small segments on the end of the legs reveals that the first and second segments (proceeding away from the body) are about equal in length (C). The claws on the end of the legs do not count as a segment. The most common kinds in the East have a single, finger-like, segmented gill on the first leg segment closest to the body (coxa) on all legs (A). — Page 323

Plate 48

Small Winter Stoneflies

Whole body, top view. (Insecta: Plecoptera: Capniidae)

Distinguishing Features of Larvae — Body length usually 5–10 mm, range up to 20 mm (mature larvae, not including antennae or tails). The body is elongate, especially the abdomen, and often dark in color. The middle portion of the end of the lower lip, between the two finger-like projections, is mostly blunt on the end with three small notches, which is best seen in bottom view (Plate 47B). There are no gills in the neck region between the head and thorax. Both pairs of wing pads do not diverge from an imaginary straight line running from the front to the back of the body. The hind wing pads are as wide, or wider, than they are long. If the hind legs are pulled straight back, the ends of the legs do not come close to the tip of the abdomen. In top view, the sides of the abdomen are zigzagged, and the abdomen is wider at mid-length than it is farther forward where it joins the thorax. — Page 322

Plate 49

Rolledwinged Stoneflies

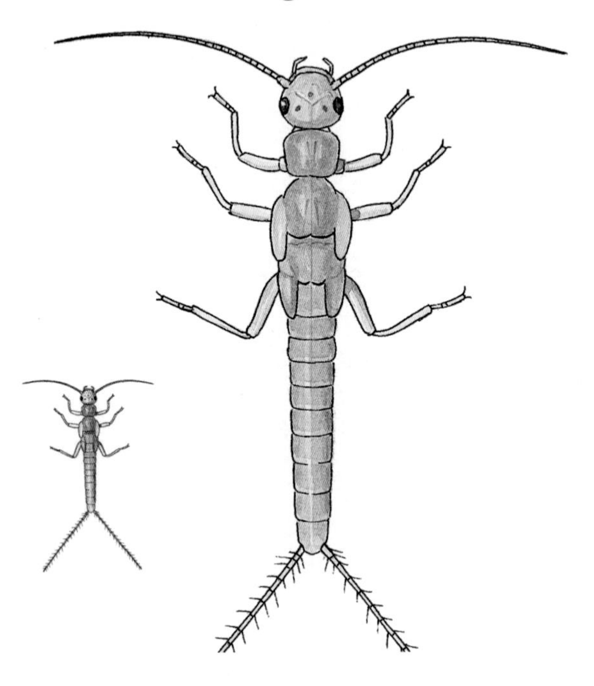

Whole body, top view. (Insecta: Plecoptera: Leuctridae)

Distinguishing Features of Larvae — Body length usually 6–10 mm, range up to 20 mm (mature larvae, not including antennae or tails). The body is slender and elongate, especially the abdomen. The middle portion of the end of the lower lip, between the two finger-like projections, is mostly blunt on the end with three small notches, which is best seen in bottom view (Plate 47B). There are no gills in the neck region between the head and thorax. Both pairs of wing pads do not diverge from an imaginary straight line running from the front to the back of the body. The hind wing pads are usually longer than they are wide. If the hind legs are pulled straight back, the ends of the legs do not come close to the tip of the abdomen. In top view, the sides of the abdomen are even, and the abdomen is about the same width from front to back. — Page 321

Plate 50

Common Burrower Mayflies

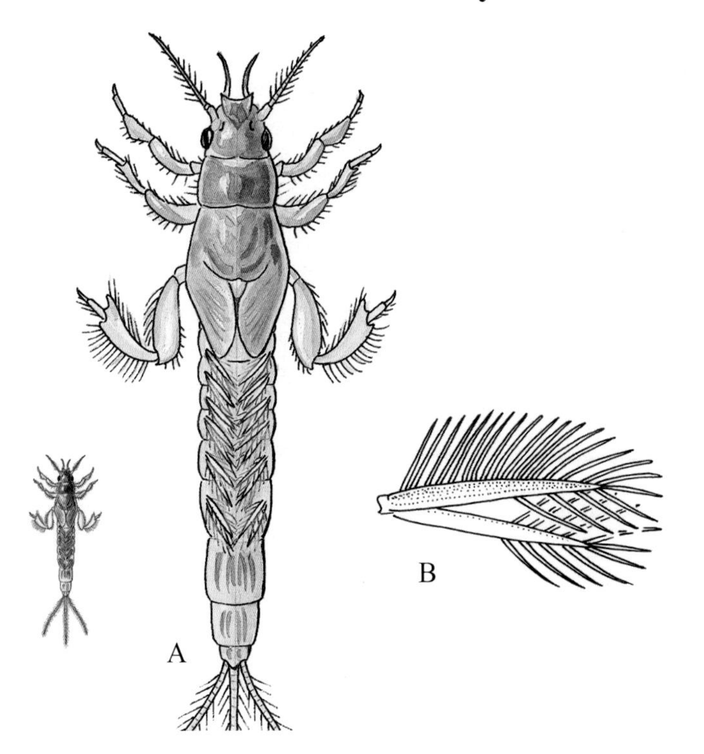

(A) Whole body, top view. **(B)** Gill, top view, enlarged. (Insecta: Ephemeroptera: Ephemeridae)

Distinguishing Features of Larvae — Body length 10–32 mm (mature larvae, not including antennae and tails). There are two tusks that originate on the mouthparts underneath the head, project forward in front of the head, and curve up at the end. The tusks are visible in top view. The front legs are widened and have sharp, pointed extensions for digging (fossorial). Most of the abdomen segments have elongate, forked gills that are attached on the top at the sides and curve over the top of the abdomen (B). The gills have fringed margins that give them a feathery appearance. — Page 276

Plate 51

Brushlegged Mayflies

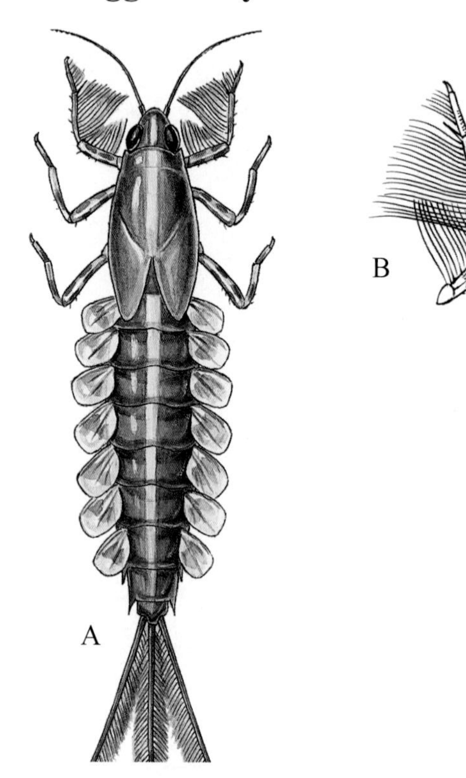

(A) Whole body, top view. **(B)** Front leg, enlarged. (Insecta: Ephemeroptera: Isonychiidae)

Distinguishing Features of Larvae — Body length 8–17 mm (mature larvae, not including antennae and tails). Most kinds are different shades of brown, with a white or cream stripe running from front to back on the top of the body. However, the stripe does not always extend the entire length of the body, and some do not have the stripe. The front legs have a dense fringe of long hairs that project toward the inside (B). Gills are present on most of the abdomen segments. Each gill is a single flat disk with an oval shape. — Page 275

Plate 52

Little Stout Crawler Mayflies

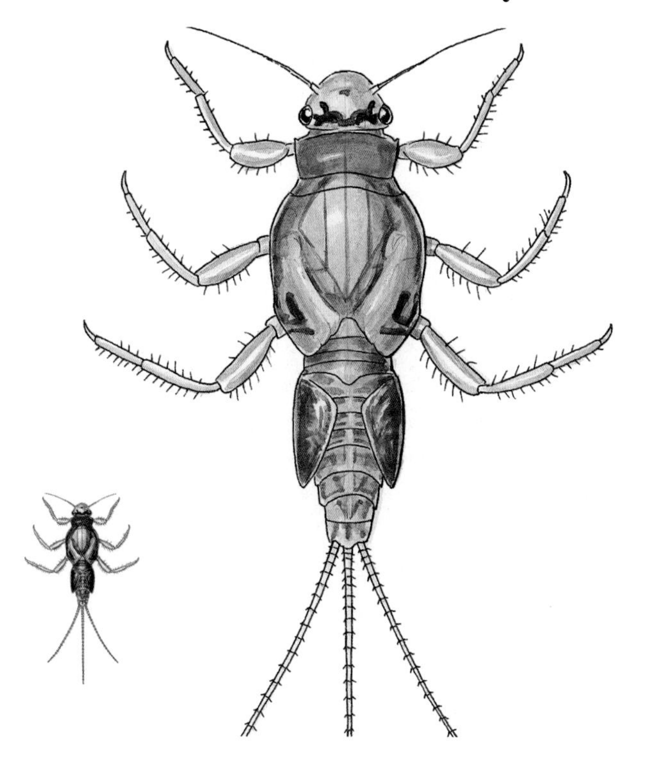

Whole body, top view. (Insecta: Ephemeroptera: Leptohyphidae)

Distinguishing Features of Larvae — Body length 3–10 mm (mature larvae, not including antennae and tails). In top view, the only structures that are visible on the abdomen other than the tails are a pair of large, thick, movable plates. These plates are attached to the rear of abdomen segment two, and they cover and protect four pairs of gills on abdomen segments three to six. The plate-like gill protectors are shaped nearly like triangles, and there is a space between them in the middle. The gills underneath the protective plates do not have any fringe on their edges. Several segments at the rear of the abdomen do not have any gills. — Page 280

Plate 53

Small Squaregill Mayflies

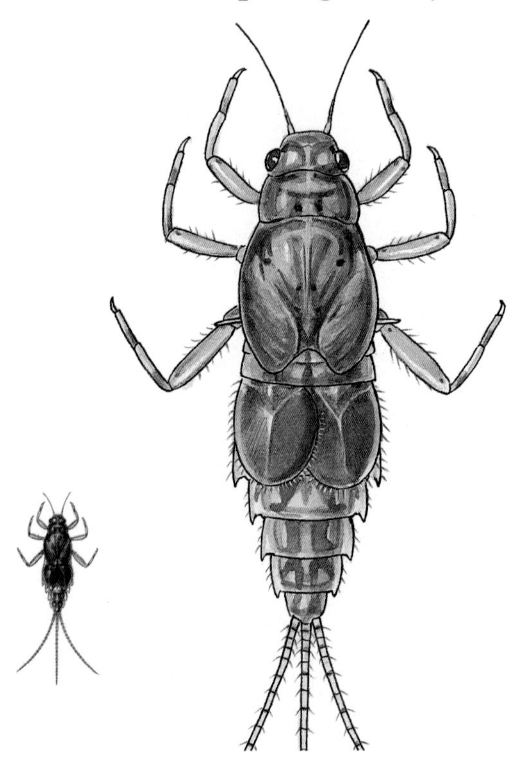

Whole body, top view. (Insecta: Ephemeroptera: Caenidae)

Distinguishing Features of Larvae — Body length 2–8 mm (mature larvae, not including antennae and tails). In top view, the only structures that are visible on the abdomen other than the tails are a pair of large, thick, movable plates. These plates are attached to the rear of abdomen segment two, and they cover and protect four pairs of gills on abdomen segments three to six. The plate-like gill protectors are shaped nearly like squares or rectangles, and they overlap in the middle. The gills underneath the protective plates have a fringe on their edges. Several segments at the rear of the abdomen do not have any gills. — Page 285

Plate 54

Spiny Crawler Mayflies

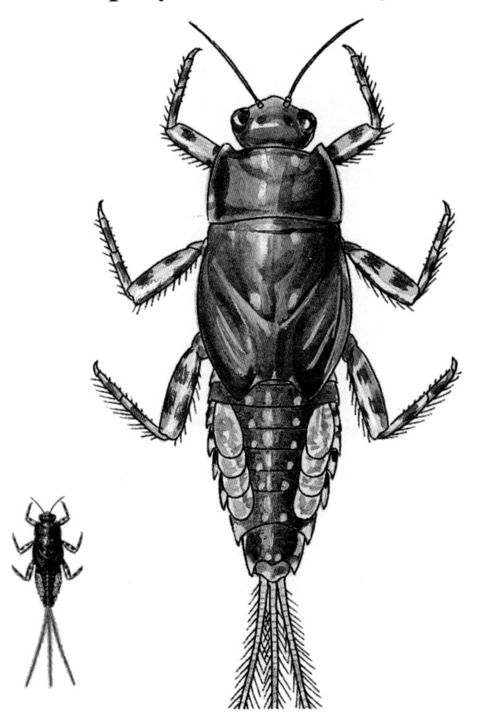

Whole body, top view. (Insecta: Ephemeroptera: Ephemerellidae)

Distinguishing Features of Larvae — Body length 4–15 mm (mature larvae, not including antennae and tails). There are no thick, plate-like gill protectors on the abdomen. In top view, the abdomen has a close group of several thin, flat, oval, disk-like gills on the middle segments. The gills are attached to the rear of abdomen segments three to seven or four to seven. Thus, you see segments without gills toward the front of the abdomen and segments without gills toward the rear of the abdomen. The gills are held tightly against the top of the abdomen. Pairs of thumb-like projections often occur in a row down the center of the tops of the abdomen segments. — Page 287

Plate 55

Flatheaded Mayflies

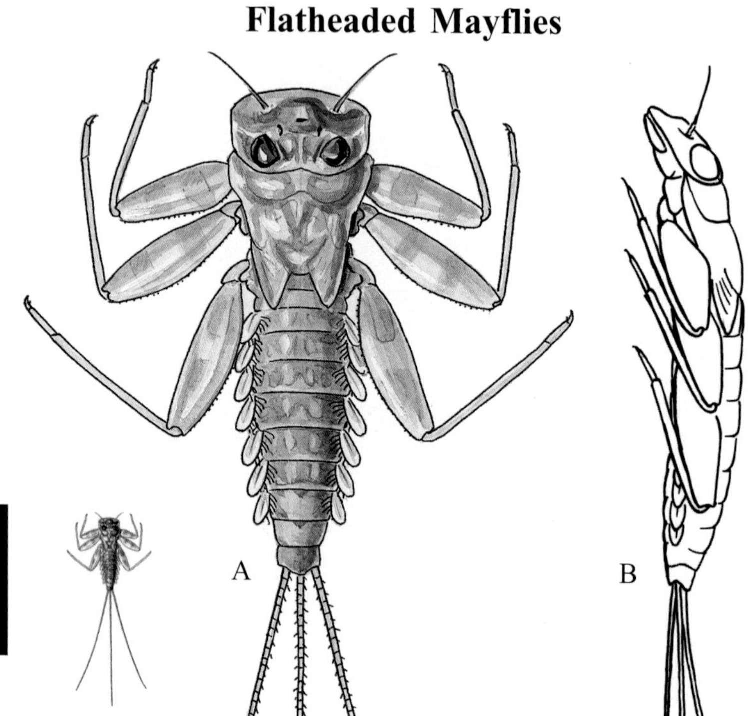

(A) Whole body, top view. **(B)** Whole body, side view. (Insecta: Ephemeroptera: Heptageniidae)

Distinguishing Features of Larvae — Body length 5–20 mm (mature larvae, not including antennae and tails). There are no thick, plate-like gill protectors on the abdomen. The body and legs are distinctly flattened (B). The head is especially flat, with the eyes on top and the mouthparts out of sight on the bottom (in top view). There are gills on top of most of the abdomen segments at the sides. The gills are thin, flat disks that are round to oval and never have any forks or long pointed filaments. The gills have a tuft of fine filaments on the underside at the base, where the gills attach to the body. — Page 278

Plate 56

Pronggilled Mayflies

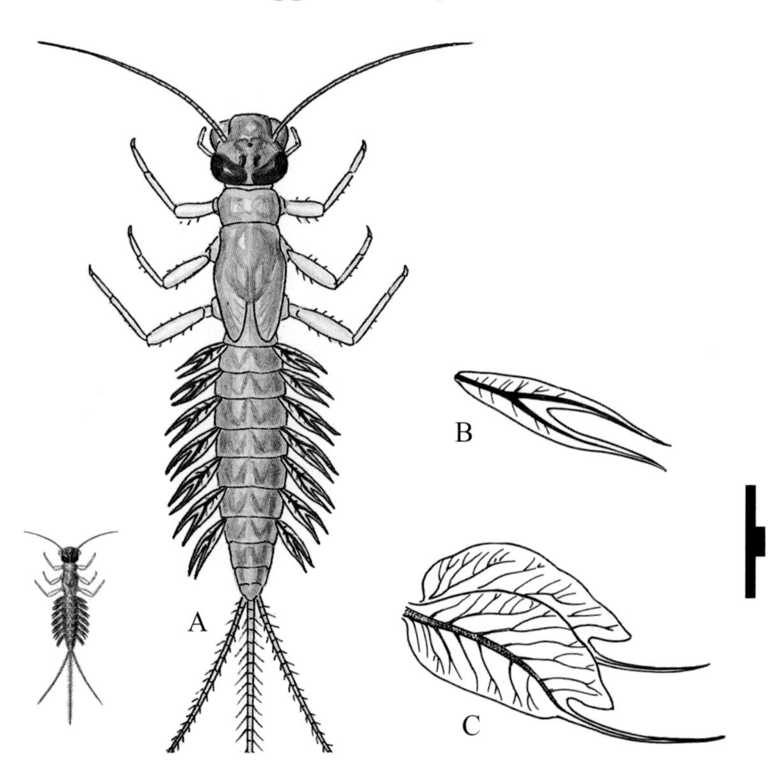

(A) Whole body, top view. **(B)** gill, top view, enlarged. **(C)** Gill found on other kinds, top view, enlarged. (Insecta: Ephemeroptera: Leptophlebiidae)

Distinguishing Features of Larvae — Body length 4–15 mm (mature larvae, not including antennae and tails). There are no thick, plate-like gill protectors on the abdomen. Gills are present on the top of most abdomen segments at the sides. Different kinds within this family have three types of gills: elongate and forked without any fringe (B), round to oval double layers with one long pointed filament (C), or tufts of small filaments. — Page 282

Plate 57

Small Minnow Mayflies

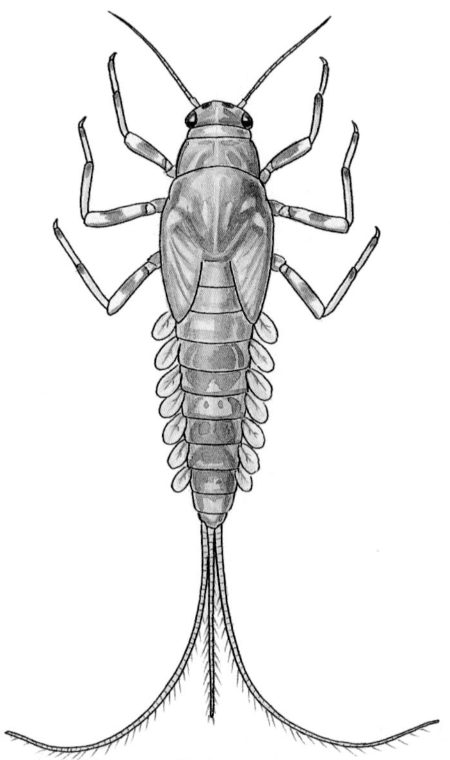

Whole body, top view. (Insecta: Ephemeroptera: Baetidae)

Distinguishing Features of Larvae — Body length 3–12 mm (mature larvae, not including antennae and tails). Gills are present on the top of most abdomen segments at the sides. The gills are thin, flat disks that are round to oval and have no forks or pointed filaments. The antennae are usually two or three times the width of the head. The upper lip usually has a notch in the middle of the edge (best seen in front view). Some kinds have only two tails or a much shorter middle tail. The last two segments of the abdomen before the tails do not have sharp spines on the outside rear corners. Small minnow mayflies closely resemble ameletid minnow mayflies (Plate 58) and primitive minnow mayflies (Plate 59). — Page 284

Plate 58

Ameletid Minnow Mayflies

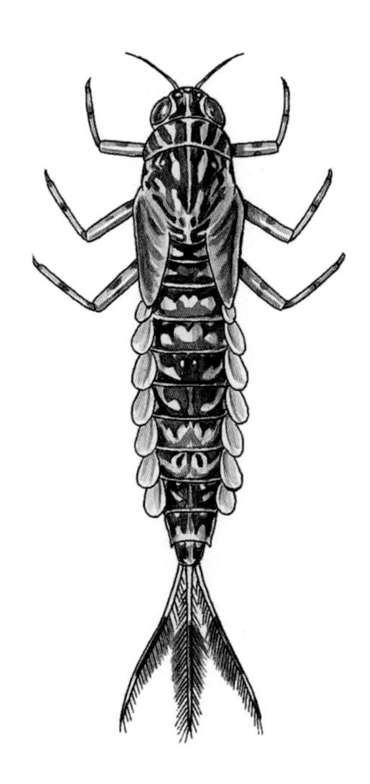

Whole body, top view. (Insecta: Ephemeroptera: Ameletidae)

Distinguishing Features of Larvae — Body length 6–14 mm (mature larvae, not including antennae and tails). Ameletid minnow mayflies closely resemble small minnow mayflies (Plate 57) and primitive minnow mayflies (Plate 59), and can only be distinguished by small features. The antennae are less than twice the width of the head. The upper lip does not have a notch on the edge (best seen in front view). The ends of the mouthparts have two dense clusters of dark, sharp spines (best seen in front or bottom view). There are always three tails, all of which are about the same length. All three tails have a dark band about midway along their length. The last two segments of the abdomen before the tails have sharp spines on the outside rear corners. — Page 274

Plate 59

Primitive Minnow Mayflies

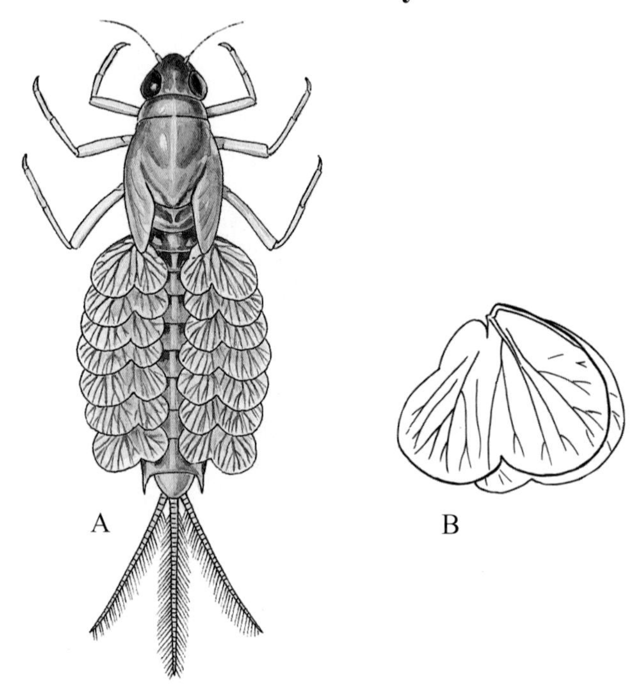

(A) Whole body, top view. **(B)** Gill, top view, enlarged. (Insecta: Ephemeroptera: Siphlonuridae)

Distinguishing Features of Larvae — Body length 6–20 mm (mature larvae, not including antennae and tails). Primitive minnow mayflies closely resemble small minnow mayflies (Plate 57) and ameletid minnow mayflies (Plate 58), and can only be distinguished by small features. The gills are often large and have two layers (B). The antennae are less than twice the width of the head. The upper lip does not have a notch on the edge (best seen in front view). There are no spines on the ends of the mouthparts (best seen in front or bottom view). There are always three tails that are about the same length. The tails do not have dark bands. The last two segments of the abdomen before the tails have sharp spines on the outside rear corners. — Page 281

Plate 60

Dobsonflies, Fishflies, Hellgrammites

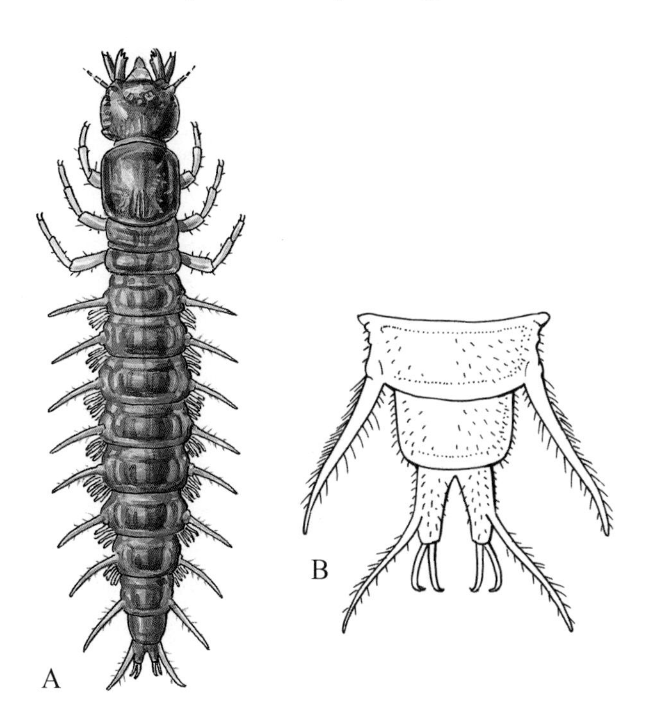

(A) Whole body, top view. **(B)** Last three segments on end of abdomen, top view, enlarged. (Insecta: Megaloptera: Corydalidae)

Distinguishing Features of Larvae — Body length 25–90 mm (mature larvae). The body is elongate and slightly flattened. The head is large, with robust, toothed jaws that project forward. The head and three segments of the thorax are hardened, but the abdomen is soft. The first eight abdomen segments and number 10 each have a pair of stout, flexible, pointed filaments that project out to the side. The end of the body (abdomen segment ten) has a pair of short, fleshy, unsegmented appendages (prolegs) that project to the rear (B). There are two claws on each of these prolegs. — Page 350

Plate 61

Alderflies

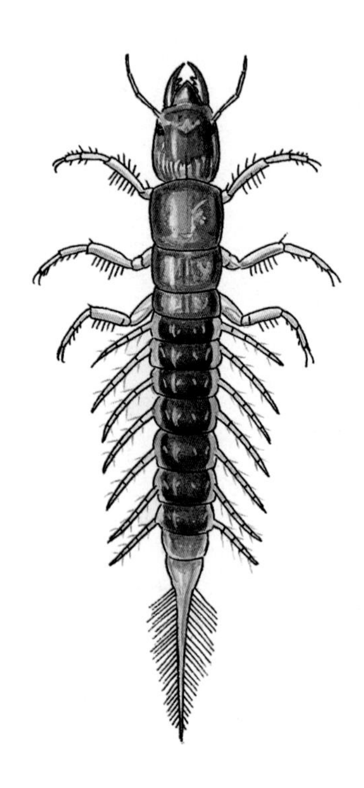

Whole body, top view. (Insecta: Megaloptera: Sialidae)

Distinguishing Features of Larvae — Body length 10–25 mm (mature larvae, not including tail). The body is elongate and slightly flattened. The head is large, with robust, toothed jaws that project forward. The head and three segments of the thorax are hardened, but the abdomen is soft. The first seven abdomen segments each have a pair of stout, flexible, pointed filaments that project out to the side. Each of these filaments consists of four or five segments. The end of the body (abdomen segment ten) has a single long, tapering filament, fringed with fine hairs, that projects to the rear. — Page 349

Plate 62

Snailcase Maker Caddisflies

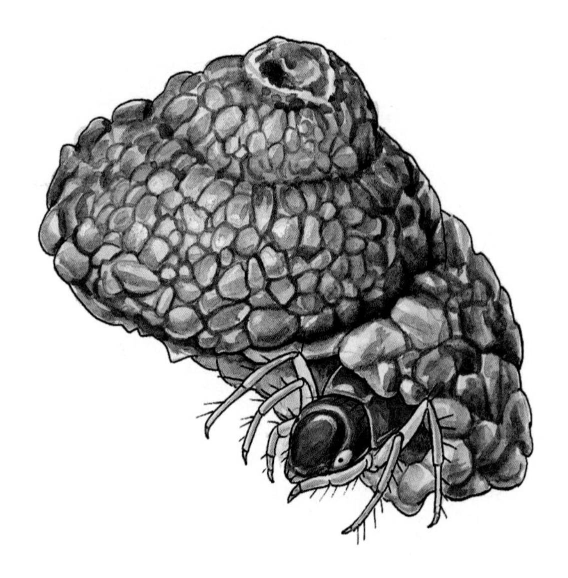

Whole body in portable case, front/side view. (Insecta: Trichoptera: Helicopsychidae)

Distinguishing Features of Larvae — Body length about 8 mm (mature larvae). Case diameter about 7 mm. They construct a unique portable case that closely resembles a snail shell. The case is a tube composed of sand grains or small rock particles formed in a helical coil. Larvae are seldom collected without their case. The body is highly curved. The head, thorax plates, and legs are very dark. The prolegs on the last abdomen segment (number ten) have numerous small teeth like a comb. — Page 392

Plate 63

Micro Caddisflies

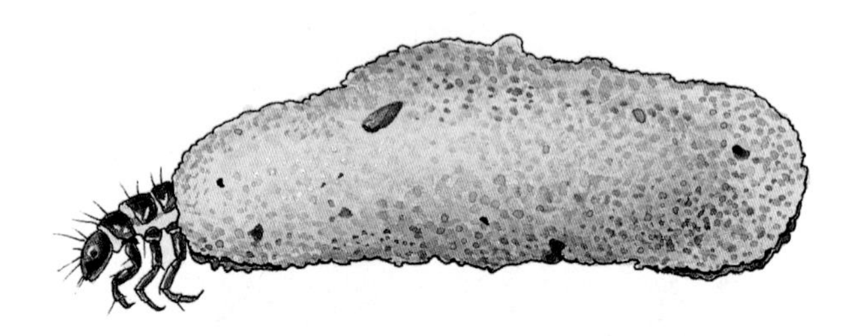

Whole body in portable case, side view. (Insecta: Trichoptera: Hydroptilidae)

Distinguishing Features of Larvae — Body length usually 2–4 mm, range up to 7 mm (mature larvae). Case length usually 3–6 mm, range up to 8 mm. Almost all kinds construct a portable case, but there is considerable variation in cases among the different kinds in this family. Many common kinds construct a case that has two similar halves and is compressed from side to side, somewhat resembling a purse. Fine sand is the most common building material, but algae, fine detritus, or just silk are also used in this shape. Other shapes vary from straight and cylindrical to tapering and flat. One common kind makes a flat, oval case that it attaches tightly to solid substrates. Usually there are similar openings at both ends of the case, and the case is often about the same length as the larva. These small larvae have hardened plates on the tops of all three thorax segments. There are no gills on any of the abdomen segments. — Page 388

Plate 64

Longhorned Case Maker Caddisflies

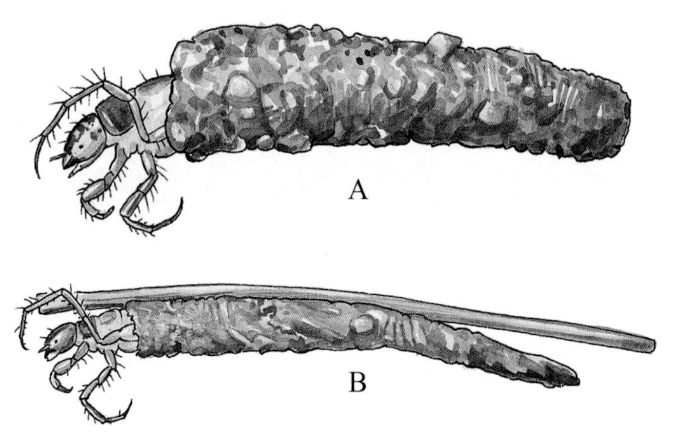

(A, B) Two different kinds, whole body in portable case, side view. (Insecta: Trichoptera: Leptoceridae)

Distinguishing Features of Larvae — Body length 7–15 mm (mature larvae). Case length 8–33 mm. Larvae construct portable cases, but there is a wide variety of cases among the different kinds (A, B). The building materials include pieces of plants, small rock particles, fine sand, and pure silk. The shape of most cases is an elongate, straight, tapering tube, but some are slightly curved. Some common kinds build their cases attached to a thin plant stem that is almost as long as the case and extends in front of the case. In most kinds the materials are glued together randomly, but some kinds attach uniform, slender pieces of vegetation in a spiral. A few kinds build a sand case with a flange on the sides and in front of the tube. The top of thorax segment two has light colored plates, and a pair of dark curved lines on the rear half. The top of thorax segment three is mostly soft and fleshy, without a continuous hardened plate. The antennae are long, at least six times their width, which is long for caddisfly larvae. It is not possible to see the antennae of other caddisfly larvae without high magnification under a microscope, so if you can make out antennae, it is probably a longhorned case maker. The hind legs are much longer than the other legs and are often held over the thorax and head. — Page 387

Plate 65

Lepidostomatid Case Maker Caddisflies

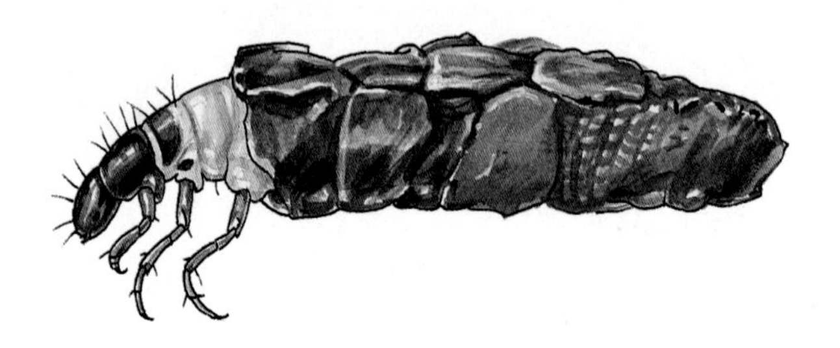

Whole body in portable case, side view. (Insecta: Trichoptera: Lepidostomatidae)

Distinguishing Features of Larvae — Body length 7–13 mm (mature larvae). Case length 7–15 mm. Common kinds in the East construct a four-sided, tapering case that is composed of coarse pieces of dead leaves or bark (illustrated). The leaves or bark are cut into pieces that are nearly square or rectangular. Cases of common kinds in the West are round and constructed of mineral particles or thinner pieces of dead vegetation. The top of thorax segment two is more than half covered with hardened plates that are usually dark, and there is not a pair of curved lines on the rear half. The top of thorax segment three is mostly soft and fleshy, without a continuous hardened plate. Abdomen segment one has a fleshy hump on each side but not on top. Refer to Plate 69B to see what these humps look like. — Page 386

Plate 66

Humpless Case Maker Caddisflies

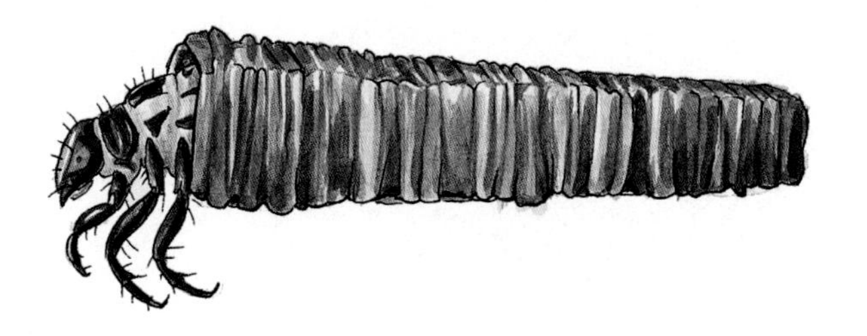

Whole body in portable case, side view. (Insecta: Trichoptera: Brachycentridae)

Distinguishing Features of Larvae — Body length 8–11 mm (mature larvae). Case length 10–17 mm. The kinds that are most commonly encountered construct two types of portable cases. One type is a four-sided, straight, tapering case that is composed of thin pieces of vegetation stacked sideways. The sides of the case resemble a log cabin. The other type of case that is common is a tapering, round cylinder that is composed of thin ribbons of plant material wrapped around the circumference. This type of case can be straight or curved. The top of thorax segment two is more than half covered with hardened plates that are usually dark, and there is not a pair of curved lines on the rear half. The top of thorax segment three is mostly soft and fleshy, without a continuous hardened plate. Abdomen segment one has no fleshy humps on the sides or the top. Refer to Plate 69B to see what these humps would look like. — Page 385

Plate 67

Saddlecase Maker Caddisflies

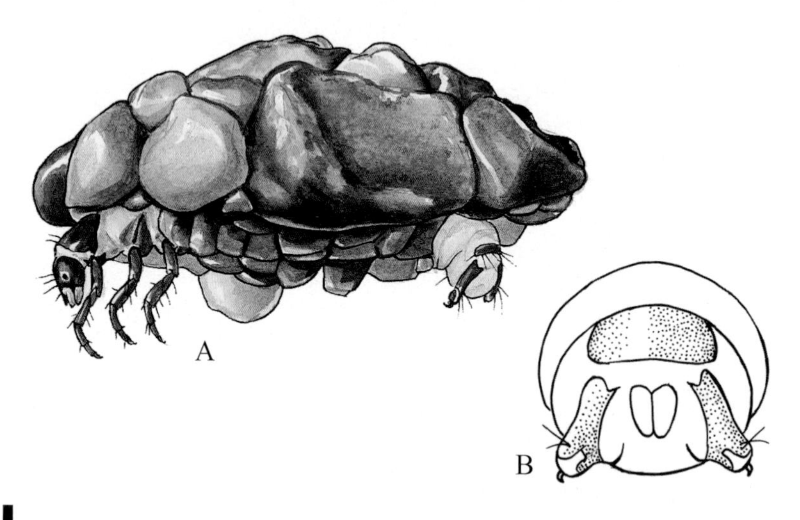

(A) Whole body in portable case, side/bottom view. **(B)** Last three segments on end of abdomen, rear/top view, enlarged. (Insecta: Trichoptera: Glossosomatidae)

Distinguishing Features of Larvae — Body length 3–10 mm (mature larvae). Case length 3–12 mm. They construct a unique portable case that has been called a saddlecase, but a turtle shell is a better comparison. The case is a dome of small rock particles that completely covers the larva in top view. On the underside of the case there is a strap of smaller rock particles or sand across the middle. The larva hangs over the strap, the head and thorax on one side and the end of the abdomen on the other. The tops of the second and third thorax segments are more than half soft and fleshy, without continuous hardened plates. There is not an elongate, sharp horn on the bottom of thorax segment one, and there are no fleshy humps on abdomen segment one. Refer to Plate 69B to see what these structures would look like. Abdomen segment nine has a hardened plate on top (B). The bottom half of the prolegs at the end of the body is fused with abdomen segment nine, so they are not completely loose and movable (B). — Page 391

Plate 68

Uenoid Case Maker Caddisflies

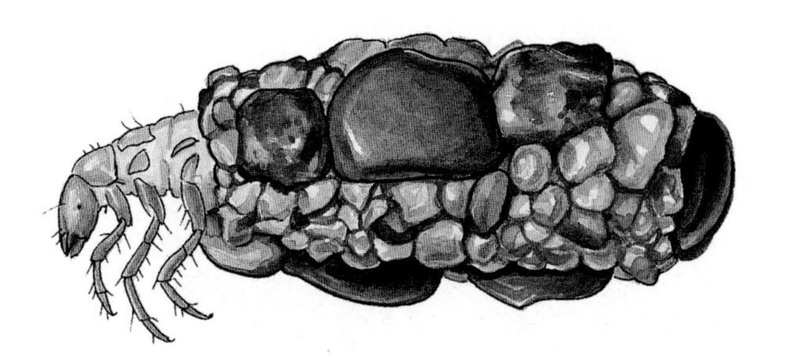

Whole body in portable case, side view. (Insecta: Trichoptera: Uenoidae)

Distinguishing Features of Larvae — Body length 6–15 mm (mature larvae). Case length 8–15 mm. The portable case is always made of mineral particles. Most common kinds in the West are thin, tapered, curved tubes made of small rock particles or sand. The common kinds in the East construct a short, stout tube made out of sand or small rock particles with three or four larger stones attached to each side. Some kinds of northern case makers (Limnephilidae) make a similar case, but there are two larger stones on each side. The top of thorax segment two is more than half covered with hardened plates that are usually dark, and there is not a pair of curved lines on the rear half. The top of thorax segment three is mostly soft and fleshy, without a continuous hardened plate. Abdomen segment one has a fleshy hump on each side and on top, and an elongate, sharp horn is usually present on the bottom of thorax segment one. Refer to Plate 69B to see what these structures look like. On the top of thorax segment two, the front edge of the plate has a rounded notch or square indentation in the middle, which is best seen in top view. — Page 394

Plate 69

Northern Case Maker Caddisflies

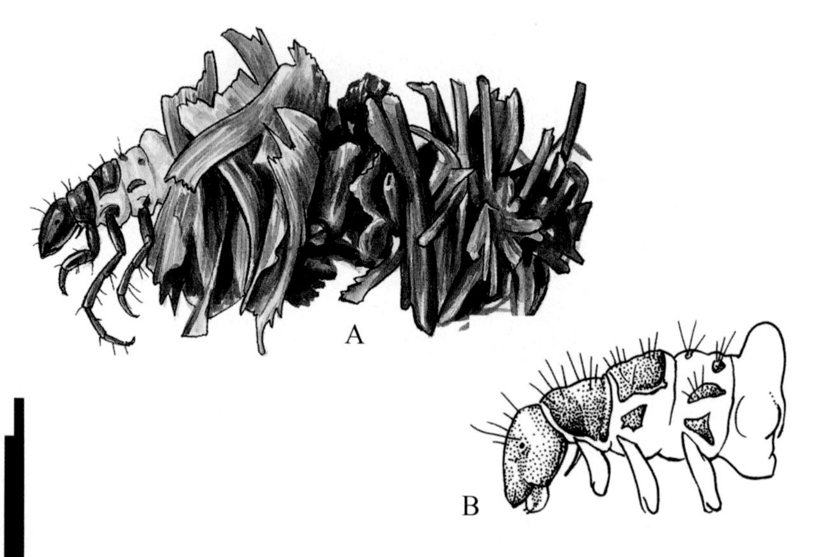

(A) Whole body in portable case, side view. **(B)** Front half of body, side view, enlarged. (Insecta: Trichoptera: Limnephilidae)

Distinguishing Features of Larvae — Body length usually 20–30 mm, range 8–35 mm (mature larvae). Case length usually 25–50 mm, range 12–76 mm. Different kinds in this family construct a wide variety of portable cases out of plant or mineral materials. Many of the common kinds use large pieces of plant material arranged either radiating away from the center (A) or lengthwise. See Plate 70 for other common kinds. The top of thorax segment two is more than half covered with hardened plates that are usually dark (B). There is not a pair of curved lines on the rear half of segment two. On the top of thorax segment two, the plate is straight across the front edge, without a rounded notch or square indentation in the middle (best seen in top view). The top of thorax segment three is mostly soft and fleshy, without a continuous hardened plate (B). Abdomen segment one has a fleshy hump on each side and on top (B). An elongate, sharp horn is usually present on the bottom of thorax segment one (B). — Page 390

Plate 70

Northern Case Maker Caddisflies

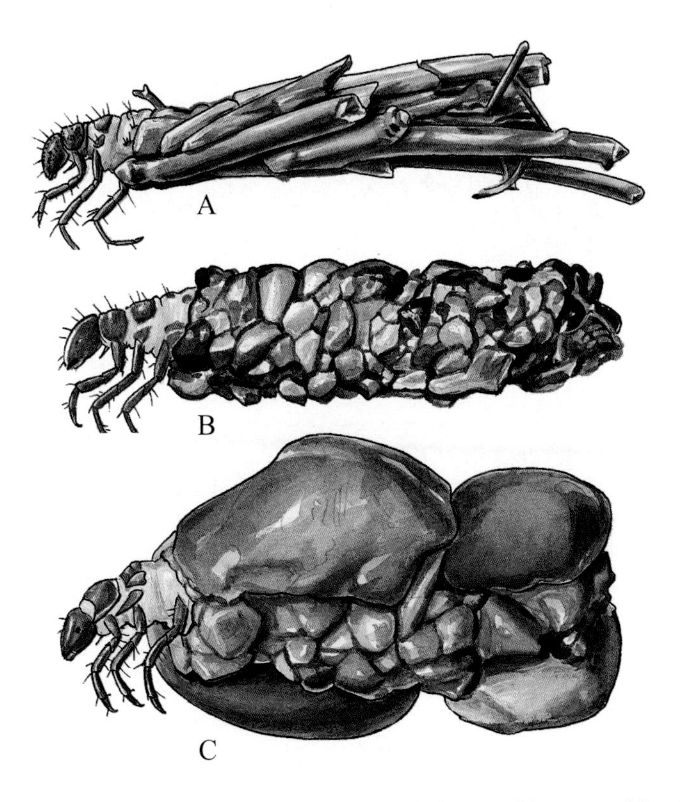

(A, B, C) Three different kinds, whole body in portable case, side view. (Insecta: Trichoptera: Limnephilidae)

Distinguishing Features of Larvae—Same as described for Plate 69. Other kinds in this family build plant cases with two or three long twigs incorporated lengthwise and trailing behind the tube (A). These have the common name "stick bait." Some kinds build mineral cases composed of particles of rocks (B), or a tube of small particles of rocks with two larger stones on each side (C). Some kinds of uenoid case makers (Plate 68) make a case similar to C, but there are three or four larger stones on each side, and the tube has a rougher texture. — Page 390

Plate 71

Strongcase Maker Caddisflies

Whole body in portable case, side view. (Insecta: Trichoptera: Odontoceridae)

Distinguishing Features of Larvae — Body length 9–20 mm (mature larvae). Case length 10–30 mm. The portable case is a round tube composed of sand or small rock particles that are very uniform in size. It is always curved a little and may have a slight taper. The top of thorax segment two is more than half covered with hardened plates that are usually dark, and there is not a pair of curved lines on the rear half. The top of thorax segment three is mostly soft and fleshy, without a continuous hardened plate. Abdomen segment one has a fleshy hump on each side and on top (Plate 69B). An elongate, sharp horn is not present on the bottom of thorax segment one. — Page 392

Plate 72

Giant Case Maker Caddisflies

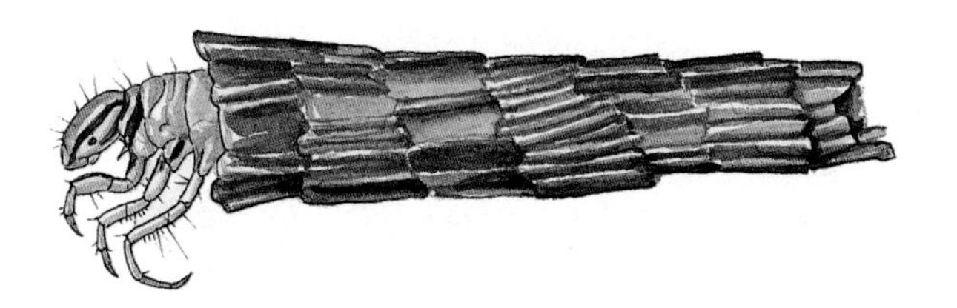

Whole body in portable case, side view. (Insecta: Trichoptera: Phryganeidae)

Distinguishing Features of Larvae — Body length 22–43 mm (mature larvae). Case length 45–60 mm. Larvae construct very long tubes out of plant materials. The common kinds cut large, rectangular pieces of vegetation, most often dead leaves or bark, and they glue them together either in a spiral or successive rings. Larvae quickly abandon their cases when disturbed, so collecting often produces free larvae and empty cases. The tops of the second and third thorax segments are more than half soft and fleshy, without continuous hardened plates. The head and top of thorax segment one are conspicuously marked with dark stripes on a yellow background. An elongate, sharp horn is present on the bottom of thorax segment one, and there are fleshy humps on the top and each side of abdomen segment one (Plate 69B). Abdomen segment nine has a hardened plate on top (Plate 77B). — Page 384

Plate 73

Common Netspinner Caddisflies

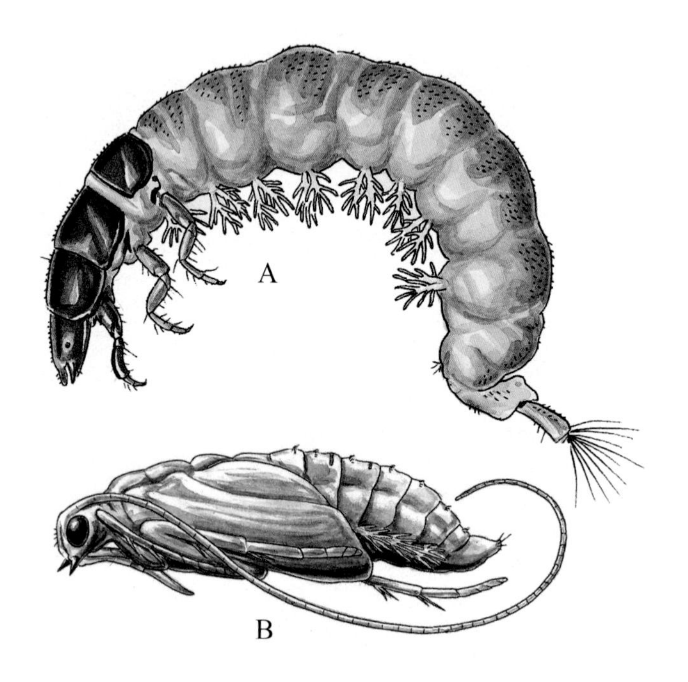

(A) Whole body, side view. **(B)** Pupa, whole body, side view. (Insecta: Trichoptera: Hydropsychidae)

Distinguishing Features of Larvae — Body length usually 13–18 mm, range 9–30 mm (mature larvae). This is one of the three families that make an attached retreat. The other families are fingernet caddisflies (Philopotamidae) and trumpetnet and tubemaking caddisflies (Polycentropodidae). Larvae are almost always collected free from their dwelling. In a pan of water, larvae usually lay on their sides in a "C" shape. They have hardened plates on the tops of all three thorax segments. There are branched filamentous gills on the bottom of most abdomen segments. The prolegs on the last abdomen segment have a prominent brush of long hairs. — Page 380

Plate 74

Common Netspinner Caddisflies

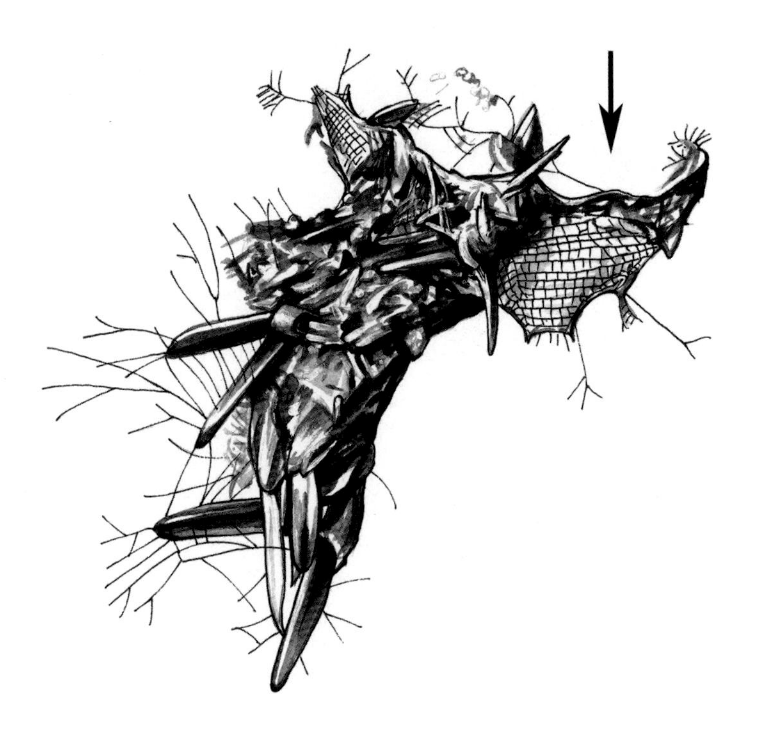

Attached tubular retreat, top view, arrow shows direction of water current. (Insecta: Trichoptera: Hydropsychidae)

Distinguishing Features of Retreat: Larvae live in tubular retreats that they glue down to solid substrates with silk. Different kinds incorporate gravel, sand, plant detritus, or combinations of these materials into the structure. There is a mesh net for filter feeding. Larvae quickly abandon the retreats when disturbed. The fragile retreats usually come apart during collecting, but sometimes they can be observed on rocks or logs in shallow, clear water. It may be possible to observe the retreat out of the water if the substrate is picked up carefully. — Page 380

Plate 75

Fingernet Caddisflies

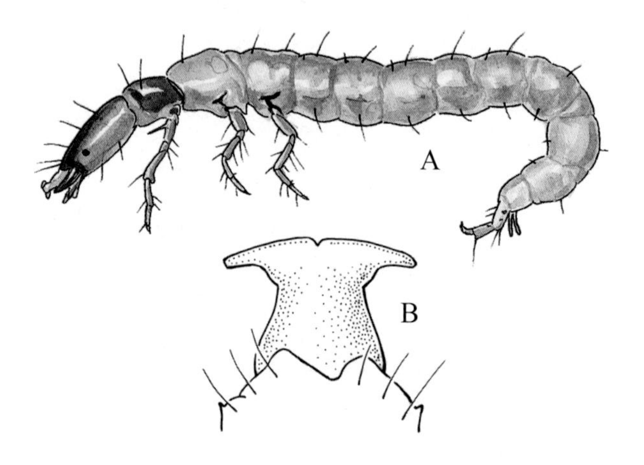

(A) Whole body, side view. **(B)** Upper lip, top view, enlarged. (Insecta: Trichoptera: Philopotamidae)

Distinguishing Features of Larvae — Body length 13–17 mm (mature larvae). Retreat length 25–60 mm, diameter 3–5 mm. This is one of the three families that make an attached retreat. The other families are common netspinners (Hydropsychidae) and trumpetnet and tubemaking caddisflies (Polycentropodidae). Larvae are almost always collected free from their dwelling. They live in a very fine, soft, elongate pouch that resembles a finger when held expanded by mild current. However, these fragile retreats come apart during collecting. Many kinds are yellow-orange while alive, but they quickly turn white or cream colored when placed in alcohol. The head is long and projects forward. The tops of the second and third thorax segments are completely soft and fleshy, without hardened plates. The top of abdomen segment nine is soft and fleshy, without a hardened plate. Close examination reveals that the upper lip (labrum) is soft and fleshy, and shaped like a wide letter "T" (B). This upper lip is light colored, or almost clear, in contrast to the orange colored head. Unfortunately, the upper lip is small and is often withdrawn into the mouth in preserved specimens. — Page 382

Plate 76

Trumpetnet and Tubemaker Caddisflies

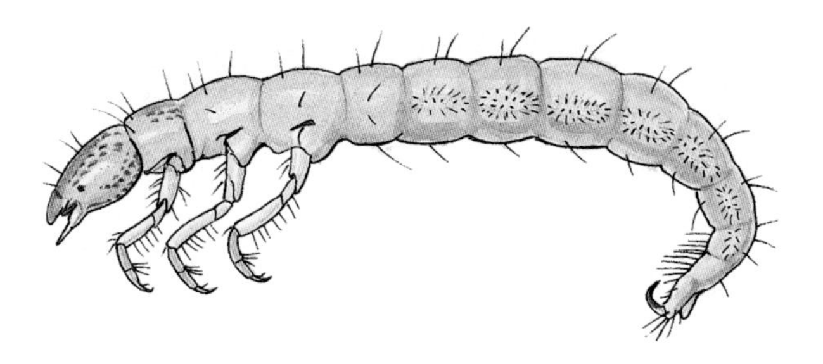

Whole body, side view. (Insecta: Trichoptera: Polycentropodidae)

Distinguishing Features of Larvae — Body length 8–25 mm (mature larvae). This is one of the three families that make an attached retreat. The other families are common netspinners (Hydropsychidae) and fingernet caddisflies (Philopotamidae). Larvae are usually collected free from their dwelling. They construct a variety of attached retreats, including wide pouches held open by current, long curved funnels resembling trumpets, shapeless flat coverings, and loose tubes with wide openings on both ends. However, some of these retreats are fragile and come apart during collecting. The head projects forward. The tops of the second and third thorax segments are completely soft and fleshy, without hardened plates. The top of abdomen segment nine is soft and fleshy, without a hardened plate. Close examination reveals that the upper lip is normal, resembling that of other caddisfly larvae, and not membranous and T-shaped, such as Plate 75B. The upper lip is hardened, curved and oval in shape, and the same color as the rest of the head. — Page 393

Plate 77

Freeliving Caddisflies

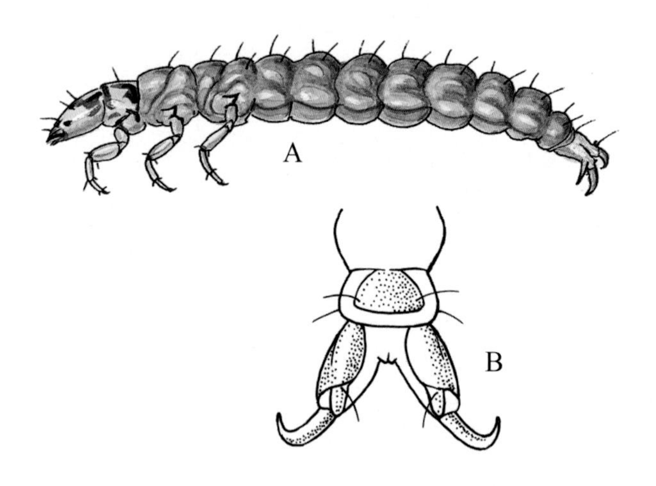

(A) Whole body, side view. **(B)** Last three segments on end of abdomen, rear/top view, enlarged. (Insecta: Trichoptera: Rhyacophilidae)

Distinguishing Features of Larvae — Body length usually 11–23 mm, range up to 32 mm (mature larvae). These roam freely and never reside in a case or retreat as larvae. The body is somewhat flattened, and the head projects forward. Many kinds are bright green while alive, but they quickly turn a purplish color when placed in alcohol. They are sometimes called "green caddisflies." The tops of the second and third thorax segments are completely soft and fleshy, without hardened plates. The abdomen is long and has deep constrictions at each segment (best seen in top view). Abdomen segment nine has a hardened plate on top. Almost the entire length of the prolegs at the end of the body is free from abdomen segment nine, so they are loose and movable (B). The claws on the prolegs are long, stout, curved, and sharply pointed. — Page 383

Plate 78

Water Pennies

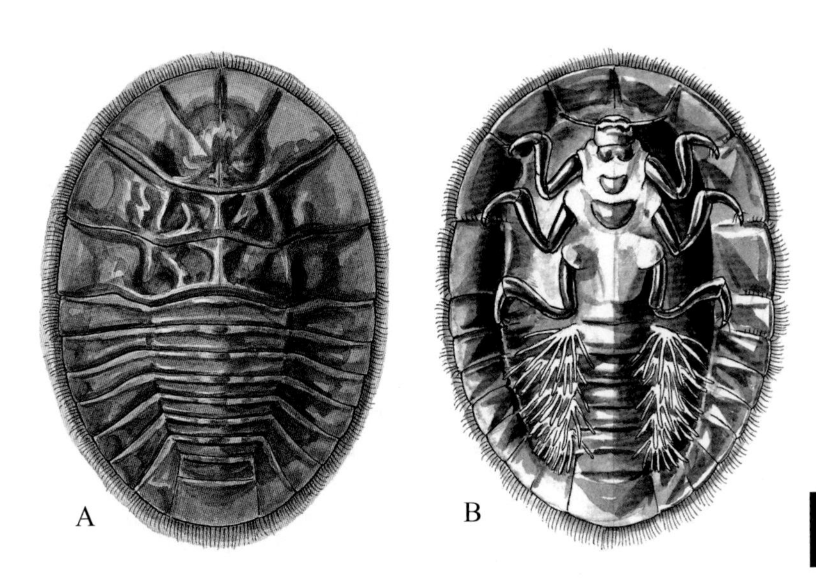

(A) Whole body, top view. **(B)** Whole body, bottom view. (Insecta: Coleoptera: Psephenidae)

Distinguishing Features of Larvae — Body length 3–10 mm (mature larvae). Water pennies are very easy to recognize, and their common name gives a good description of their appearance. They have conspicuously flattened bodies that are oval to almost circular. Most of the thorax and abdomen segments have thin flat plates that extend away from the body, hiding the head and legs in top view (B). In some kinds of water pennies (not illustrated), there are spaces between the plates, giving the appearance of saw teeth on the outer edge. The legs have four segments (not counting the claws). There is one claw on the end of each leg. — Page 365

Plate 79

Crawling Water Beetles

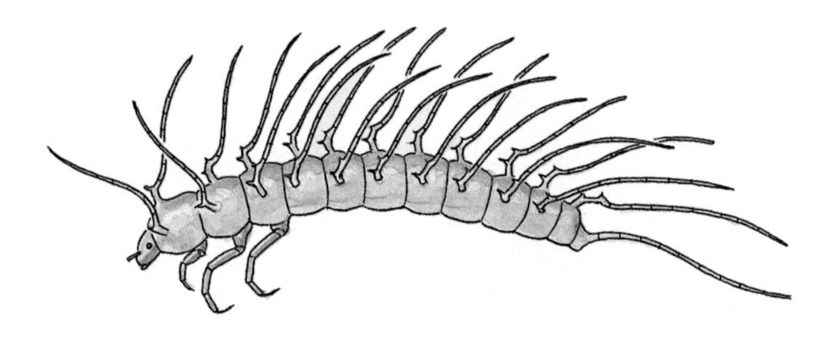

Whole body, side view. (Insecta: Coleoptera: Haliplidae)

Distinguishing Features of Larvae — Body length 5–12 mm (mature larvae). Some kinds within this family look very different, and it is not possible to distinguish all crawling water beetles with one set of simple characteristics. This illustration and description will distinguish the kinds that are most likely to be found. The body is cylindrical and somewhat stout. The legs have five segments (not counting the claws). There is one claw on the ends of the legs. There are no stout, pointed filaments on the sides of the abdomen or hooks on the end of the abdomen. The top of the body has numerous long, thin, erect projections that make the larvae look like small porcupines. — Page 359

Plate 80

Whirligig Beetles

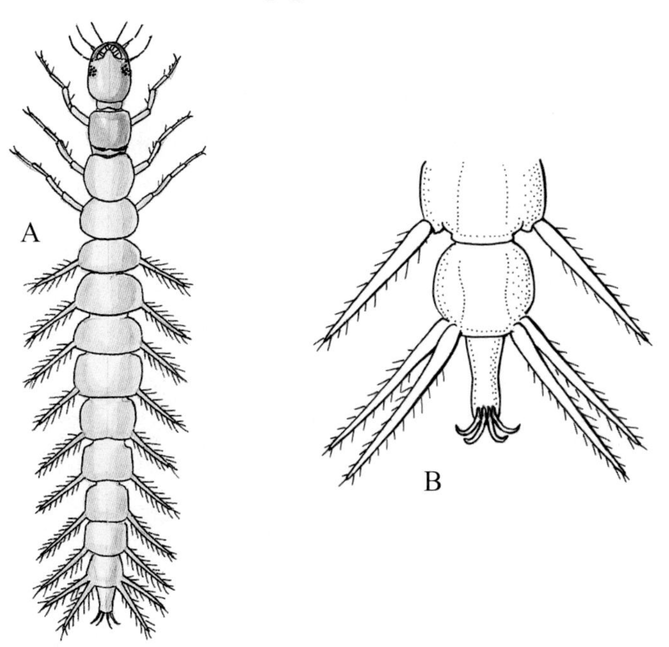

(A) Whole body, top view. **(B)** Last three segments of abdomen, top view. (Insecta: Coleoptera: Gyrinidae)

Distinguishing Features of Larvae — Body length 6–30 mm (mature larvae). The body is elongate, narrow, and somewhat flattened. The color of the soft skin is creamy white, and the hardened parts are yellow or yellow-brown with brown marks on the head and top of thorax segment one. The legs have five segments (not counting the claws). There are two claws on the ends of all legs. The abdomen has ten segments. The first nine abdomen segments have stout, pointed filaments that project out to the side. There is one pair of these filaments on the first eight abdomen segments and two pairs on abdomen segment nine (B). The last segment of the abdomen has four hooks on the end. At first glance, these might look like hellgrammites or alderfly larvae, but close examination of the last abdomen segment will distinguish whirligig beetle larvae. — Page 368

Plate 81

Riffle Beetles

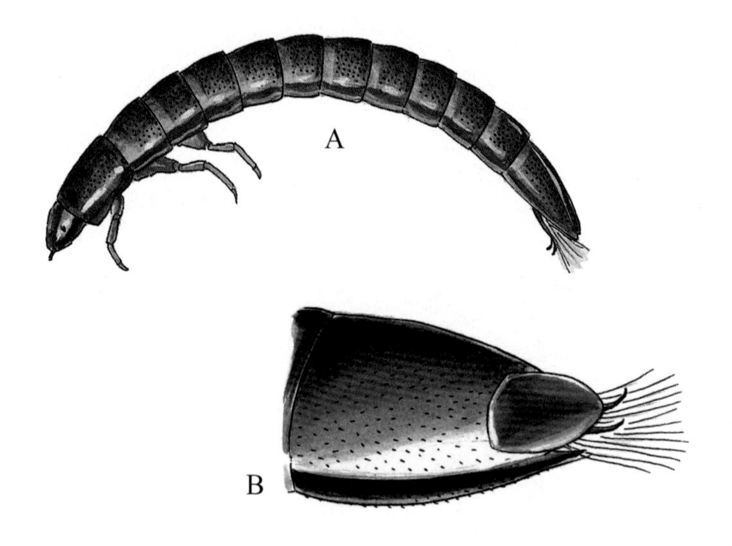

(A) Whole body, side view. **(B)** End of abdomen, bottom/side view, enlarged. (Insecta: Coleoptera: Elmidae)

Distinguishing Features of Larvae — Body length usually 3–8 mm; range up to 16 mm (mature larvae). The body is elongate, cylindrical, and hard. They are usually dark brown or red-brown. The legs have four segments (not counting the claws). There is one claw on the end of each leg. The abdomen has nine segments. Abdomen segment nine has a cavity that is covered by a hinged lid, and there is a tuft of filamentous gills that can be withdrawn into this cavity (B). — Page 364

Plate 82

Predaceous Diving Beetles

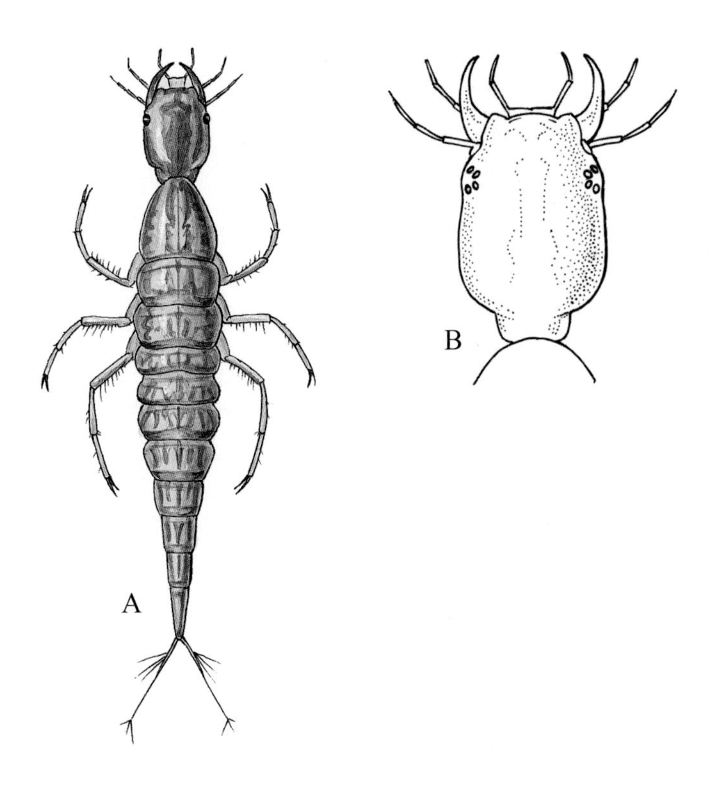

(A) Whole body, top view. **(B)** Head, top view, enlarged. (Insecta: Coleoptera: Dytiscidae)

Distinguishing Features of Larvae — Body length 2–70 mm (mature larvae, not including antennae and tails). There are long, slender jaws, without any teeth, that curve smoothly to a sharp point (B). The jaws project in front of the head. The legs have five segments (not counting the claws) and are long and slender for swimming. There are two claws on the end of each leg. The abdomen narrows distinctly to a point at the end. Larvae of predaceous diving beetles and water scavenger beetles (Plate 83) look a lot alike. — Page 362

Plate 83

Water Scavenger Beetles

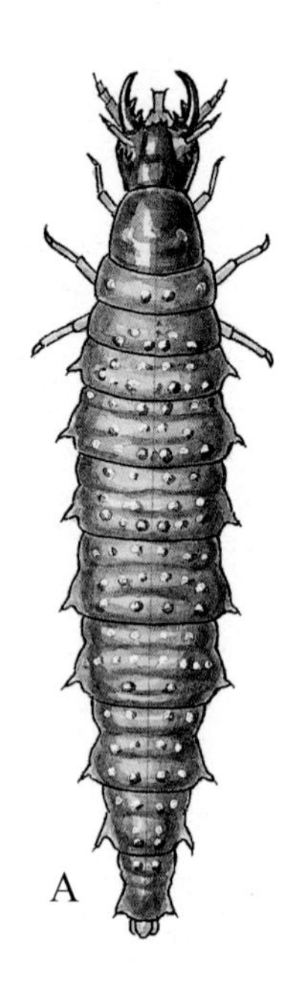

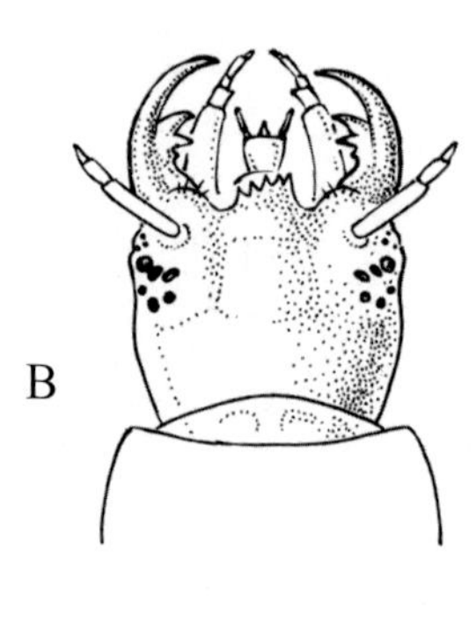

(A) Whole body, top view. **(B)** Head, top view, enlarged. (Insecta: Coleoptera: Hydrophilidae)

Distinguishing Features of Larvae — Body length 2–60 mm (mature larvae). There are curved, stout jaws with one to three teeth on the inside margins (B). The jaws project in front of the head. The legs have four segments (not counting the claws) and are short for crawling. There is one claw on the end of each leg. The abdomen does not taper and is bluntly rounded on the end. Larvae of water scavenger beetles and predaceous diving beetles (Plate 82) look a lot alike. — Page 367

Plate 84

Net-Winged Midges

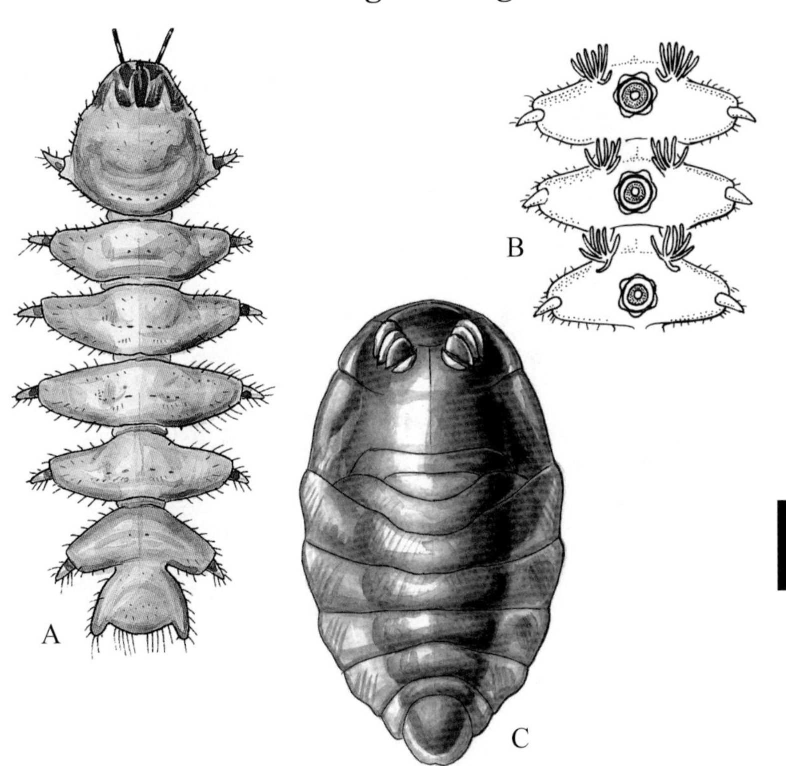

(A) Whole body, top view. **(B)** Three segments in middle of body, bottom view, enlarged. **(C)** Pupa, top view. (Insecta: Diptera: Blephariceridae)

Distinguishing Features of Larvae — Body length 4–12 mm (mature larvae). The body is flattened from top to bottom. There are six deep constrictions on the sides that separate the body into seven major divisions. The first division is larger than the others because it is made up of the head, all three thorax segments, and the first abdomen segment, which are all fused into a single section. The head is barely distinguishable from the rest of the first division. Each of the first six body divisions has a sucker on the bottom side in the middle of the division (B). — Page 418

Plate 85

Mountain Midges

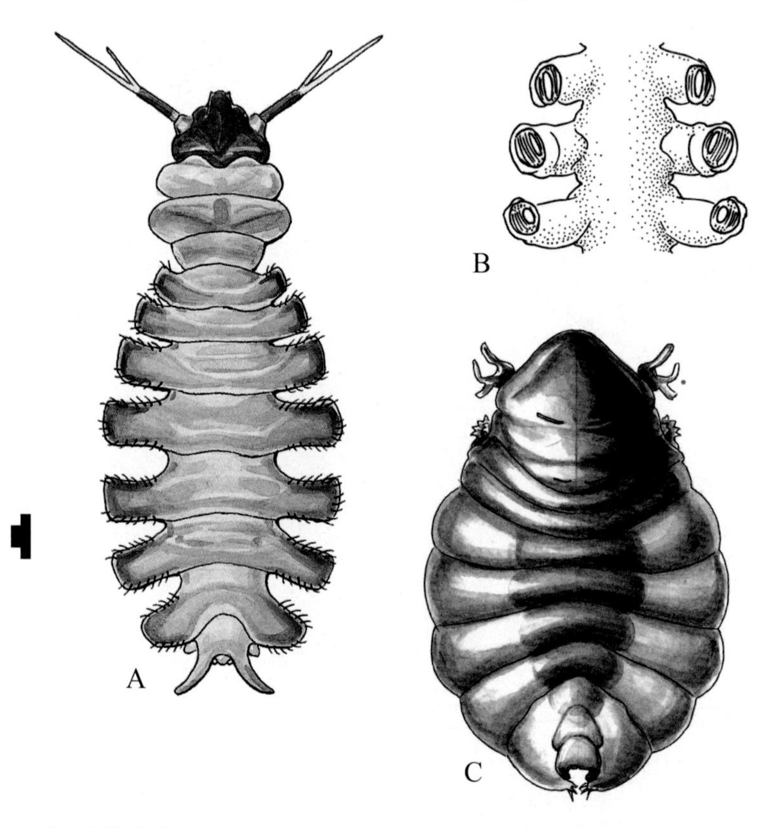

(A) Whole body, top view. **(B)** Three segments in middle of body, bottom view, enlarged. **(C)** Pupa, top view. (Insecta: Diptera: Deuterophlebiidae)

Distinguishing Features of Larvae — Body length 3–6 mm (mature larvae). The body is flattened from top to bottom. The head and three segments of the thorax are each distinct. The antennae are forked and much longer than the head. There are seven pairs of elongate, rounded, fleshy lobes on the sides of the abdomen. The lobes on the abdomen project to the side and downward. Several rows of tiny, hooked spines encircle the ends of the lobes (B). — Page 417

Plate 86

Phantom Crane Flies

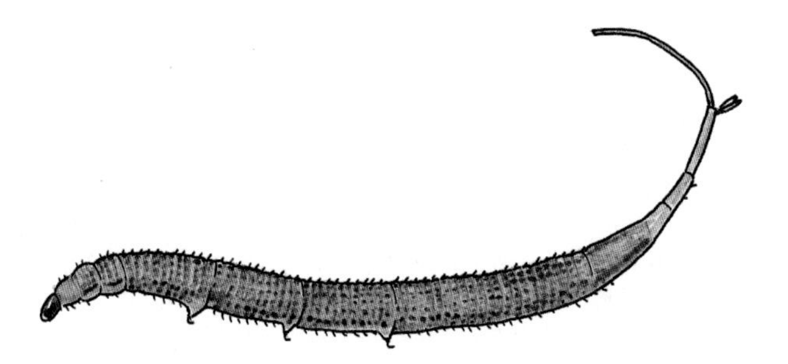

Whole body, side view. (Insecta: Diptera: Ptychopteridae)

Distinguishing Features of Larvae — Body length 10–25 mm (mature larvae, not including breathing tube); range up to 60 mm with breathing tube extended. There is a capsule-like head that is distinct from the thorax. All of the body segments have multiple ridges or rows with tiny hairs or small, soft bumps running across the body. Each of the three thorax segments is distinct, and none are enlarged or have any prolegs on them. The first three abdomen segments have a pair of small prolegs on the bottom, each with one slender, curved claw. At the end of the body, there is a long, thin, telescoping breathing tube. — Page 423

Plate 87

Mosquitoes

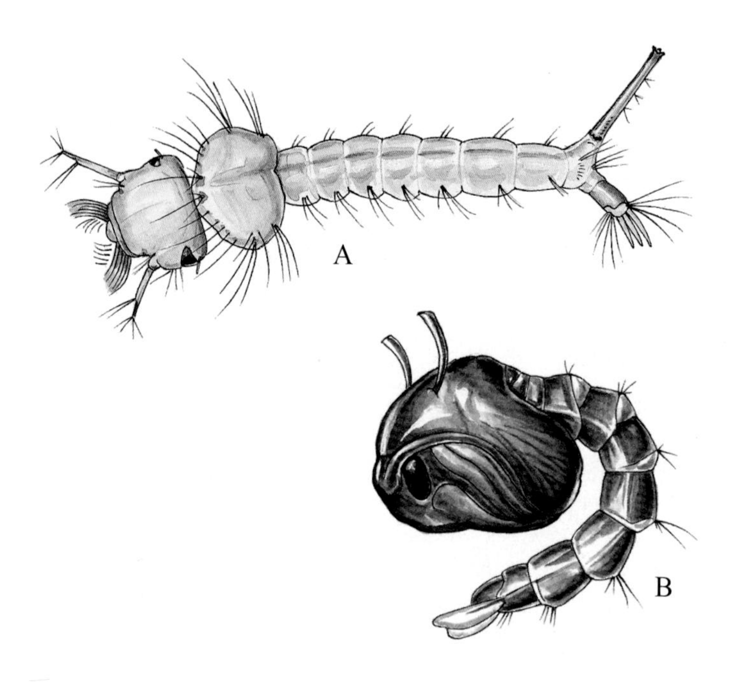

(A) Whole body, side/top view. **(B)** Pupa. (Insecta: Diptera: Culicidae)

Distinguishing Features of Larvae — Body length 4–18 mm (mature larvae). There is a capsule-like head that is distinct from the thorax. The head is rather large in relation to the body. There are prominent brushes of long hairs at the front of the head on the sides of the mouth. The three thorax segments are fused into one swollen segment that is much wider than the abdomen. There are no prolegs on the thorax or abdomen. Most species have a short breathing tube on the top of the last abdomen segment. — Page 415

Plate 88

Phantom Midges

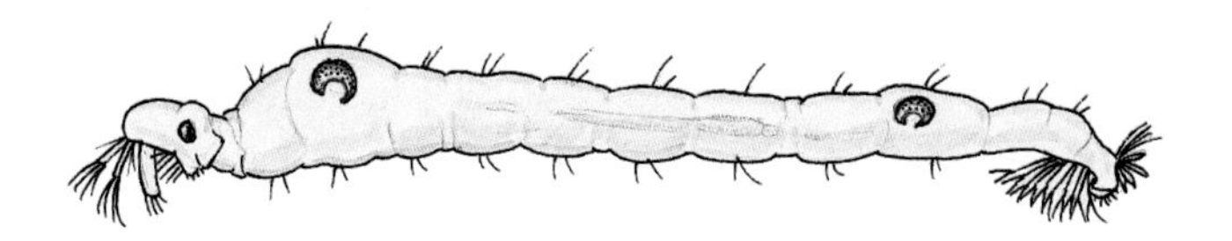

Whole body, side view. (Insecta: Diptera: Chaoboridae)

Distinguishing Features of Larvae — Body length 6–12 mm (mature larvae). They resemble mosquito larvae (Culicidae), but the common species of phantom midges are nearly transparent when they are alive. There is a capsule-like head that is distinct from the thorax. There are no prominent brushes of long hairs at the front of the head on the sides of the mouth. Stout antennae, with conspicuous long bristles on the ends, project down from the front of the cone-shaped head. The three thorax segments are fused into one swollen segment that is wider than the abdomen. There are no prolegs on the thorax or abdomen. Common species have two dark air sacs that are visible through the skin in the thorax and near the end of the abdomen. There is no breathing tube at the end of the abdomen. — Page 424

Plate 89

Dixid Midges

Whole body, side view. (Insecta: Diptera: Dixidae)

Distinguishing Features of Larvae — Body length 3–25 mm (mature larvae). There is a capsule-like head that is distinct from the thorax. The head is usually held tilted up. Each of the three thorax segments is distinct, and none are enlarged or have any prolegs on them. There is a pair of prolegs on the bottom of abdomen segment one and usually abdomen segment two. The prolegs have tiny hooks on the ends. There is no breathing tube on the end of the abdomen. Just before the end of the abdomen, there are two flattened lobes located on the top and toward the outside. There are fine hairs on the edges of these lobes. The abdomen ends with a segment that tapers to a blunt point. — Page 412

Plate 90

Black Flies

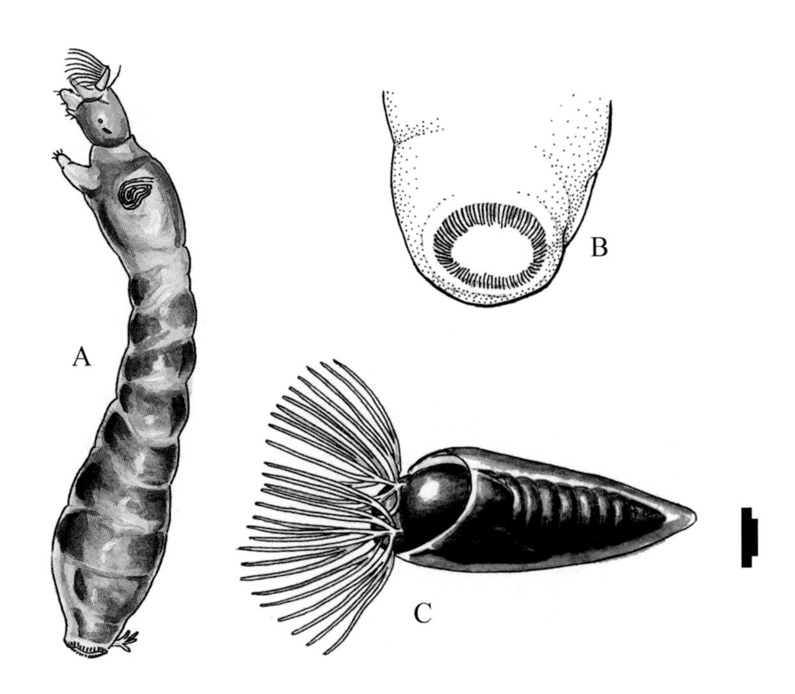

(A) Whole body, side view. **(B)** End of abdomen, rear/side view, enlarged. **(C)** Pupa in cocoon, side/top view. (Insecta: Diptera: Simuliidae)

Distinguishing Features of Larvae — Body length usually 5–8 mm, range 3–15 mm (mature larvae). There is a capsule-like head that is distinct from the thorax. There are two conspicuous clumps of long hairs on the top of the head on each side of the mouth. These clumps of hairs can be opened and closed, like a folding fan. The three separate thorax segments are hard to distinguish, making the thorax look like one long segment that is not enlarged very much in width. The first thorax segment has one proleg on the bottom. The rear one-third of the body is enlarged, giving the body a shape that resembles a vase. The rear segment of the body has a ring of minute hooks on the end (B). — Page 407

Plate 91

Non-Biting Midges, Midges

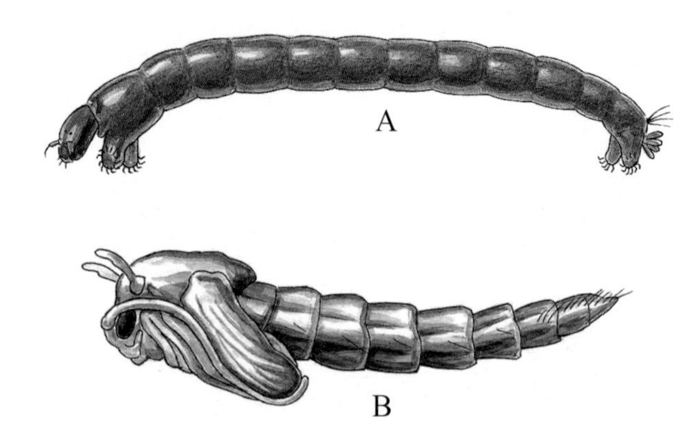

(A) Whole body, side view. **(B)** Pupa, side view. (Insecta: Diptera: Chironomidae)

Distinguishing Features of Larvae — Body length usually 2–20 mm, range up to 30 mm (mature larvae). These very common insect larvae are narrow and elongate. They somewhat resemble aquatic earthworms (Oligochaeta), but there is a capsule-like head that is distinct from the thorax. There are no brushes of long hairs on the head. Each of the three thorax segments is distinct, and none are enlarged. There are two pairs of prolegs on the bottom of the body, one pair on the first segment of the thorax segment and one pair on the last segment of the abdomen. There are tiny hooks on the ends of the prolegs. Some kinds are bright red throughout their body when they are alive. Red color is a reliable characteristic for only a portion of the family. Many other kinds of non-biting midges range from nearly white to dark green. — Page 420

Plate 92

Biting Midges, No-See-Ums, Punkies

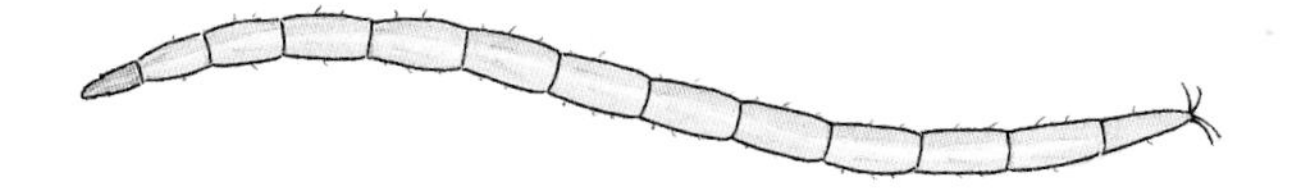

Whole body, side/top view. (Insecta: Diptera: Ceratopogonidae)

Distinguishing Features of Larvae — Body length 2–15 mm (mature larvae). Some kinds within this family look very different, and it is not possible to distinguish all biting midge larvae with one set of simple characteristics. This illustration and description will distinguish the kinds that are most likely to be found. The most common members of this family are recognized primarily by the absence of structures. They resemble non-biting midges (Chironomidae), but their bodies are even simpler. Their narrow, elongate bodies have a snake-like appearance. There is a capsule-like head that is distinct from the thorax. The head is small and narrow and does not have any brushes of long hairs. Each of the three thorax segments is distinct, and none are enlarged. The common kinds have no prolegs on the thorax or abdomen. The skin is smooth, shiny, and creamy white and lacks all surface features except for a few hairs that may be noticeable at the tip of the last segment of the abdomen. — Page 406

Plate 93

Moth Flies

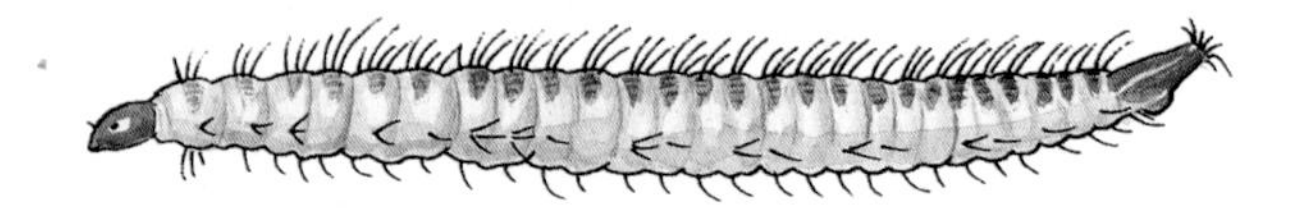

Whole body, side view. (Insecta: Diptera: Psychodidae)

Distinguishing Features of Larvae — Body length 3–6 mm (mature larvae). Some kinds within this family look very different, and it is not possible to distinguish all moth fly larvae with one set of simple characteristics. This illustration and description will distinguish the kinds that are most likely to be found. There is a capsule-like head that is distinct from the thorax. There are no brushes of long hairs on the head. Each of the three thorax segments is distinct, and none are enlarged. There are no prolegs on the thorax or abdomen. All of the body segments are secondarily divided into two or three subdivisions, giving the appearance of more than twenty-five segments behind the head. Some of the subdivisions have hardened plates on the top. The rest of the skin has numerous dark spots. Live larvae are gray-brown. There is a short, cone-shaped breathing tube with two spiracles at the end of the abdomen. — Page 416

Plate 94

Crane Flies

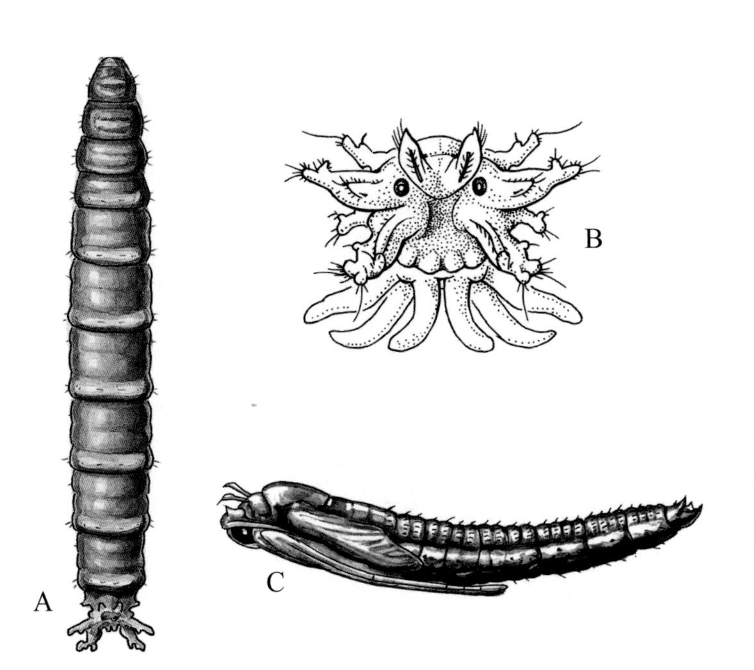

(A) Whole body, top view. **(B)** End of abdomen, rear view, enlarged. **(C)** Pupa, side view. (Insecta: Diptera: Tipulidae)

Distinguishing Features of Larvae — Body length usually 10–25 mm, range 3–100 mm (mature larvae). The body is cylindrical and usually rather stout. The skin often has fleshy lumps or short, fine hairs. The head is not visible. There is a normal capsule-like head, but it is withdrawn in the thorax. At the end of the abdomen, there are two spiracles situated on a somewhat flat area that is surrounded by several pairs of short, fleshy lobes (B). There can be one to seven pairs of these lobes, and they often have a fringe of hair. — Page 410

Plate 95

Soldier Flies

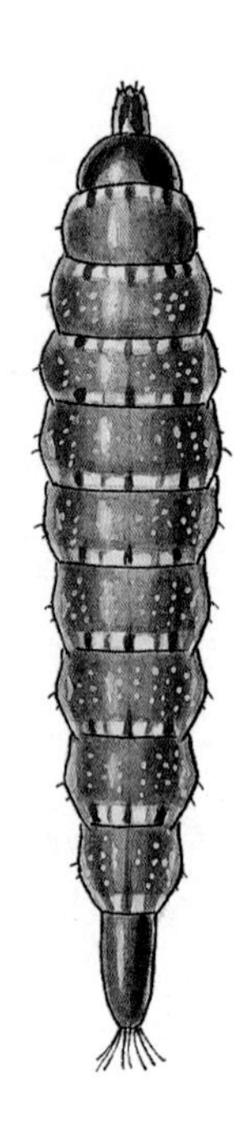

Whole body, top view. (Insecta: Diptera: Stratiomyidae)

Distinguishing Features of Larvae — Body length usually 5–35 mm, range 3–50 mm (mature larvae). The body is somewhat flattened from top to bottom. The skin has a tough, leathery, grainy texture. The head is not a distinct capsule. The head is reduced toward the rear and is partially or almost completely withdrawn into the thorax. The major mouthparts are two parallel hooks that move vertically, something like snake fangs. Usually the eyes bulge out on the sides of the head. The thorax is much wider than the head, making the larvae look like they have broad shoulders. A cleft, surrounded by hairs, runs across the end of the abdomen and contains two spiracles. — Page 428

Plate 96

Horse Flies, Deer Flies

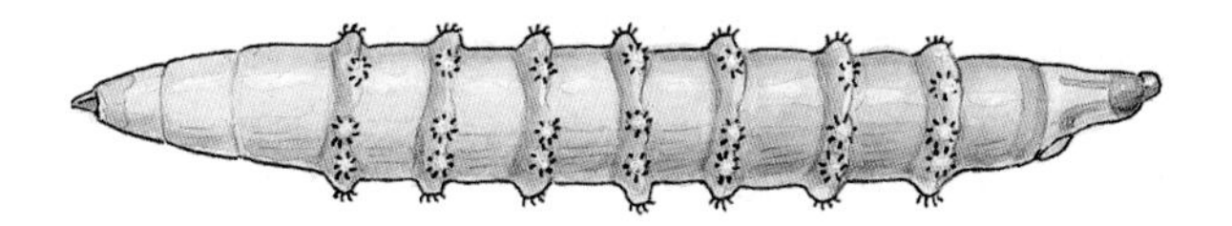

Whole body, side view. (Insecta: Diptera: Tabanidae)

Distinguishing Features of Larvae — Body length 11–60 mm (mature larvae). The body is cylindrical, with both ends tapering to a cone-shaped point. The head is not a distinct capsule. The head is partially reduced toward the rear, but some portions of a hardened head are visible sticking out in front of the thorax. The major mouthparts are two parallel hooks that move vertically, something like snake fangs. There are no structures on any of the three segments of the thorax. The first seven abdomen segments have three or four pairs of fleshy welts or prolegs arranged in a circle around each segment. There are tiny hooks on the ends of these structures. There are no tails, breathing tubes, lobes, or prolegs on the end of the abdomen. — Page 413

Plate 97

Aquatic Snipe Flies

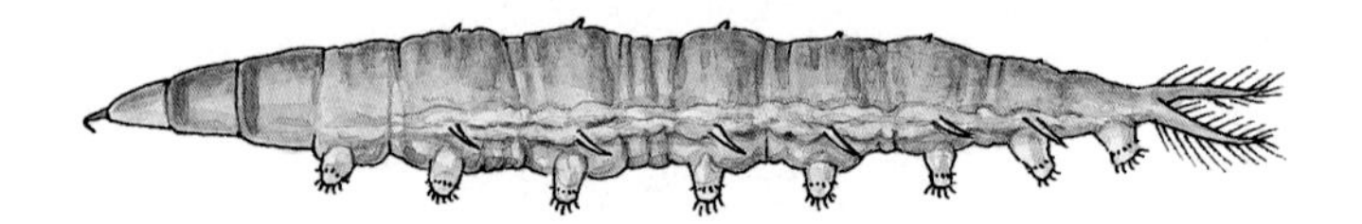

Whole body, side view. (Insecta: Diptera: Athericidae)

Distinguishing Features of Larvae — Body length 10–18 mm (mature larvae). The body is mostly cylindrical, with the front end tapering to a cone-shaped point. The head is not a distinct capsule. The head is partially reduced toward the rear, but some portions of a hardened head are visible sticking out in front of the thorax. The major mouthparts are two parallel hooks that move vertically, something like snake fangs. There are no structures on any of the three segments of the thorax. The first seven segments of the eight-segmented abdomen have a pair of prolegs on the bottom, while abdomen segment eight has only a single proleg. All prolegs have small hooks on the ends. Abdomen segments one to seven have two pairs of pointed, fleshy outgrowths, one pair near the top and one pair on the sides. These structures get progressively larger from front to back. There are two stout, pointed tails at the end of the abdomen. These tails are longer than the proleg on the last segment, and they have a fringe of hairs on their sides. — Page 405

Plate 98

Dance Flies

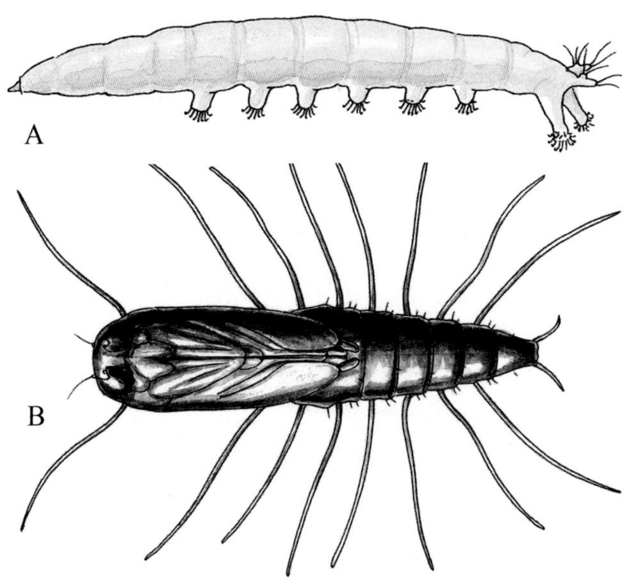

(A) Whole body, side view. **(B)** Pupa, bottom view. (Insecta: Diptera: Empididae)

Distinguishing Features of Larvae — Body length 2–20 mm (mature larvae). There is much variation in the appearance of the different kinds within this family, but the following features will distinguish many of the common kinds. The body is cylindrical, with the front end tapering to a cone-shaped point. The head is not a distinct capsule. The head is partially reduced toward the rear, but some portions of a hardened head are visible sticking out in front of the thorax. The major mouthparts are two parallel hooks that move vertically, something like snake fangs. Each of the eight abdomen segments has a pair of prolegs on the bottom (some kinds do not have prolegs on abdomen segment one). All prolegs have small hooks on the ends. The prolegs on the last segment are much longer than the others. The last segment of the abdomen has one to four rounded lobes coming off the rear, near the top. These lobes are shorter than the prolegs on the last segment of the abdomen. — Page 411

Plate 99

Rat-Tailed Maggots, Flower Flies

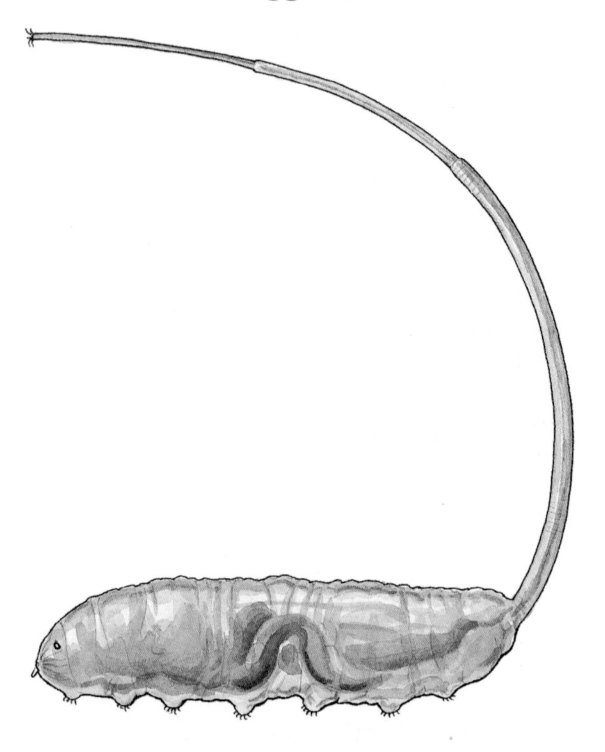

Whole body, side view. (Insecta: Diptera: Syrphidae)

Distinguishing Features of Larvae — Body length 4–16 mm (mature larvae, not including breathing tube); range up to 70 mm with breathing tube extended. The body is cylindrical and chubby, with the front end blunt and slightly rounded. The soft skin is wrinkly all over the body. The head is almost completely reduced, without any portion of a hardened head or mouthparts sticking out in front of the thorax. There are short prolegs on thorax segment two and abdomen segments one to six. On the last segment of the abdomen there is a single, long, thin, breathing tube. This tube moves in and out like a telescope, and can be several times longer than the body when fully extended. — Page 426

Plate 100

Shore Flies, Brine Flies

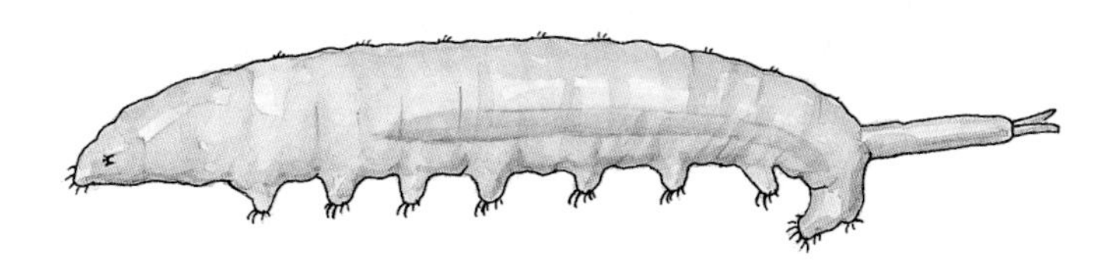

Whole body, side view. (Insecta: Diptera: Ephydridae)

Distinguishing Features of Larvae — Body length 1–14 mm (mature larvae). Some kinds within this family look very different, and it is not possible to distinguish all shore fly larvae with one set of simple characteristics. This illustration and description will distinguish the kinds that are most likely to be found. The body is nearly cylindrical, with the front end tapering to a cone-shaped point. The head is almost completely reduced, without any portion of a hardened head sticking out in front of the thorax. Usually, some hardened parts of the head can be seen inside of the first thorax segment, looking something like an internal skeleton. The major mouthparts are two parallel hooks that move vertically, resembling snake fangs. The tips of these hooks barely protrude from the front of the thorax. Each of the eight abdomen segments usually has a pair of prolegs on the bottom, with the prolegs on the last segment larger than the others. All prolegs have small hooks on the ends. The last segment of the abdomen usually has a stout, tail-like process on the end with a breathing tube that can be withdrawn. The breathing tube can be single or forked.— Page 426

References on Identification of Different Kinds of Freshwater Invertebrates

Brigham, A. R., W. U. Brigham, and A. Gnilka, editors. 1982. *Aquatic Insects and Oligochaetes of North and South Carolina.* Midwest Aquatic Enterprises, Mahomet, Illinois. 722 pages.

Burch, J. B. 1975a. *Freshwater Sphaeriacean Clams (Mollusca: Pelecypoda) of North America.* Malacological Publications, Hamburg, Michigan. 99 pages.

___________. 1975b. *Freshwater Unionacean Clams (Mollusca: Pelecypoda) of North America.* Malacological Publications, Hamburg, Michigan. 204 pages.

___________. 1989. *North American Freshwater Snails.* Malacological Publications, Hamburg, Michigan. 365 pages.

Edmunds, G. F., Jr., S. L. Jensen, and L. Berner. 1976. *The Mayflies of North and Central America.* University of Minnesota Press, Minneapolis, Minnesota. 330 pages.

Kathman, D. D., and R. O. Brinkhurst. 1999. *Guide to the Freshwater Oligochaetes of North America.* Aquatic Resources Center, College Grove, Tennessee. 264 pages.

McAlpine, J. F., B. V. Peterson, G. E. Shewell, H. J. Teskey, J. R. Vockeroth, and D. M. Wood, Coordinators. 1987. *Manual of Nearctic Diptera.* Volume 2. Research Branch Agriculture Canada Monograph 28. Canadian Government Publishing Centre, Supply and Services Canada, Hull, Quebec, Canada. Pages 1–674.

McAlpine, J. F. (editor), B. V. Peterson, G. E. Shewell, H. J. Teskey, J. R. Vockeroth, and D. M. Wood, Coordinators. 1981. *Manual of Nearctic Diptera.* Volume 1. Research Branch Agriculture Canada Monograph 27. Canadian Government Publishing Centre, Supply and Services Canada, Hull, Quebec, Canada. Pages 675–1332.

McCafferty, W. P. 1981. *Aquatic Entomology. The Fisherman's and Ecologists' Illustrated Guide to Insects and Their Relatives.* Science Books International, Boston, Massachusetts. 448 pages.

Merritt, R. W., and K. W. Cummins, editors. 1996. *An Introduction to the Aquatic Insects of North America.* 3rd edition. Kendall/Hunt Publishing Company, Dubuque, Iowa. 862 pages.

Needham, J. G., M. J. Westfall, Jr., and M. L. May. 2000. *Dragonflies of North America.* Scientific Publishers, Gainesville, Florida. 939 pages.

Peckarsky, B. L., P. R. Fraissinet, M. A. Penton, and D. J. Conklin, Jr. 1990. *Freshwater Macroinvertebrates of Northeastern North America.* Cornell University Press, Ithaca, New York. 442 pages.

Smith, D. G. 2001. *Pennak's Freshwater Invertebrates of the United States: Porifera to Crustacea,* 4th edition. John Wiley & Sons, Inc., New York. 648 pages.

Stehr, F. W., editor. 1987. *Immature Insects.* Kendall/Hunt Publishing Company, Dubuque, Iowa. 754 pages.

__________. 1991. *Immature Insects.* Volume 2. Kendall/Hunt Publishing Company, Dubuque, Iowa. 974 pages.

Stewart, K. W., and B. P. Stark. 1988. *Nymphs of North American Stonefly Genera (Plecoptera).* Entomological Society of America, College Park, Maryland. 460 pages.

Thorp, J. H., and A. P. Covich, editors. 2001. *Ecology and Classification of North American Freshwater Invertebrates.* 2nd edition. Academic Press, San Diego, California. 1056 pages.

Ward, J. V., B. C. Kondratieff, and R. Zuellig. 2002. *An Illustrated Guide to the Mountain Stream Insects of Colorado.* 2nd edition. University Press of Colorado, Niwot, Colorado. 248 pages.

Westfall, M. J., Jr. and M. L. May. 1996. *Damselflies of North America.* Scientific Publishers, Gainesville, Florida. 649 pages.

Wiggins, G. B. 1996. *Larvae of the North American Caddisfly Genera (Trichoptera).* 2nd edition. University of Toronto Press, Toronto, Ontario, Canada. 457 pages.

Section 3

Information about Different Kinds

Information about Different Kinds

Section 3 is arranged in evolutionary sequence, from the most primitive to the most advanced major groups of freshwater invertebrates. The different kinds (usually families) within the major groups are arranged alphabetically according to common names.

Flatworms

Phylum Platyhelminthes, Class Turbellaria; Plate 1

The flatworms are the free-living members of their phylum. Most of the other kinds are internal parasites that are not of interest in freshwater ecology. There are approximately 200 species of free-living flatworms in the freshwaters of North America. There are many more marine species and a few terrestrial species. More than half of the freshwater species are microscopic in size (< 1 mm) and not likely to be encountered. Many of the larger flatworms that are commonly found in freshwaters are triclads (Tricladida), which is classified as either an order or suborder within the class Turbellaria. The triclads are also commonly called planarians. The different kinds of flatworms are difficult to distinguish because it involves examining very small structures, which are sometimes located internally. This guide will only cover them collectively as a class.

Distinguishing Features — The length of mature flatworms in different species ranges from <1 to 30 mm, but most common kinds are 5–20 mm. Most are dark shades of gray, brown, or black on the top side, often with stripes, spots, or mottling. The bottom side is usually light and without patterns. Flatworms do not undergo any discernible metamorphosis as they grow and develop into adults. Immature flatworms look essentially like small adults. Flatworms can be distinguished from other aquatic invertebrates by the following combination of characteristics (Figure 9):

- Body is soft, elongate, and flattened from top to bottom.
- Body has no individual segments.
- Sides of body are slightly constricted near front, forming a head that is somewhat triangular, resembling an arrowhead.
- Two eyespots are usually situated on the top of the head, making some common kinds look "cross-eyed."
- Digestive tract has only one opening, which is called the mouth but functions as mouth and anus.
- Mouth is usually on the bottom side, positioned about one-fifth to three-fifths the length of the body.
- Mouth is often on an extendible tube (pharynx) that can be pushed out almost one-half the length of the body.
- No segmented appendages are present.

Note: It is essential to examine flatworms while they are alive. When flatworms are preserved in alcohol, they shrivel up and become distorted, resembling small pieces of chewed gum.

Explanation of Names — The most widely used common name, flatworm, obviously originated from the general appearance of the organism. The other common name that is sometimes used, planarian, comes from one of the families of flatworms, Planariidae,

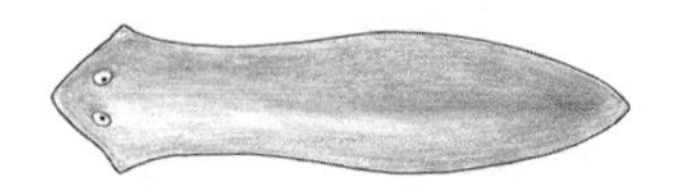

Figure 9. Representative flatworm.

which was derived from the Latin word *planus*, meaning level. The scientific name Turbellaria came from the Latin word *turba*, meaning confusion, crowd, bustle, or stir. This refers to the minute currents created in the water as flatworms wave the very small hairs on the bottom of their bodies to glide across the substrate.

Habitat — Most kinds of flatworms live in lentic-littoral or lotic-depositional habitats, but some will be found in slow-moving lotic-erosional habitats. Specific types of freshwater habitats for flatworms include springs, streams, ditches, marshes, pools, ponds, lakes, and caves, but they also occur in temporary habitats, such as puddles. Some kinds are semi-aquatic, inhabiting such places as water sitting on fallen leaves in a forest or damp soil in a meadow. They are usually found only in shallow waters, but some kinds live on the bottom of lakes as deep as 100 m. Most live only on solid substrate, but different kinds live on all types of objects: mineral, live plant, or detritus. There are also some kinds that reside on top of soft sediment. Some kinds have narrow habitat requirements and only live in running or standing water, in cool or warm temperature, or on a particular type of substrate. Other kinds have very broad requirements and live practically everywhere. The most common place to find flatworms is on the surfaces of stones in still or slow-moving water. Also, some kinds are particularly common in limestone springs and live on the plants that grow there, primarily cress. Almost all kinds are photonegative, which means that they move away from light and inhabit the undersides of objects during the daytime.

Movement — The common kinds of flatworms are clingers, gliding slowly on the substrate. They accomplish this by secreting a thin layer of mucous on the substrate then moving very fine hairs (cilia) on the bottom of their body against the mucous layer. There may also be slight waves of muscular contraction from the front to the back of the body.

Feeding — Flatworms demonstrate a variety of ways of obtaining food, including piercer-predators, engulfer-predators, and collector-gatherers. They are primarily predators of various soft-bodied invertebrates. They have no teeth or digestive enzymes to break down their prey. Most kinds simply position themselves on top of their prey, then use their feeding tube and suction to take in fluids from the body of their prey, remove chunks of their prey, or consume small organisms whole. Some can secrete neurotoxins to immobilize their prey. All kinds do some scavenging, particularly of dead animal matter. They are attracted to injured or recently dead invertebrates, possibly by some means of chemical detection. Some kinds produce a thread, layer, or mass of sticky mucous, then roll the mucous into a ball and eat it along with whatever is stuck. The food gathered in this manner usually consists of detritus, bacteria, algae, or protozoans. A few kinds feed on diatoms from the substrate.

Breathing — Flatworms obtain dissolved oxygen over the entire surface of their body.

Life History — Different species of flatworms reproduce sexually, asexually by dividing, or alternate between these methods according to ecological conditions. For example, some kinds reproduce sexually in the spring when water temperature is cool, but in the summer the reproductive organs degenerate and they switch to asexual division when water temperatures rise. All kinds are hermaphroditic. However, when they reproduce sexually they almost always mate with another individual of their species. Eggs are usually laid singly and are enclosed in a capsule that may be on a stalk. Other kinds put several eggs in a cocoon. Eggs either hatch in about 2 weeks or remain dormant through the winter. Species that reproduce sexually usually live from a few weeks to a few months. Some live longer by forming a cyst during unfavorable conditions. The number of generations per year varies from one to many in different species. This is also variable within the same species according to ecological conditions. Species that reproduce asexually may live indefinitely. Most large species divide every 5–10 days, while small species divide more frequently. Asexual flatworms have been kept in laboratory cultures for many years. Flatworms are also capable of regeneration. If an individual is cut into several pieces, most of the pieces will grow into a complete flatworm.

Significance — The free-living flatworms are not particularly significant. They are eaten by some other invertebrates, but not to a great extent. They contribute some useful information in pollution studies.

Stress Tolerance — Most common kinds of flatworms are somewhat tolerant, but this ranges to somewhat sensitive for others. Some tolerant kinds live where the decay of organic matter produces very low dissolved oxygen. These can withstand conditions without dissolved oxygen for awhile. Usually, flatworms occur in low to moderate numbers. If flatworms represent a high proportion of the invertebrates at a particular location, that is usually a reliable indicator of the presence of pollution from organic matter or nutrients.

Segmented Worms, Annelids

Phylum Annelida

Many kinds of this phylum live in terrestrial and marine environments. Two major groups of annelids, aquatic earthworms and leeches, are commonly encountered in freshwater environments, and these are covered in the following sections.

Distinguishing Features — The body wall of annelids is soft, muscular, and covered with a thin skin. An important distinguishing feature of this phylum is that the bodies of annelids are arranged in a series of many segments that are easily seen and all look essentially alike. The other distinguishing features have to do with things that are absent from annelids. There are no groups of segments, or regions, that are specialized for particular functions. Most conspicuously, there is no head at the front of the body. There are no mouthparts or antennae. Lastly, there are no segmented appendages of any kind anywhere on the body. Segmented worms do not undergo any discernible metamorphosis as they grow and develop into adults. Immature worms look essentially like small adults.

Explanation of Name — The scientific name for the phylum comes from the Latin word *annellus*, which means little ring. This refers to the body consisting of many small, round segments.

Aquatic Earthworms

Class Oligochaeta; Plate 3

One group of annelids that is very familiar to most people is the earthworms (class Oligochaeta). Among the approximately 3,500 species of earthworms in North America, about 170 species live in freshwater environments. The different kinds of aquatic earthworms are difficult to distinguish because doing so involves examining very small structures, some of which are located internally. This guide covers aquatic earthworms collectively as a class.

Distinguishing Features — The length of the body is usually 1–30 mm, but may be as long 150 mm. Aquatic earthworms can be distinguished from other freshwater segmented worms by the following combination of characteristics (Figure 10):

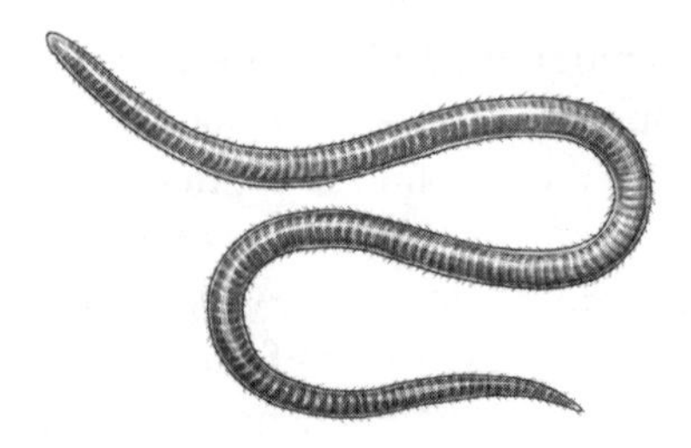

Figure 10. Representative aquatic earthworm.

- Body is soft, moderately muscular, elongate, and cylindrical in shape.
- Body consists of many similar, round, ring-like segments arranged in a row.
- Number of body segments is commonly 40–200, with a range of 7–500 in some kinds.
- Each segment after the first has bundles of tiny hairs (chaetae).
- No suckers or eyespots are present.

Explanation of Names — The common name worm is derived from the Latin word for these organisms, *vermis*. This became *wurm* in Old German and *wyrm* in Old English, both of which referred to serpents as well as worms. The prefix earth was obviously added to reflect their habit of burrowing in the soil. The scientific name for earthworms comes from two Greek words: *olig*, meaning long and *chaite*, meaning hair. This refers to the bundles of hairs that occur on the body segments.

Habitat — The kinds of earthworms that are most commonly found live in lentic-littoral, lentic-profundal, or lotic-depositional habitats. Some kinds are abundant in lotic-erosional habitats, but these are very small and are usually overlooked by beginning aquatic biologists. Most common kinds live in silt and mud in ponds, lakes, stagnant pools, and slow-moving sections of streams. Some live within accumulations of coarse detritus, on live aquatic plants, in mats of algae, or damp areas at the margins of aquatic habitats. In ponds and lakes, they are most common in the shallows down to a depth of 1 m, but some kinds live in the deepest parts of lakes. The ones that live in riffles are mostly in areas of gravel or sand.

Movement — For the most part, aquatic earthworms are burrowers, moving in the same way as terrestrial earthworms. They live in the upper layers of soft, fine sediment and move through this material by contractions of the muscular body wall and pushing on the sediment with the bundles of hairs. Some common kinds live in vertical mud tubes, with their head down and rear end protruding and waving around.

Feeding — The majority of aquatic earthworms are collector-gatherers, but a few are engulfer-predators. Most common kinds eat mud as they burrow through the sediment and digest the organic component (fine detritus) as it passes through their digestive tract. Some kinds graze on bacteria, protozoa, and algae. These kinds usually acquire their food by extending a muscular tube (pharynx) from their mouth and picking up particles with its lower sticky surface. The tube is then retracted into the mouth, along with the particles of food it has picked up. A few are predaceous on small invertebrates.

Breathing — Aquatic earthworms acquire dissolved oxygen all over the surface of their body. This may be enhanced by pumping water in and out of their anus and acquiring dissolved oxygen through their digestive tract. A few kinds have gills that are richly supplied with blood vessels. The kinds that live upside down in vertical mud tubes wave their body to circulate water and improve breathing. These kinds also have hemoglobin in their blood to help transport oxygen within their body.

Life History — Aquatic earthworms can reproduce either sexually or asexually by dividing. They are hermaphroditic, but sexual reproduction almost always involves mating with another individual of the same species. Self-fertilization is rare. Cocoons containing embryos are placed on rocks, vegetation, or debris. Mating usually takes place in late summer or early autumn, and most of the resulting cocoons are eaten by small fish. For asexual reproduction, cells in a particular section of the body, known as the budding zone, grow into a new worm that separates from the original body. Some kinds can enclose themselves in a cyst to endure adverse conditions, such as cold water temperature during winter. They tightly coil the body, then produce a coat of mucous that hardens around them. They emerge from the cyst when the water warms up. Regeneration occurs among some of the common kinds. They can regenerate front or rear portions of the body that are damaged or lost. The length of life in different kinds ranges from several weeks to several years.

Significance — Because aquatic earthworms are abundant and easy to find, they are eaten by many other invertebrates and also fish. However, the most important role of aquatic earthworms is probably their influence on the physical and chemical properties of the bottom sediment, especially in lakes. When they feed, they ingest fine particles that are a few centimeters deep in the bottom sediments. After they have digested the organic component, the rest of the material is deposited as pellets in a layer covering the bottom. Aquatic earthworms are usually very abundant in lakes, especially in the deep parts. Thus, their feeding results in the continuous vertical mixing of the top 5–10 cm of bottom sediment. Only the uppermost layer of bottom sediment is exposed to dissolved oxygen in the water, so anaerobic conditions can develop in the underlying sediment if it remains sealed off from the water. Some microbes are the only living things that can exist permanently in anaerobic conditions. The vertical mixing of the sediment by aquatic earthworms exposes the layer of pellets to dissolved oxygen and keeps the sediment in an aerobic condition that is conducive to a diverse assemblage of organisms, including invertebrates. Aquatic earthworms are also widely used in biomonitoring studies.

Stress Tolerance — The best known kinds of aquatic earthworms are the long red ones that are very tolerant of pollution. These may develop tremendous populations in places where all other organisms have been eliminated by pollution or natural sources of stress. These kinds are especially tolerant of the low dissolved oxygen conditions that occur in bodies of water that have been polluted with organic wastes, such as sewage. They also become exceptionally abundant in the deepest parts of lakes that have little or no dissolved oxygen for part of the year because of natural reasons. These kinds of aquatic earthworms can build up densities as high as 8,000 organisms per square meter of bottom. As the dissolved oxygen concentration becomes lower, the worms extend their bodies a greater distance out of their tubes, and they wave around more. When the dissolved oxygen becomes practically nonexistent, they become very still. There are no aquatic earthworms that can withstand the complete absence of dissolved oxygen indefinitely. Studies have shown that about one-third of the population will survive for 48 days in water without dissolved oxygen, but only a very small percentage will survive for 120 days.

However, it is incorrect to think that all aquatic earthworms are very tolerant and indicative of pollution. Many are only facultative to pollution and a few are somewhat sensitive to very sensitive, especially to heavy metals. It is not unusual to collect a few aquatic earthworms in a pond, lake, stream, or river. It is only when aquatic earthworms, particularly the long red kinds, represent a majority of the invertebrate assemblage that they are reliable indicators of pollution from organic wastes. Sometimes, red midge larvae (Diptera: Chironomidae) also occur in high numbers along with aquatic earthworms under these polluted conditions.

Leeches

Class Hirudinea; Plate 2

The leeches (Hirudinea) are a class of segmented worms that are closely related to earthworms (Class Oligochaeta). They are primarily freshwater organisms, with only a few species that are marine or terrestrial. There are about 69 species of leeches in North America's inland waters. More species of leeches occur in the northern United States and Canada than in the southern part of the continent. The different kinds of leeches are difficult to distinguish because key features are small and require a microscope to be seen. Some structures that are important for identifying leeches are located internally and require dissection. Therefore, this guide will only cover leeches collectively as a class.

Distinguishing Features — The length of different kinds of leeches ranges from 4 to 450 mm (body fully extended). Almost all kinds have some pattern on the top of their body, and many are brightly colored. Color patterns include red, yellow, orange, and green arranged in stripes, zigzags, spots, or splotches. Background colors are usually various shades of tan, brown, gray, or black. Colors must be observed while leeches are alive. If they are placed in a preservative such as alcohol, red, brown, and black hold up while green, yellow, and orange quickly fade. Leeches can be distinguished from other freshwater segmented worms by the following combination of characteristics (Figure 11):

- Body is somewhat soft, but very muscular, and flattened from top to bottom.
- Body is composed of 34 actual segments arranged in a row.
- Body appears to have many more segments because a lot of the segments have fine lines (annuli) running across them, creating secondary subdivisions that look like actual segments.
- Actual segments in middle portion of body commonly have 3–6 secondary subdivisions; with as many as 14 in some kinds.
- No bundles of tiny hairs are present on any segments.
- First few segments of the body have a number of small eyespots on top.
- Two distinct suckers are situated on the bottom of the body, one at the front and one at the rear.
- Mouth is on the bottom of the body, within the front sucker.
- Anus is on the top of the body at the rear, just in front of the rear sucker.

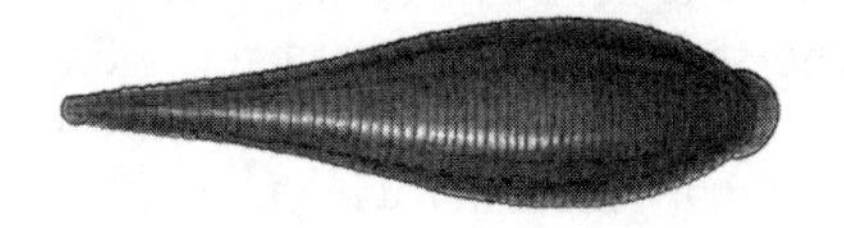

Figure 11. Representative leech.

Explanation of Names — The common name leech comes from the Old English word *laece*, meaning worm. This became *leche* in Middle English. Leech has also been a synonym for physician because of the widespread use of leeches in medicine throughout history. The scientific name simply comes from the Latin word for these organisms, *hirudo*.

Habitat — Most leeches live in lentic-littoral and lotic-depositional habitats, but a few are lotic-erosional. Different kinds of leeches either live freely on solid substrates or attached to hosts in a variety of habitats. Most of the common kinds are found in quiet, protected, shallow areas of ponds, lakes, marshes, and slow sections of springs and streams, where plants, plant debris, or stones are plentiful. Free-living kinds avoid light by hiding under objects during the day, then becoming active at night. Most live in water that is less than 2 m deep, but they may occur as deep as 50 m. Leeches do not live on substrates with fine sediment because they cannot attach their suckers. In suitable habitats, it is common to have as many as 700 leeches per square meter of bottom. Densities have been reported as high as 50,000 per square meter in exceptional circumstances. There are a few kinds that live in riffles and a few others that are semi-aquatic, roaming around on damp shores.

Movement — All leeches are clingers by means of the two suckers on the bottom of their body. They move, like an inchworm caterpillar, in a looping motion by alternately gripping the substrate with their suckers. While holding on with the front sucker, they contract the body muscles, bring the rear sucker close to the front sucker, then attach the rear sucker This forms a loop above the substrate. Then they release the front sucker and extend the body muscles as far forward as possible before grabbing on to the substrate with the front sucker. The shape of the body changes dramatically during these contractions and extensions. Some kinds of leeches are excellent swimmers that gracefully wiggle their bodies in the water. Some of the large species look like small snakes swimming through the water.

Feeding — All leeches feed on the fluids of other organisms, but different kinds of leeches obtain their food by two mechanisms. They are either piercer-predators that kill small prey, or external parasites that suck the blood from larger prey but seldom kill them. Although leeches are notorious for their blood-sucking habits, most of the leeches that are collected by persons looking for invertebrates with a net or picking up material on the bottom are piercer-predators that feed on a variety of invertebrates. Most of these have a muscular tube (proboscis) that can be extended from the mouth, which lies within the front sucker. They either use this tube to suck all of the fluids from their prey or they swallow small prey whole. If prey are swallowed whole, leeches still only feed on the fluid contents because they quickly get rid of the hard body parts through the mouth or anus. Common prey organisms include insect larvae, aquatic earthworms, snails, and other leeches. Some kinds have a distinct preference for specific types of invertebrate prey, while others are generalists. Some of these kinds also obtain part of their food by scavenging dead animal matter.

Only a few kinds of leeches are blood-sucking external parasites. Most of the blood suckers have three sharp-toothed jaws behind large mouths. They cut the skin of their hosts and introduce an anticoagulant chemical (hirudin) to keep the blood flowing while they suck it into their mouth. Some of the blood suckers do not have teeth and instead release digestive enzymes that dissolve layers of their host's skin until blood is released. Most of the blood suckers are temporary external parasites on frogs, turtles, and fish. A few attack warm-blooded animals in the water such as waterfowl, mammals, and occasionally humans. They take a large meal of blood, increasing their weight up to five times, then leave the host until their next meal. The wound bleeds for a long time after the leech has detached because of the anticoagulant that it put into the wound. Blood suckers use their food slowly. Some do not feed again for as long as 2 years after they have filled themselves with blood. The turtle leeches are probably the most commonly encountered blood suckers. One species called the horse leech lives in livestock watering troughs and attacks cattle and horses when they come to drink. Some kinds of blood-sucking leeches are nearly permanent external parasites of fish and turtles. These do not leave the host except to mate.

Breathing — Leeches obtain dissolved oxygen all over the surface of their body. They sometimes increase the rate of oxygen transfer by attaching one or both suckers and wiggling their body to create more water flow over their body. None of the freshwater species have gills.

Life History — Leeches always reproduce sexually. They are hermaphroditic, but sexual reproduction almost always involves mating with another individual of the same species. Mating takes place in the spring. Some kinds put their eggs in a cocoon, which they deposit in soft mud or along the shoreline. Certain kinds of these leeches lie on the cocoon after depositing it. Other kinds carry their eggs in a membranous sac on the bottom side of their body. The young leeches of these kinds remain attached to the parent after they hatch from the eggs. While they remain attached, they feed on the remains of the membranous sac. Most leeches spend the winter in a dormant state, buried in the substrate. Particular kinds can exist in temporary habitats by burrowing into the mud and constructing a mucous-lined cell to protect them from drying out while there is no water. Some kinds of leeches have two generations per year, but many require several years to mature. There have been instances in which leeches kept in captivity have lived 10–15 years.

Significance — Leeches have been used for medicinal purposes throughout most of recorded history. The belief was that illness could be cured by getting rid of "bad blood." The use of leeches in medicine reached its peak during the eighteenth and nineteenth centuries, when raising and selling leeches was a thriving business. The use of leeches in medicine practically disappeared in North America and western Europe in the twentieth century. However, the early physicians were not completely wrong about the usefulness of leeches, and recently there has been a minor resurgence in their use. They are occasionally used to remove blood from areas where tissue has been transplanted or reattached.

Leeches are both beneficial and a nuisance in the natural aquatic environment. On the positive side, they serve as food for quite a few vertebrates and invertebrates. Vertebrates that eat leeches include fish, newts, salamanders, snakes, and birds. Because fish readily consume them, they are raised and sold as live bait, particularly in the North. Invertebrates prey upon them to a limited extent. These are primarily aquatic insect larvae, such as dragonflies, true bugs, dobsonflies, beetles, and caddisflies. Some of the blood-suckers become an annoyance when they seek a meal from bathers. This is primarily a problem at northern lake resorts. Leeches are attracted to disturbances in the water, so they congregate in the areas where people are swimming and splashing. The kinds of leeches that are permanent parasites of fish cause considerable losses in hatcheries and in nature.

Stress Tolerance — The ability of leeches to tolerate pollution has not been studied very much. A few kinds can withstand at least moderate pollution. For example, some can survive several days without any dissolved oxygen. Collecting a few leeches at a location does not mean that pollution exists. It is when leeches form a high proportion of the invertebrates at a site that it may indicate some form of pollution has appreciably reduced the dissolved oxygen.

Mollusks

Phylum Mollusca

Mollusks (Phylum Mollusca) include the animals known as snails, limpets, clams, and mussels. The most obvious characteristic of mollusks is their hard shell, which can enclose their entire body. Inside the shell, they all have soft bodies that are not arranged in a series of similar segments. Instead, the body is organized into several irregular sections. These include a muscular foot for locomotion, a head region, the main portion of the body called the visceral mass, and a thin layer of fleshy tissue called the mantle that covers the body. The mantle secretes the shell, which is made of calcium carbonate and protein.

There are about 110,000 species of mollusks in the world, but most of them are marine or terrestrial. In freshwater environments of North America, there are less than 800 species of mollusks. Mollusks are divided into two major groups. Snails and limpets make up the group known as gastropods (Class Gastropoda). Mussels and clams, known as bivalves (Class Bivalvia), are the other major group of mollusks.

Explanation of Names — The scientific name for the phylum comes from the Latin word *molluscus*, which means thin-shelled, such as the shell of a nut.

Snails

Class Gastropoda; Plates 4 –11

Snails are the freshwater mollusks with one shell, and the various kinds of these are referred to collectively as gastropods. Snails, which are also called gastropods, account for almost three-fourths of the 110,000 species of mollusks in the world, but most of those are marine or terrestrial. Freshwater gastropods are moderately diverse in North America, with 15 families and about 500 species. This guide covers the eight families that are most likely to be encountered.

Gastropods are divided into two major groups, mostly according to how they breathe. Some of the families use gills to breathe oxygen dissolved in the water, and, accordingly, they are called gilled snails or prosobranchs (Subclass Prosobranchia). The other families obtain oxygen from the air by means of a structure that works something like a lung, so they are called the lunged snails or pulmonates (Subclass Pulmonata). Another common name for pulmonates is pouch snails. In addition to differences in breathing, there are other biological aspects that are different between gilled snails and lunged snails, so they are frequently discussed according to these two subclasses. Of the eight families covered in this guide, four are in each subclass.

Distinguishing Features — Snails have a single shell (Figure 12). In most kinds, the shell is coiled and elongate, like a spiral staircase, but in one family the shell is coiled flat and in one family the shell is a simple cone with no coiling. Undisturbed freshwater snails show a prominent head, which has a mouth with a ribbon of rasping teeth and a pair of elongate structures called tentacles that superficially resemble the antennae of arthropods. (Land snails have two pairs of tentacles.) The muscular foot that projects from the shell is inconspicuously colored grayish, brownish, or blackish, often with speckles or blotches of yellow or white (Figures 1T, 2J). The gilled snails have a flat, hard plate, called an operculum, on the top of the rear end of the foot. When the foot is withdrawn into the shell, the round to oval operculum fits snugly in the opening of the shell and seals the snail inside. The shells of adult freshwater gastropods range in size from about 2 to 70 mm in their maximum

Figure 12. Representative snail.

dimension. Freshwater gastropods do not undergo any discernible metamorphosis as they grow and develop into adults. Immature gastropods look essentially like small adults. You can usually tell if a snail is an adult by counting the coils. Adults usually have four or more coils. Most of the shells are dull, nondescript shades of gray, tan, black, or brown. Some have faint bands or blotches of a slightly different shade of the background color. Only a very few freshwater snails have bright tinges of green, yellow, or red.

The shells of different kinds of snails vary considerably in size, shape, and thickness, and these features are used to distinguish the families. Snails with elongate, coiled shells are described as being either left-handed (sinistral) or right-handed (dextral). You determine this feature by holding the shell with the narrow end up and the opening facing you. The shell is left-handed if the opening is on the left, and right-handed if the opening is on the right. Another feature of the shell that is used to distinguish different families is the extent to which the individual coils bulge out on the sides. In some families, the individual coils (whorls) bulge out distinctly, making the sides of the shell a series of curves with deep incisions between the coils. In other families, the individual coils hardly bulge out, making the sides of the shell almost straight lines with only shallow incisions between the coils. The pattern of the lines on the operculum is used to distinguish the families of gilled snails.

Explanation of Names — The common name snail is derived from the Middle English word *snahhan*, which was derived from the Old English *snaegl* and Old High German *snecko*. These words mean to creep and refer to the slow, gliding movement of these organisms. The explanation of the scientific name for the class of snails lies in the fact that part of the digestive system begins in the muscular foot projecting from the shell. Hence, Gastropoda, comes from two Greek words: *gaster*, meaning stomach, and *pod*, meaning foot.

Habitat — Snails are common in lentic-littoral, lotic-depositional, and lotic-erosional habitats, but not in lotic-erosional habitats that are exceptionally swift or turbulent. Specific types of habitats include ponds, lakes, ditches, swamps, and slow reaches of springs, streams, and rivers. The types of substrates inhabited by different kinds of gastropods include rocks, sand, mud, vegetation, and plant debris. Many kinds of snails require a specific type of substrate. Some kinds have very narrow habitat requirements, such as plant species that grow in thick beds versus plants that grow in loose accumulations. Snails usually do not occur on shifting sand or gravel in swift portions of streams or wave-swept shores of lakes. Most kinds live in shallow water, less than 3 m deep, regardless of the type of habitat. Natural water chemistry has a distinct effect on snails. There are usually more kinds of gastropods in hard water, which is the term used for water that has a lot of calcium carbonate dissolved in the water from limestone in the underlying rocks. Hard water is probably a better environment in which to live because gastropods require calcium carbonate for manufacturing their shells. Water hardness does not matter to some kinds that live in a wide variety of habitats. Snails rarely occur in waters that are naturally acidic, most notably bogs. There are usually more kinds of gastropods in larger bodies of water, presumably because there is a greater variety of specific habitat types. Gastropods are usually not common in waters that are continuously cold, such as lakes and streams high in the mountains, or in waters that are continuously very warm (above 30° C). Some lake species of snails migrate seasonally to different depths in order to find suitable temperatures. Some of the lunged snails are well adapted to survive in intermittent habitats, whereas gilled snails only live in permanent habitats.

Movement — Snails exist as either clingers or sprawlers, according to the type of substrate that they inhabit. They all glide on the substrate on a film of mucus secreted from the bottom of the foot. The mucus lubricates the surface and allows them to ease forward by muscular movements of the foot and the action of tiny cilia on the bottom of the foot. This mechanism of locomotion loosely seals some kinds to the surface of firm substrates, thereby making them effective clingers. The broad, flat foot also prevents some kinds from sinking in soft sediment, thus allowing them to exist as sprawlers.

Feeding — Most freshwater snails are scrapers, but various kinds also use quite a few other methods to acquire their food. These other methods include being collector-gatherers, shredder-herbivores, shredder-detritivores, or collector-filterers. Snails have a long, strap-like structure called a radula, which is a little bit like a human tongue, inside their mouth. The radula is very hard and is covered with many (sometimes thousands) small, sharp teeth. They push the radula out of their mouth until it touches the substrate, then they drag it along the substrate as they pull it back into their mouth. The teeth on the radula act like a rasp or file to loosen small pieces of food, which then accompanies the radula back into the mouth. Most snails remove living algae, especially filaments of green algae, from submerged firm surfaces. However, some kinds, particularly the ones that live on soft substrate, gather in fine detritus that is lying on the substrate. Some kinds shred bits of living aquatic plants, or they break down leaves and other coarse detritus that has entered the water and begun to decay. A few snails use their gill to set up currents of water, from which they filter out algae and fine detritus suspended in the water. Gastropods occasionally eat small invertebrates, but probably only when they accidentally ingest them while they are scraping algae or gathering detritus.

Breathing — Gilled snails rely on oxygen dissolved in the water, which they obtain mostly by means of a gill on the body inside of the shell. Water enters the shell and comes in contact with the gill. This internal gill (ctenidium) has many flat leaflets arranged like the teeth on a comb. These leaflets provide a great deal of surface area for transferring oxygen from the water into their body. The lunged snails obtain oxygen primarily from the air. They have a cavity in their body, and they periodically come to the surface to fill it with air. Oxygen enters the body through the thin tissue that lines the air cavity. However, some of the smaller lunged snails do not carry air pockets, and are evidently able to obtain sufficient oxygen through lungs filled with water. Some of these types never have to come to the surface of the water to obtain air.

Life History — Some kinds of snails have separate sexes, while others are hermaphroditic. As a general rule, most gilled snails have separate sexes, while most lunged snails are hermaphroditic. Most of the hermaphroditic kind mate with another individual of the same species, even though some self-fertilization is possible. Eggs of all kinds are usually laid in the spring. The number of eggs varies from a few to hundreds. Lunged snails produce more eggs than gilled snails. Snails enclose the entire batch of eggs in a mass, which they attach to various objects in the water. Snail egg masses often resemble a drop of mucus and may be easy to see when collecting aquatic invertebrates. Freshwater gastropods develop within the egg mass for awhile, so that upon hatching they look like tiny snails with only one or two coils. Most species of lunged snails complete their life cycle in a year or less, with some producing two or three generations per year. Most kinds of gilled snails live for 2–5 years. Lunged snails are able to hibernate in the mud during adverse conditions, such as when shallow ponds freeze solid or intermittent ponds and streams dry up. They are more successful at hibernating because they do not obtain their oxygen by means of a gill that must be in contact with water. Some lunged snails make up for the lack of an operculum by secreting a sheet of mucus in the opening of their shell. This hardens and seals off the snail and helps protect it from dry conditions.

Significance — Some snails are intermediate hosts of parasites that cause diseases in humans, but this is not much of a problem in North America. Freshwater snails are generally regarded as beneficial organisms. They serve as food for many vertebrates and invertebrates. Fish, amphibians, and waterfowl are the primary vertebrates that consume appreciable numbers of snails. Snail-eating invertebrates include leeches, crayfish, water beetle larvae, true bugs, true flies, and dragonfly larvae. The vertebrate predators and crayfish crush the shells to consume the snails. The various invertebrate predators invade the intact shells and eat the soft tissue of the snails. Snails that graze on attached algae affect the assemblage of algae, as well as other components of the ecosystem, in a positive way. Moderate grazing by snails stimulates more vigorous growth of algae by getting rid of the old cells that are past their peak of production. This prevents a thick layer of dead algae from building up, which would shade out the vigorous, young algae cells. Snails are also beneficial to large, rooted submerged plants because they keep a thick coating of algae from building up on the leaves and twigs. Reducing the layer of algae allows more sunlight to reach the plant tissues where photosynthesis takes place, thereby leading to lusher beds of submerged aquatic plants. Thus, the net result of snail grazing is more high quality food and habitat for other organisms in the ecosystem.

Stress Tolerance — Snails provide useful information in studies of pollution in freshwater environments. Most gilled snails are somewhat sensitive, ranging from very sensitive to facultative. They require high concentrations of dissolved oxygen, so their absence might be an indication of pollution from organic wastes or nutrients. On the other hand, almost all lunged snails are somewhat tolerant, with only a few facultative kinds. Because they are air breathers, some can live in severe organic or nutrient pollution that uses up all of the oxygen in the water. Having a few lunged snails present at a site is not meaningful, but if they make up a high proportion of the invertebrate assemblage that is a reliable indication of pollution.

Biology of Common Families

There is not much about the biology of snails that is unique or consistent for entire families. Many snail genera, and even individual species, have very broad biological features, and they are best described as opportunistic generalists. Many of the common species can be found in running or standing waters, where they move around on plants, rocks, or soft sediment, and thrive on a diet of microscopic algae, rooted aquatic plants, or detritus. Because most of the families contain quite a few species that are not particular about where they live or what they eat, it is meaningless for the intended purpose of this guide to summarize biological information about families. The families usually reflect all of the biology described above for the entire class of snails. Therefore, this section only includes a few biological features that are fairly consistent within a family. Unfortunately, the families of snails do not have accepted common names that apply to all members of the family. I have converted the scientific names to common names by dropping –ae from the end of the scientific name. In the Different Kinds section, I have alphabetically listed common names that are used for various individual kinds within the family.

Bithyniid Snails

Subclass Prosobranchia, Family Bithyniidae; Plate 4

Different Kinds in North America — Few; 1 genus, 1 species (*Bithynia tentaculata*). The genus of the single species in this family is also used as a common name, bithynia.

Distribution in North America — Bithynia was introduced from northern Europe to the Great Lakes region in the late 1870s. It has spread throughout the Great Lakes drainage from Wisconsin to Pennsylvania and New York.

Habitat — Usually live on aquatic vegetation.

Feeding — Scrapers, collector-filterers. This species grazes on attached algae in the normal manner for snails, but it also uses its internal gills to filter algae that lives suspended in the water. The ability to feed by filtering may explain why this species becomes very abundant in lakes that have become enriched with nutrients because of human activities.

Stress Tolerance — Facultative.

Hydrobiid Snails

Subclass Prosobranchia, Family Hydrobiidae; Plate 6

Different Kinds in North America — Moderate; 28 genera, 152 species. Common names for various kinds in this family include hydrobes, pebblesnails, siltsnails, and springsnails.

Distribution in North America — Throughout.

Habitat — Usually found on aquatic plants. Sometimes these small snails occur in enormous numbers.

Feeding — Mostly scrapers.

Stress Tolerance — Somewhat sensitive.

Pleurocerid Snails

Subclass Prosobranchia, Family Pleuroceridae; Plate 7

Different Kinds in North America — Moderate; 7 genera, 153 species. Common names for various kinds in this family include elimias, hornsnails, mudalias, and rocksnails.

Distribution in North America — Throughout, but the greatest diversity occurs in the southeastern United States.

Habitat — Mostly lotic-erosional. Pleurocerids usually occur in rocky riffles or sandy shoals. They are often conspicuously abundant on stones in shallow, slow-moving riffles in medium-sized rivers.

Feeding — Mostly scrapers.

Stress Tolerance — Somewhat sensitive to facultative.

Viviparid Snails

Subclass Prosobranchia, Family Viviparidae; Plate 5

Different Kinds in North America — Few; 5 genera, 19 species. Common names for various kinds in this family include campelomas and mysterysnails.

Distribution in North America — Throughout, but more kinds in southeastern and central United States.

Habitat — Most common on sandy bottoms.
Feeding — Mostly collector-gatherers.
Stress Tolerance — Facultative.

Ancylid Snails

Subclass Pulmonata, Family Ancylidae; Plate 8

Different Kinds in North America — Few; 4 genera, 11 species. Limpet is frequently used as a common name for members of this family.

Distribution in North America — Throughout.

Habitat — Lotic-erosional. Ancylids are more common on firm substrates in current.

Feeding — Mostly scrapers.

Other Biology — Even though ancylids are in the subclass of lunged snails, they do not have an air cavity in their body, and they do not have an internal gill like the subclass of gilled snails. Instead, they have a projection from the foot that acts as an accessory gill, and they probably obtain a lot of dissolved oxygen all over the surface of their body.

Stress Tolerance — Somewhat tolerant.

Lymnaeid Snails

Subclass Pulmonata, Family Lymnaeidae; Plate 10

Different Kinds in North America — Moderate; 9 genera, 57 species. Common names for various kinds in this family include fossarias, lymnaeas, and pondsnails.

Distribution in North America — Throughout, but more kinds in the North.

Habitat — Occur on firm surfaces, such as rocks, woody debris, and aquatic plants, as well as on soft, silty substrate.

Feeding — Scrapers and collector-gatherers.

Stress Tolerance — Somewhat tolerant.

Physid Snails

Subclass Pulmonata, Family Physidae; Plate 11

Different Kinds in North America — Few; 4 genera, 37 species. The genus of the most common species in this family is also used as a common name for the family, physas.

Distribution in North America — Throughout.

Habitat — Most common on soft, silty substrate, but also found on any solid objects that protrude above the bottom.

Feeding — Mostly collector-gatherers, some scrapers.

Stress Tolerance — Somewhat tolerant.

Planorbid Snails

Subclass Pulmonata, Family Planorbidae; Plate 9

Different Kinds in North America — Moderate; 11 genera, 44 species. Common names for various kinds in this family include gyros, orbsnails, rams-horns, and sprites.

Distribution in North America — Throughout.

Habitat — Most common on soft, silty substrate, but also found on any solid objects that protrude above the bottom.

Feeding — Mostly collector-gatherers, some scrapers.

Other Biology — Planorbids have hemoglobin in their blood. This allows them to live in some habitats with very low concentrations of dissolved oxygen, such as small, warm ponds.

Stress Tolerance — Facultative to somewhat tolerant.

Mussels and Clams

Class Bivalvia; Plates 12 –15

Mussels and clams are the mollusks that have two shells. Mussels and clams, which are also called bivalves, account for about one-fourth of the 110,000 species of mollusks in the world, but most of those species occur in marine environments. There are about 270 species of freshwater bivalves in North America. They are often the largest invertebrates present in freshwater environments, in terms of body mass. Excluding those that occur in estuaries and brackish waters, there are only five families of bivalves in freshwater environments in North America: Margaritiferidae, Unionidae, Sphaeriidae, Corbiculidae, and Dreisseniidae. The Margaritiferidae family contains only a few species that amateur aquatic biologists will have a hard time distinguishing from Unionidae. In this guide, these two families are described and discussed together as the superfamily Unionacea because both families have the same overall appearance and biological features. The other three families are described separately. Two of the families, Corbiculidae and Dreisseniidae, have only one freshwater species in each, and both of these are introduced pests.

Distinguishing Features — Mussels and clams have two shells that are opposite each other and strongly connected on top by an elastic hinge ligament (Figure 13). Under natural conditions, the shells are held slightly open with the foot and two tubular structures, called siphons, protruding (Figure 1V). The foot is located on the bottom of the organism at the front. It is a thick, muscular structure shaped like an axe. The two tube-like siphons are on the top of the body at the rear margin. Their purpose is to bring water in and out of the body. The foot is pushed down in loose substrate, while the siphons are held in the water. The soft body tissue is pinkish to

Figure 13. Representative bivalve.

grayish. Bivalves have no distinct head, and hence no tentacles or eyes, like the gastropods. The length of shells ranges from 2 to 250 mm.

There is much variation in the shape of shells among different species of mussels and clams, including elongate, oval, almost circular, or with three or four sides that are almost straight. The umbo is a raised area on the top of the shell, just in front of the hinge ligament. This is where shell growth begins as a juvenile clam or mussel. The concentric lines that surround the umbo are growth rings. As bivalves grow, the shell is enlarged by adding layers to the outside. Some kinds have bumps, wrinkles, or ridges on their shells. The color of the shell also varies among different kinds of bivalves, ranging from whitish or light tan or light yellow-green to dark green, brown, or blackish. Some have pigmented rays or blotches on their shell. The shells of large, old bivalves usually have some of the dark outer protective layer worn away near the umbo, exposing white calcium carbonate underneath (Plate 12). The interior of bivalve shells is lined with several layers, called mother of pearl, that reflect light in colors ranging from silvery white to pink or dark purple. Some bivalves undergo dramatic metamorphosis during their growth and development. These kinds have parasitic or free-swimming larvae that have no shells and look nothing like clams or mussels. Larvae transform into juveniles that do look like adults while they are growing and developing. The descriptions given in this guide pertain to juvenile and adult bivalves.

Explanation of Names — The common name clam dates back before the twelfth century. It is derived from the Old High German *klamma*, meaning constriction. In Old English, this became *clamm* and was used to mean bond. The current spelling and use of clam began in Middle English and refers to the ability of these organisms to hold their shells tightly together. The other common name for some members of this group, mussel, is derived from the Latin word for muscle, *musculus*. This is a reference to the strength of these organisms at holding their shells together. The scientific name for the class Bivalvia was derived in the seventeenth century. In

Latin, *bi* was a word for two, and *valva* was used for the hinged and movable pieces of several shell-bearing animals. *Valva* came from another Latin word, *volvere*, meaning to roll or wrap, in reference to the function of the shells enclosing the body of these animals.

Habitat — Mussels and clams live throughout all sizes of lentic and lotic habitats, but they are most abundant and diverse in lotic-erosional habitats with moderate current in medium to large rivers. They also live in natural lakes, with large lakes containing more kinds than small ones. Some kinds live in reservoirs constructed by damming rivers, but there are always many less kinds in an impoundment than in the original free-flowing river. Most live in shallow water, 2–7 m deep, where there is enough current to keep the bottom free of fine sediment and the dissolved oxygen concentration is high. Because they position themselves at least partially below the bottom, different kinds have distinct preferences for substrate composition. Most kinds are found where the substrate is stable and consists of gravel or a combination of gravel and sand. A few kinds of bivalves live in other, less accommodating natural habitats, such as springs and small streams, shifting sand or mud substrates, solid rock substrate, and at depths greater than 30 m in lakes with little dissolved oxygen.

Movement — Almost all mussels and clams are burrowers. They situate themselves upright, with the bottom half or more of their shells buried in the substrate and their foot extended into crevices. Bivalves are much less mobile than gastropods, but they may move slowly through the substrate by contractions of the large, muscular foot. They elongate their foot and push it forward. The front part of the foot swells and wedges in the substrate, then the foot muscles contract, which pulls the whole organism forward. Tracks may be visible on the bottom where they have dragged their shells through the substrate. Thin-shelled species are more active than thick-shelled ones. Some kinds lie entirely buried except for their siphons, while others are completely under the substrate. Most kinds position themselves deeper in the substrate during the winter.

The only exception to the burrowing habit is the zebra mussel, which is a clinger.

Feeding — All mussels and clams are collector-filterers. They create a one-way current of water that enters their body through one of the tubular siphons and then exits through the other siphon. Minute hair-like processes, called cilia, on the mantle, gills, and body beat in unison to make the water move. Small suspended particles of food are filtered out of the water by cilia and a mucous coating on the gill filaments. There are grooves on the gills with other cilia that move the captured particles to the mouth. The particles of food that they filter out of the water consist of microscopic algae (phytoplankton), bacteria, and detritus. Live algae is the most nutritious of these foods, and most bivalves grow faster when it is abundant in the water. Bivalves themselves also put some of the food that they filter into suspension. As the foot moves through the substrate it stirs up fine detritus lying on the bottom, and the clam or mussel filters this material out of the water before it settles back to the bottom. Even the bivalves that bury themselves completely in the bottom are filter feeders. They create a momentary suspension of detritus in the small space around their shells.

Breathing — Bivalves rely on dissolved oxygen that they obtain primarily with their gills and mantle. The gills consist of many long, thin filaments inside the mantle cavity. Bivalves have very large gills, and there appears to be more surface area than is needed to obtain the amount of dissolved oxygen that they need. The large size of their gills probably has more to do with filter feeding than breathing. Some dissolved oxygen may be obtained all over the body surface.

Life History — Almost all freshwater bivalves hold the eggs in the gills and incubate them until they have developed into young larvae. However, there are distinct differences among the families as to how the immature stages develop once they have left the parent.

Mussels are unique among mollusks because they have a parasitic larval stage. Fertile eggs are produced in midsummer. These are held within the gills of the female for 1–10 months while the embryos develop. Females may incubate several thousand to several million embryos. The offspring are released through the outgoing siphon as tiny (0.05–0.5 mm) specialized larvae called glochidia. These larvae usually must become attached to a fish host. Often there are just a few species of fish, or perhaps only one, that will serve as a suitable host for any given species of mussel. Glochidia cannot move on their own. They simply float or lie on the bottom until they encounter a fish, at which time they attach themselves to the fish's gill filaments, skin, fins, or scales. Fish become parasitized by brushing against the bottom, stirring up bottom sediments, or taking in water through their gills. Glochidia can only live unattached for a few days after being released from the mother. They detach if they accidentally become attached to the wrong species of fish. Many die without finding a suitable host, which is why female mussels produce so many glochidia. Some mussels have interesting mechanisms for luring host fish nearby, so the glochidia can attach to them. Some species release their glochidia in masses of mucous that mimic food items. In other kinds, the adults have muscular projections on the edges of their mantle that look like prey to fish. Both types of lures are often brightly colored and look like flatworms, leeches, aquatic earthworms, insect larvae, or small fish. For glochidia that successfully attach, the fish tissue grows over them and seals them in a cyst. They remain attached to the fish in this manner for 1–10 weeks and apparently do little harm to the host. The main benefit they obtain from the fish host is dispersal, not nutrition. During this time the larvae transform into juvenile mussels, but their size does not increase. The juvenile mussels break out of the cyst, drop to the bottom, and begin their growth and development to the adult stage.

Fingernail clams also retain the eggs and incubate the developing embryos in chambers within the gills. However, the young organisms that are eventually released through the outgoing siphon are not parasitic larvae. They are juvenile clams that look like miniature adults. The unique feature of the life history of fingernail clams is the size of the eggs and juveniles that are incubated. Even though this family has the smallest adults among the bivalves, the eggs and juveniles are the largest. The juveniles may become one-fourth to one-third the size of the adults before the adults release them. As a result, the number of offspring is the lowest among the bivalves, ranging from 1 to 135 per parent.

The Asian clam, which is an introduced species, shares some life history traits with the fingernail clams and the mussels. Like the fingernail clams, Asian clams retain the eggs in a brood chamber and release juvenile clams that are not parasitic. The difference is the size of the offspring. The eggs that are incubated by Asian clams are very small, only 0.2 mm long. The juveniles that are released from the outgoing siphon are only about 0.3 mm long. Up to 1 million juveniles may be released. The exceptionally high number of offspring is similar to the mussels. Juvenile Asian clams look like miniature adults, and they are ready to begin their development to the adult stage as soon as they find a suitable location. Although they do not swim, the newly released juveniles are small enough to remain suspended by current, so they disperse great distances before they settle to the bottom.

The zebra mussel, which is another introduced species, is the one exception among the bivalves that does not retain the eggs and incubate them. Zebra mussels have high reproductive rates, with females producing 10,000–1 million eggs per year. The eggs are released right away into the water, where they are fertilized and develop quickly into a specialized free-swimming larva called a veliger. These microscopic larvae swim and float for 10–30 days without needing a fish host. Thus, they are able to disperse great distances by means of currents. The veligers transform into juveniles, acquire their shells, and begin to settle out of the water onto any firm surface. They are still very small at this stage, about the size of sand grains. There are so many of them that underwater objects often feel like sandpaper when zebra mussels are reproducing.

The different life histories of these four families account for their very different abilities to disperse and the resulting distribution patterns. The mussels depend on certain species of fish for their larval stage, so they can only disperse and occur where their host fishes occur. Thus, many kinds of mussels have very restricted distributions, and their existence has been threatened by pollution and habitat modification. Fingernail clams, the Asian clam, and the zebra mussel, do not have a larval stage that is dependent on fish hosts, hence, they occur widely. The two major introduced pest species, first the Asian clam and more recently the zebra mussel, have dispersed rapidly because the former produces small juveniles that float freely for a long time, and the latter produces tiny, free-swimming larvae that move around with currents for as long as a month.

In addition to the normal mechanisms of dispersing in the water, juveniles of fingernail clams and the two introduced pest species are known to hitchhike to other bodies of water. They become temporarily attached to filamentous algae and rooted aquatic plants as juveniles. These plants get tangled on the feet or feathers of water birds that transport them to other drainage systems. Fingernail clams and the Asian clam are known to clamp their small juvenile shells on the body parts of highly mobile animals such as adult aquatic insects, salamanders, and water birds. Some species of fingernail clams are eaten by ducks, but not digested. The ducks usually regurgitate them alive, but sometimes not until they have transported them great distances during the migratory season. Dispersal of the zebra mussel is assisted by the recreational activities of humans. Juveniles of this species readily attach to underwater objects, such as boats, outboard motors, and scuba gear, which are transported and used in other bodies of water, where the hitchhiking pests drop off. The larvae can be transported in this manner also, because they can withstand drying for several days. Zebra mussel larvae can also be transported in water contained in bilges, live wells, or bait buckets, if this water is dumped elsewhere.

Significance — Mussels and clams play an important role in freshwater ecosystems by virtue of their filter feeding. They have high filtration rates and occur in high densities under natural conditions. Thus, they are able to remove excess algae and suspended particles of organic matter from the water and make it available to other organisms residing in the sediment, while purifying the water in the process. However, introduced species, especially the zebra mussel, build up such large populations that they upset the balance in freshwater ecosystems (see Zebra Mussel).

Bivalves serve as food for a variety of animals. Shore birds, ducks, turtles, frogs, mud puppy salamanders, fish, and crayfish eat fingernail clams, zebra mussels, and young Asian clams and mussels. Some invertebrates eat the glochidia of mussels before they attach to their fish hosts. Most fish that eat bivalves, ingest relatively soft immatures, but a few fish, primarily catfish, can crack the shells of larger, older individuals. Raccoons, muskrats, mink, and otter also eat larger mussels.

About 40 species of bivalves have been harvested for economic enterprises. Beginning in the late nineteenth century and continuing until the 1940s, there was a sizeable industry in manufacturing buttons from mussel shells. They were harvested primarily from the large rivers of the midwestern and southern United States. Fortunately, plastics eliminated the market for buttons made from mussel shells. However, in the 1950s the Japanese discovered that small fragments of freshwater mussel shells were very effective for culturing marine pearls. Another mussel harvesting industry developed to meet this demand and persisted through the 1960s. Today, over harvesting, government regulations, and economic factors have greatly reduced commercial uses of mussels, but there is still an active harvest in the rivers of the midwestern United States.

As a result of pollution, modifying habitats to build dams and dredge navigation channels, and commercial harvesting, the populations of mussels have declined dramatically over the last century. Some species are now endangered or extinct because of these activities. Freshwater bivalves are some of the few invertebrates for which conservation programs have been established. At least 25 species have been included on the US government's endangered species list. Research and restoration programs are in progress.

Almost all kinds of bivalves, even including the introduced Asian clam, are sensitive to most types of pollution and environmental stress. Thus, they are useful as biological monitors for assessing the health of freshwater environments.

Biology of Common Families

Asian Clam

Family Corbiculidae; Plate 14

Different Kinds in North America — Few; 1 genus, 1 species (*Corbicula fluminea*).

Distribution in North America — The Asian clam was not native to North America but was introduced from southeastern Asia. It was first observed in the Columbia River in the Pacific Northwest of the United States in 1938. Since then it has become widespread. The Asian clam now occurs in the drainages of the west coast, all along the southern tier of states from California to Florida, all along the Mississippi River and eastward to the Atlantic from Illinois to New Jersey. It does not occur in the most northern states. Apparently it cannot tolerate low winter temperatures, so its spread may be about finished.

Habitat — Lotic-erosional, lotic-depositional, lentic-littoral. The Asian clam prefers slow moving, well oxygenated water at depths of 0.5–1.5 m in all sizes of streams or rivers. It sometimes lives in fast water. It also occurs in reservoirs, occasionally in much deeper water if there is plenty of dissolved oxygen. The Asian clam is usually found on sand and mud bottom, but it can live on any substrate, including bedrock.

Movement — Burrower.

Feeding — Collector-filterer.

Other Biology — The Asian clam has become the dominant species of invertebrate on the bottom of many freshwater habitats. It is capable of building up very high densities, sometimes as many as 10,000 per square meter. The empty shells often accumulate in enormous numbers on sand and gravel bars and become the most abundant component of the substrate. Asian clams grow very rapidly because they have a high filtration rate for obtaining food. The tiny juveniles become 15–30 mm long by the end of their first year. The life span is 3 or 4 years. The Asian clam grows larger in reservoirs than it does in free-flowing rivers. They breed whenever the water temperature is above 18–19 °C. This introduced species is considered a pest because it multiplies rapidly, possibly to the detriment of native mollusks. Sometimes Asian clams colonize and clog the water intakes of factories, power plants, and municipal water supplies. This causes costly removal and prevention procedures, but the pest status of the Asian clam is being surpassed by the more recently introduced zebra mussel (see below).

Stress Tolerance — Somewhat sensitive to facultative. This species is sensitive enough to some types of pollution, especially metals, to be used as the subject for toxicity tests.

Fingernail or Pea Clams

Family Sphaeriidae; Plate 13

Different Kinds in North America — Few; 5 genera, 36 species. Pill clam is another common name that is sometimes used for members of this family.

Distribution in North America — Throughout.

Habitat — Lentic-littoral, lentic-profundal, lotic-depositional, lotic-erosional. Most kinds of fingernail clams are more common in standing waters. Fingernail clams are the dominant bivalves in temporary habitats, ponds, and the deep areas of lakes. However, some members of this family live in all freshwater habitats, including springs and small streams. They occur in all water velocities, from fast to still, and at depths ranging from 0.5 to 30 m. They can be found on any type of substrate, except bedrock, but the greatest diversity of fingernail clams occurs where the bottom is composed of fine sand, silt, and clay. Fingernail clams are not restricted to waters with high concentrations of calcium carbonate, like the mussels.

Movement — Burrowers. Fingernail clams sometimes burrow as deep as 25 cm in soft sediment when they go into a resting stage.

Feeding — Collector-filterers.

Other Biology — Densities of fingernail clams can be as high as 10,000 per square meter. Most kinds live for 1–2 years, but some live as long as 3 years. One species completes its development in 1 month or less. Some can withstand drought for as long as several months by burrowing into the substrate, which explains why fingernail clams sometimes inhabit intermittent ponds.

Stress Tolerance — Facultative to somewhat tolerant.

Mussels

Superfamily Unionacea, Families Unionidae and Margaritiferidae; Plate 12

Different Kinds in North America — Many; 44 genera, 227 species. Each species of mussel has its own individual common name composed of two words. Some of them share the same root words, which may be used regionally as common names for all mussels. Some of the more interesting ones are creekshells, elktoes, floaters, heelsplitters, lances, moccasinshells, pearlshells, pearlymussels, pigtoes, pocketbooks, and spikes.

Distribution in North America — Throughout, but many species are restricted to individual drainage basins.

Habitat — Lotic-erosional, lotic-depositional, lentic-littoral. The greatest number of mussel species live in the shallow waters of larger rivers. A high concentration of calcium carbonate in the water favors more kinds and higher abundance, because it is easier for them to construct their shells. They prefer depths of 2–6 m, sometimes up to 10 m. Many also do well in the shallows of large natural lakes, and some are successful in impoundments. Members of this group do not live in springs or small streams, and they are not common in ponds. The substrate where they live is usually stable gravel or a mix of gravel and sand, although a few live in shifting sand or mud. They prefer slow to moderate current rather than swift current or still water. They usually do not occur where there is very much rooted vegetation.

Movement — Burrowers.

Feeding — Collector-filterers.

Other Biology — Densities of mussels in good habitat with no pollution can be as high as 50 per square meter. For most kinds, it takes 1–8 years to become an adult capable of reproducing. Large species of mussels live 10–20 years or more, and one species is known to sometimes live for a century.

Stress Tolerance — Somewhat sensitive to facultative. Mussels do best in unpolluted streams.

Zebra Mussel

Family Dreisseniidae; Plate 15

Different Kinds in North America — Few; 1 genus, 1 species (*Dreissena polymorpha*).

Distribution in North America — The zebra mussel was not native to North America but was introduced from the Caspian Sea in eastern Europe. It was first discovered near Detroit, Michigan in 1988. It probably came from a freighter discharging ballast water that it had taken on in Europe. Based on the length of time that it takes for development, it was probably first introduced in 1985 or 1986. Since then, it has spread throughout the Great Lakes and at the time of this writing is showing up in the Arkansas, Ohio, Tennessee, Cumberland, Atchafalaya, and Mississippi river systems as far south as New Orleans. The zebra mussel is certain to extend its range farther, just as the Asian clam did in the twentieth century.

Habitat — Lotic-depositional, lotic-erosional, lentic-littoral, lentic-profundal. The zebra mussel is usually found in slow moving or still waters that are rich in algae and well oxygenated. It generally lives at depths of 2–10 m, but it can occur in lakes and reservoirs as deep as 60 m if there is plenty of dissolved oxygen. The zebra mussel is unique among the North American bivalves in that it prefers solid substrates, such as bedrock, cobble, gravel, wood, metal, plastic, and rubber. Its unique ability for attachment (below) also allows this species to colonize areas with fast current.

Movement — Clingers. The zebra mussel is the only bivalve found in North America that attaches firmly to solid objects. They accomplish this with tufts of fibers, called byssal threads. Natural attachment sites are rocks and logs, but these mussels will attach to any object, including other species of mussels, crayfish, turtles, buoys, boat hulls, and industrial pipes and condensers. They attach in clusters, something like marine barnacles. Zebra mussels can move again if their initial attachment site is not satisfactory. Relocation is more common with juveniles and young developing mussels. The byssal threads are shed, then the foot is used in the normal manner to move to another place, where the organisms make new byssal threads. In habitats with little hard substrate, zebra mussels aggregate on soft bottom. Juveniles settle out of the water on small pebbles or other small hard objects. Then other juveniles attach themselves to the same group, forming masses up to 10 cm in diameter. These masses of living zebra mussels, called druses, are loose and can roll freely on the bottom.

Feeding — Collector-filterers.

Other Biology — Zebra mussels live up to 5 years. As described above, they disperse widely, reproduce in vast numbers, and grow quickly. Zebra mussels out compete native mussels for food, and physically smother native mussels when they use them for attachment sites. The spread of this recently introduced species is already causing economic problems and may lead to dramatic ecological changes. The immediate problems are caused by their habit of attaching to any firm substrate. Unfortunately, this often includes the inside of water pipes where they restrict, or even block, the flow of water. This presents a formidable problem for industries, power companies, and municipal water supply agencies. They can also clog the cooling systems of engines on inland commercial vessels and recreational boats. This pest poses a multi-billion dollar threat in the United States, and managing the problem is going to be a major nuisance to shippers, recreational boaters, commercial fishermen, anglers, and persons who use inland beaches. Diving ducks and some fish, such as freshwater drum, do eat zebra mussels, but not enough to control them.

In addition, the zebra mussel's role in food webs could cause far reaching effects on freshwater ecosystems. Each individual mussel can filter about 1 liter of water per day. They are very efficient at filtering, even though they do not eat everything that they filter from the water. What they do not eat, they mix with mucous and discharge the material onto the bottom. Because they have such high populations, they can remove almost all of the algae and fine detritus from the water. Without algae and fine detritus in the water, other organisms, especially the microscopic animals that live suspended in the water (zooplankton), cannot get enough food. This essential change in the food web could lead to reductions of other organisms, including fish. On the other hand, zebra mussels may help some other bottom dwelling invertebrates. Loose aggregates of these mussels and the abundant unused food that they deposit on the bottom provide additional habitat and food for flatworms, aquatic earthworms, scuds, and midge larvae. The net ecological effect of the zebra mussel will not be known until long-term research has been conducted, but the introduction of this species into North America will most likely prove to be detrimental to freshwater ecosystems that are already stressed from human activities.

Stress Tolerance — Facultative. Zebra mussels live successfully in a wide range of environmental conditions, but they must have adequate levels of dissolved oxygen.

Arthropods

Phylum Arthropoda

About 85% of all known invertebrate species are in the phylum of arthropods. This is a considerable number of species, considering that invertebrates account for 96% of the 1,032,000 species of animals. Arthropods are clearly the most successful animal phylum on land. It so happens that they are prominent in freshwater environments as well.

Distinguishing Features — The size of arthropods varies greatly, from less than 1 mm to more than 150 mm. The bodies of arthropods are segmented, but the segments may not be easy to distinguish in some kinds. The body segments are grouped into two or three distinct regions that look different from each other and perform different functions. One of these regions is usually the head, which is specialized for acquiring food and sensing information about the environment, such as vision, smell, and taste. Another body region is usually the abdomen, which is specialized for physiological functions such as digesting food, eliminating wastes, and reproduction. Some arthropods have a separate thorax region, which is specialized for moving the organism by means of legs or wings. In some arthropods, the thorax is combined with the head, in which case the region is called the cephalothorax. The most distinctive characteristic of this phylum is the paired, segmented appendages that are used for locomotion. However, many kinds of insects do not have legs when they are in their immature stages called larvae. The bodies of arthropods are always bilaterally symmetrical, which means that the left and right sides are just alike.

The bodies of arthropods are protected by a skin that acts as an external skeleton (see Introduction, Fundamentals of Invertebrate Biology, Life History). The skin varies from being hard and thick to soft and thin, and it comes in many different colors, according to different species. On the inside of their body, arthropods do not have any veins or arteries to hold the blood. Their blood flows loosely throughout the body, a condition known as an open circulatory system. In almost all arthropods, the blood does not carry oxygen to where it is needed for respiration. Instead, these animals have a network of fine, branching tubes (tracheae) that carry oxygen as a gas directly to the tissues where it is needed for respiration.

Arthropods are further classified into three major groups that are called subphyla. These arthropod subphyla include: (1) Chelicerata, water mites (page 234); (2) Crustacea, aquatic sow bugs, scuds, crayfishes, shrimps (page 242); and (3) Atelocerata, aquatic insects (page 260). The most common members of these subphyla are covered below.

Explanation of Name — The scientific name for this phylum comes from two Greek words: *arthr*, meaning joint, and *pod*, meaning foot. The name refers to the most conspicuous feature of these animals, the segmented appendages that bend and allow them to move around so effectively.

Water Mites

Subphylum Chelicerata, Class Arachnida, Order Acariformes
Hydracarina or Hydrachnida; Plate 16

Water mites are the only members of the arthropod subphylum Chelicerata that are common in freshwater environments. Most of the well-known chelicerates are arachnids (Class Arachnida) that live on land: scorpions, daddy-long-legs, spiders, ticks and mites. Chelicerates also live in marine environments, for instance the familiar horseshoe crabs (Class Merostomata, Subclass Xiphosura). However, the only chelicerates that inhabit freshwater environments are some spiders (Order Araneae) and mites (Order Acariformes). A few spiders are capable of living underwater for brief periods of time with a bubble of air, but most spiders associated with freshwater environments are semi-aquatic kinds that live on land at the edge of the water or occasionally venture out on the surface of the water. Thus, spiders are not included in this guide. However, there are more than 1,500 species of mites in North America that live in freshwater environments. All of the water mites are lumped together under the scientific names of Hydracarina or Hydrachnida, but these names do not represent a unit of the official systematic naming scheme, such as an order or family. Water mites belong to the same order (Acariformes) as their arachnid relatives that live on land. In addition, the 38 families and 124 genera that water mite species belong to also contain many more terrestrial species. Hydracarina or Hydrachnida are just names that scientists made up for the convenience of talking about a group of mites that are ecologically similar. Although small, water mites swim conspicuously and are commonly collected. They are easy to recognize, once you have been introduced to the group. This guide only covers them together as one group, because the individual kinds are hard to distinguish and the biology of all kinds is about the same.

Distinguishing Features — In addition to having the features of all arthropods, water mites also have the characteristics that are typical of all chelicerates (Figure 14). They have two narrow, pointed mouthparts that move vertically, something like the fangs of snakes. These are called chelicerae, which is where the subphylum Chelicerata gets its name. Chelicerates have no antennae. Individual body segments are not distinct, but the segments are grouped

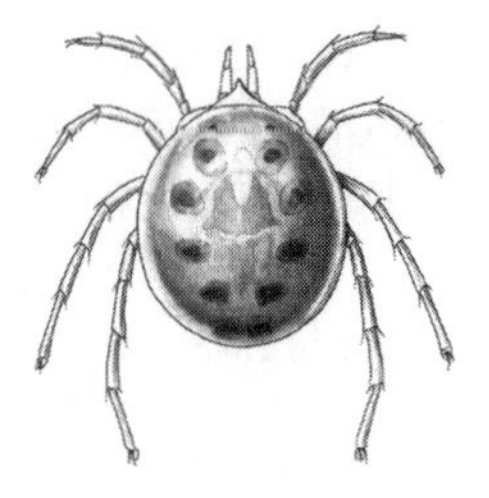

Figure 14. Representative water mite.

into two major regions, the cephalothorax and abdomen. As members of the arachnid class, the bodies of water mites are usually somewhat oval and have four pairs of legs when they are adults. The appearance of water mites changes appreciably during their metamorphosis. Larvae have only three pairs of legs, but water mites are usually only collected as adults. Therefore, the following description pertains only to adults. Water mites can be distinguished from other freshwater chelicerates by the following combination of characteristics (Figure 14):

- General appearance is similar to spiders.
- Body is very small, usually about 2–3 mm in diameter with a range of <1–7 mm.
- Body is frequently brightly colored in patterns of green, blue, orange, yellow, or red, but may be only dull shades of brown or black in some kinds.
- Body shape is usually globular or spherical, but body may be shaped like teardrops, disks, saucers, footballs, or bean bags in some kinds.
- Skin is generally soft; but may be leathery or thickened to form hard plates in some kinds.
- Cephalothorax is very small, making the abdomen the only visible body region.
- Two pairs of widely separated simple eyes occur on the front margin of the body.
- A pair of finger-like, segmented structures (pedipalps) stick out in front of the body in the middle.
- A pair of fang-like mouthparts are located on the front of the body in the middle, but these are usually withdrawn and not visible.
- Four pairs of segmented legs appear to project from the sides of the body when viewed from the top, but these legs actually originate from a plate on the bottom of the body.

- Legs may have brushes of long, thin hairs all along their length or patches of short, stout hairs only at the bases of the leg segments.
- Two claws are usually situated on the end of each leg.

Explanation of Names — The common name mite can be traced back to its use in Middle Dutch before the twelfth century. At that time the word meant a small copper coin. It came to Middle English through Middle French and by the fourteenth century mite was used as an adverb to mean a very little and as a noun to mean any small object or creature. Its use for these arachnids is probably related to their small size as well as their round shape. The scientific name Hydracarina was developed by combining two Greek words, *hydor* and *akari*. *Hydor* meant water, and *akari* was the word that the ancient Greeks used for the small organisms that we now call mites. The other scientific name, Hydrachnida, was coined from the Greek words *hydor* and *arachne*, which meant spider. Of the two scientific names, Hydracarina would seem to provide the more accurate description of these creatures.

Habitat — Different kinds of water mites live in all types of freshwater habitats, but the greatest abundance and the most different kinds are found in lentic-littoral habitats among aquatic plants that are no more than 1–2 m beneath the water surface. They are very common in the shallow waters of ponds, lakes, swamps, marshes, and bogs. In these habitats, the abundance of water mites frequently reaches 2,000 organisms per square meter. Some kinds also live in running waters, ranging in size from seeps and springs to large rivers. They inhabit both erosional and depositional zones of these lotic habitats. A few kinds live in the lentic-profundal zone of lakes as deep as 100 m. Some water mites migrate to deeper water when the plants in shallow water die back for the winter. Some inhabit temporary habitats by burrowing in the mud when the water dries up.

Movement — Most water mites are swimmers, but they do not venture far above the bottom. Many swim awkwardly by uncoordinated movements of their legs, but some swim agilely. The best swimmers have long, thin hairs on their legs, especially the rear pairs of legs, for propelling them through the water. They sink quickly if they are not moving their legs. Quite a few water mites crawl on the bottom or burrow in the upper few millimeters of the sediment, while others crawl about on plants. The kinds that are crawlers and burrowers have patches of short, stout hairs at the bases of the leg segments to help them move, in addition to two claws at the ends of the legs. Some kinds are able to inhabit riffle areas of running waters as clingers by using their claws. These kinds cling to rocks or crawl among loose pebbles and gravel. Still-water kinds are mostly globular in shape, while those inhabiting running water tend to be flattened for resistance to the current. Water mites are active only during daylight hours.

Feeding — Adults and mature larvae of water mites are mostly piercer-predators, but some kinds are external parasites, collector-gatherers, or piercer-herbivores. Young larvae of water mites are external parasites. In most kinds of water mites, the adults and mature larvae prey on larvae of small aquatic insects, especially true flies, and microscopic crustaceans. Water mites grab prey with their finger-like pedipalps, sometimes using their legs to help, and quickly pierce the bodies of these organisms with their fang-like mouthparts. Then they inject digestive enzymes produced by their salivary glands into the prey. After the enzymes have dissolved some of the tissues of the prey, the water mites suck in the fluids that are created. Indigestible internal parts and the skin of the prey are discarded after the water mites have finished feeding. Water mites that are strong swimmers hunt down active prey, such as mosquito larvae or crustaceans in the zooplankton. Crawler and burrower species of water mites feed on bottom dwelling organisms, such as midge larvae, aquatic earthworms, or the eggs of aquatic insects. As a group, adult water mites fit the description of omnivores, because a few of them eat live plants, plant detritus, and carrion, some are external parasites, and a few are even cannibalistic.

The early larval stages of all water mites are external parasites that feed on tissues of their hosts, usually without killing them. The hosts are mostly aquatic insects, especially true flies such as midges and mosquitoes. After true flies, the most common aquatic insect hosts are stoneflies, dragonflies, damselflies, predaceous diving beetles, and water boatmen. A few kinds parasitize mussels, snails, or sponges. Parasitic stages of water mites attach themselves at the joints between the body segments of their hosts, where they can insert their mouthparts into soft membranous tissue. They usually attach between segments of the thorax or abdomen, but sometimes between leg segments.

Breathing — Water mites use dissolved oxygen, which diffuses across their overall body surface into a typical arthropod tracheal system. Most kinds can survive dissolved oxygen concentrations as low as 1 part per million.

Life History — Water mites have a very complicated life history that is somewhat similar to complete metamorphosis in insects, but with additional stages. Scientists have not agreed upon a single set of terms for the various stages of water mites, but, unfortunately, most of the terms are rather long and technical. The names that are used most commonly for the stages of water mites are egg, prelarva, larva, protonymph, deutonymph, tritonymph, and adult. Most users of this guide will not need to remember these terms, but they are necessary to explain the unusually complicated life history of these organisms.

Male and female water mites reproduce sexually. Females deposit groups of 20–400 fertilized eggs on submerged rocks, aquatic vegetation, woody debris, or coarse detritus. A few kinds of water mites insert their eggs into the tissue of aquatic plants. There may be a jelly-like covering over individual eggs or groups of eggs. Most water mite eggs are red.

Inactive prelarvae develop inside the egg and emerge as active six-legged larvae in about 1–6 weeks. Larvae swim around briefly until they locate an appropriate host, then they attach themselves as parasites and feed on the body fluids of the host. They usually stay attached to the same host throughout the larval stage. Sometimes they must detach temporarily and reattach when a growing immature host sheds its skin. Larvae transform into protonymphs. The protonymph is a resting stage that takes place inside of the larval skin while it is still attached to the host. The protective structure provided by the larval skin is called a nymphochrysalis. Protonymphs develop into deutonymphs inside the nymphochrysalis, then the deutonymphs break out and become free-living organisms instead of parasites. The active deutonymphs have eight legs and generally look, move, and feed like adult water mites. The deutonymph stage lasts from 5 days to 6 months in different species. Deutonymphs must transform one more time to become adult water mites. To do this, they attach to algae filaments, rooted plants, or detritus and then become inactive tritonymphs inside the skin of the deutonymphs. The protective structure made from the skin of the deutonymph is known as an imagochrysalis. Tritonymphs transform to adult water mites inside the imagochrysalis, then they burst through that structure to begin the final stage of their lives. The greatest numbers of adults occur in late spring and early autumn, but different kinds of water mites occur as adults in all seasons. Some even remain active under ice during the winter.

Many water mites are parasitic larvae when their immature aquatic insect hosts transform into adults and leave the water. Larvae that are attached to emerging adult insects become protonymphs and deutonymphs while the adult insect hosts are out of the water. Many of these immature water mites probably perish while they are away from the water, but sufficient numbers manage to return. Some probably fall off of adult insects over water by chance. More likely explanations are that the immature water mites reenter the water either when female adult insects lay their eggs or when adult insects die on the surface of the water.

There are also water mite species that are parasites throughout their lives. Sometimes these species parasitize different kinds of hosts when they are larvae and when they are adults. For example, some water mites are parasites of adult midges as young larvae, then they switch to being parasites of mussels when they are older larvae and adults.

Significance — Water mite larvae routinely parasitize about 20–50% of the emerging adults in many populations of aquatic insects. In some populations almost all aquatic insect adults are parasitized. Even though parasitic water mites usually do not kill their hosts, if female aquatic insects are parasitized when they are larvae, they may not produce as many eggs when they become adults. Larval aquatic insect hosts may grow less and become sufficiently weakened by water mite parasites that they have lower survival rates. Because water mites occur in such high numbers in some habitats, predation by deutonymphs and adults probably causes a significant reduction in the abundance of their prey. Thus, the long-term feeding activities of water mites might be one of the important factors determining the abundance and relative proportions of other invertebrate organisms living in particular habitats.

Water mites do not appear to be particularly important as food for vertebrate or invertebrate predators. Fish, turtles, and some invertebrates occasionally eat them, but the bright red water mites are thought to taste bad. Predators learn to recognize and avoid them.

In Europe, water mites have been shown to be excellent indicators of habitat quality, when they are identified to species. The taxonomy and ecology of the different species of water mites in North America have not been studied well enough to use them in environmental protection programs.

Stress Tolerance — As a group, water mites should be considered facultative, however, individual species range from somewhat sensitive to somewhat tolerant.

Crustaceans

Subphylum Crustacea; Plates 17 – 20

Crustaceans are another of the three subphyla of arthropods. They are a highly diverse group, but much more so in marine environments. In North America, there are a total of 26 orders with about 36,000 species of crustaceans. Of that total, 13 orders containing about 1,100 species live in freshwater environments. Many of the crustaceans are microscopic in size, being less than 1 mm in length even when fully grown. The majority of these small kinds, such as the common water fleas (Cladocera), spend their lives suspended in the waters of lakes and ponds as members of the zooplankton community. Because of their size and lifestyle, the microscopic crustaceans are not collected by the same methods as the larger kinds that live on the bottom. Identifying the small kinds requires technical procedures and a microscope. The larger, bottom-dwelling kinds are part of a class of crustaceans named Malacostraca, commonly referred to as malacostracans. The various kinds of malacostracans occur in almost all types of freshwater environments, where they often account for a significant portion of the animals living on the bottom. This guide includes three common orders of malacostracan crustaceans: aquatic sow bugs (Isopoda), scuds (Amphipoda), and crayfishes and shrimps (Decapoda).

Distinguishing Features — Crustaceans do not undergo any discernible metamorphosis as they grow and develop into adults. Immature crustaceans look essentially like small adults. Crustaceans have the characteristics previously mentioned for all arthropods. They can be distinguished from the other major groups of arthropods, water mites and aquatic insects, by the following features. The length of the body varies greatly for different kinds of crustaceans, from 1 mm to 150 mm. The body shape may be either cylindrical, flattened from side to side, or flattened from top to bottom. The skin of most crustaceans is somewhat brittle because calcium carbonate is incorporated into its structure along with the protein chitin. Crustaceans continue to shed their skin and grow a new one after they become adults.

Some individual body segments of crustaceans are conspicuous. The individual body segments are grouped into either two or three generalized regions. If there are two body regions, they are a cephalothorax, consisting of the head and entire thorax, and an abdomen. If there are three body regions, they are a cephalothorax that consists of the head and first thorax segment, a thorax, and an abdomen. Most of the body segments have segmented appendages, or limbs. The numerous appendages of crustaceans are variously specialized for feeding, sensory reception, walking, swimming, defense, grooming, breathing, and reproducing. On the head, the mouthparts are two jaw-like structures that move laterally, from outside to inside. There are two pairs of antennae, sometimes called feelers, which are used for sensory reception. There are five to eight pairs of appendages on the thorax or cephalothorax. These are called pereiopods. The abdomen has six pairs of appendages, the first five of which are called pleopods and the last of which is called the uropod. The appendages on the thorax or cephalothorax are longer and more conspicuous than the ones on the abdomen. If you look closely, you will see that most of these appendages have two branches or forks. Scientists call this condition biramous. One of the branches of the appendages may be much longer than the other.

Explanation of Names — The scientific name for this subphylum is derived from the Latin word *crusta*, meaning a hard or brittle external coat or covering. The calcium carbonate incorporated into the skin of these animals makes them have a crust-like covering. The scientific name of the class of crustaceans covered in this guide, Malacostraca, is a combination of two Greek words: *malakos* (soft) and *ostrakon* (shell). This would appear to be a poorly chosen name because crustaceans have a hard, somewhat brittle covering. However, much of the skin of crustaceans is bendable because it is composed mostly of proteins, with some calcium carbonate incorporated into the structure. Malacostracan crustaceans are soft in comparison to mollusks, which have shells composed almost entirely of calcium carbonate.

Aquatic Sow Bugs

Order Isopoda, Family Asellidae; Plate 17

Aquatic sow bugs, which are also called isopods, are one of the orders in the subphylum Crustacea and class Malacostraca. There are about 130 species of freshwater isopods in North America. This represents only about 5% of the total number of species, because most isopods are marine or terrestrial. Practically all of the species of aquatic sow bugs that live freely in surface waters throughout most of the continent are in one family, Asellidae. There are three other families with a few species, but these are seldom encountered because they have very specialized lifestyles or narrowly restricted distributions. In this guide, all of the information on biology and identification of aquatic sow bugs pertains to the family Asellidae.

Distinguishing Features — The body length of aquatic sow bugs ranges from 5 to 20 mm, without the antennae and tails. Most of the common kinds are gray, but there are frequently blackish or brownish tones. Less common colors are reddish or yellowish. Some kinds have mottled markings on the body. Aquatic sow bugs can be distinguished from other freshwater crustaceans by the following combination of characteristics (Figure 15):

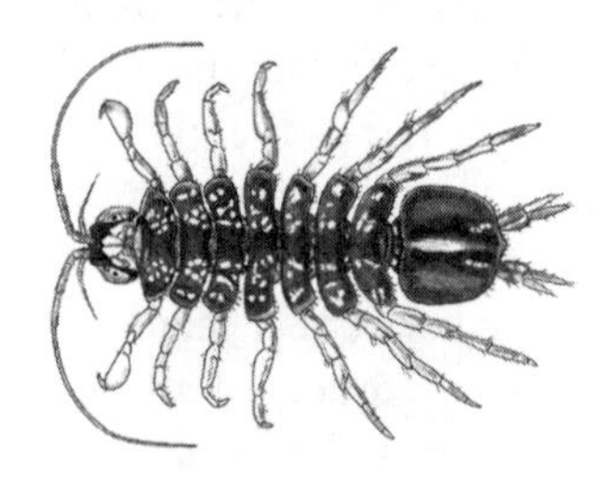

Figure 15. Representative aquatic sow bug.

- Body is strongly flattened from top to bottom.
- Body segments are grouped into three regions: cephalothorax, thorax, and abdomen.
- Cephalothorax is a combination of the head and first thorax segment.
- Thorax is composed of the next seven segments after the cephalothorax.
- Abdomen consists of all segments to the rear of the thorax, fused into a short, broad, shield-like region.
- One pair of antennae on the cephalothorax is much longer than the other pair.
- Seven pairs of long walking legs are present on the thorax, one pair per segment.
- First pair of walking legs has enlarged ends with hinged claws for grasping.
- Next six pairs of walking legs get longer proceeding toward the rear of the body, and each has a simple, sharp claw on the end.
- All thorax segments have a shelf-like projection on each side that covers the base of the legs.
- Six pairs of short appendages are present on the underside of the abdomen.
- First five pairs of appendages are hidden under the abdomen in top view, but the sixth pair protrudes behind the end of the abdomen, looking like a pair of flat tails.

Explanation of Names — Sow bugs got their common name from the way the adult females take care of their offspring when they first hatch. Female isopods hold a large number of young in a pouch on the bottom of their body. Since terrestrial isopods tend to look like fat pigs, the comparison was made to a female pig, or sow, nursing her piglets. The term has been carried over to the aquatic forms. The scientific name for this order comes from two Greek words: *iso*, meaning equal and *pod*, meaning foot. This name refers to the appendages on the bottom of the body, as seen in side view. Similar-looking, long walking legs protrude down almost the entire length of the body. All of these appendages are about alike. To understand the derivation of this name, you must compare it to the name of their close relatives the scuds (Amphipoda), which have two different kinds of appendages.

Habitat — Aquatic sow bugs live in a wide variety of habitats, including lotic-erosional, lotic-depositional, lentic-littoral, and subterranean. They are almost always in shallow waters, rarely occurring in places that are more than 1 m deep. They are most common in seeps, springs, and small spring-fed streams. If the substrate is complex with hiding places, they can be very abundant. A few kinds live in ponds or the shallow margins of lakes. There is usually only one species in a small habitat, such as a spring or pond. Some kinds of isopods are seldom found because they live deep down in the loose substrate of streams or waters flowing through caves.

Movement — They are slow-moving crawlers that are very secretive. They avoid light by hiding in crevices among rocks or under leaves, roots, or detritus.

Feeding — Aquatic sow bugs eat a wide variety of foods and belong to several functional feeding groups, including collector-gatherers, shredder-detritivores, shredder-herbivores, and engulfer-predators. They are accurately described as omnivores. Their primary source of food is fine particles of detritus lying on the bottom or trapped among rocks, roots, or accumulations of plant debris. They also scavenge dead animal matter and prey on injured animals as well as feed on live and decaying plants. The plants that they feed on include terrestrial leaves and grasses as well as aquatic plants.

Breathing — These organisms breathe dissolved oxygen by means of gills. The gills are thin, flat structures that come off the bases of the appendages on the abdomen.

Life History — Female aquatic sow bugs hold the eggs for protection in a pouch called a marsupium. The marsupium is on the bottom of the thorax. It is formed by plate-like structures that project inward at the bases of several of the walking legs. Females maintain a current of fresh, oxygenated water through the marsupium by moving the platelike structures. They brood 20–250 eggs in the marsupium, then hold the newly hatched young for 20–30 days. Young isopods shed their skin five to eight times before becoming adults. They continue to shed their skin at least seven to ten times after becoming adults, for a total of at least fifteen molts. Most kinds reproduce only once per year. Breeding occurs throughout the year in springs with relatively constant temperatures. In other waters that are subject to changes in annual temperatures, they breed mostly in spring. Most of the common kinds live for about 1 year, from the egg stage through adulthood. The exceptions are subterranean species, which may live for 4–6 years.

Significance — Some invertebrates, such as crayfish, and bottom-feeding fish eat aquatic sow bugs. However, they are usually not important in the diet of fish because isopods live primarily in very small bodies of water where fish are not common. In lakes large enough to have trout populations, aquatic sow bugs may be an important part of the fish diet. Some kinds of isopods occasionally become pests by eating watercress that is being grown commercially in springs, hence, another common name for them is cress bugs. Attempts to control cress bugs in watercress by means of chemical pesticides have caused fish kills in the streams that receive the spring waters.

Stress Tolerance — Most kinds of aquatic sowbugs are somewhat tolerant, especially of organic wastes. They are reliable indicators of the zone where streams are beginning to recover from pollution by sewage.

Scuds, Sideswimmers

Order Amphipoda; Plate 18

Scuds, which are also called sideswimmers or amphipods, are one of the orders in the subphylum Crustacea and class

Malacostraca. Like all crustaceans, most of the amphipod species are marine. There are about 150 species of freshwater scuds in North America, but many of those live in underground waters and are not likely to be encountered by beginning aquatic biologists. Most of the species of scuds that are likely to be collected in surface waters belong to three families: Hyallelidae, Gammaridae, and Crangonyctidae. This guide does not distinguish the three individual families because the important structures are small and require a microscope to see. The biology of all common scuds is similar and is thus presented collectively.

Distinguishing Features — The body length of scuds ranges from 5 to 20 mm, without the antennae and tails. The color of most live organisms is a creamy light gray or brown, and they are somewhat translucent. Some kinds have brilliant colors, including green, blue, purple, lavender, or red. The same species is often different colors in different locations. Unfortunately, the colors of all kinds of scuds fade to dull white, cream, or gray when preserved. Scuds can be distinguished from other freshwater crustaceans by the following combination of characteristics (Figure 16):

- Body is strongly flattened from side to side.
- Body segments are grouped into three regions: cephalothorax, thorax, and abdomen.
- Cephalothorax is a combination of the head and first thorax segment.
- Thorax is composed of the next seven segments after the cephalothorax.

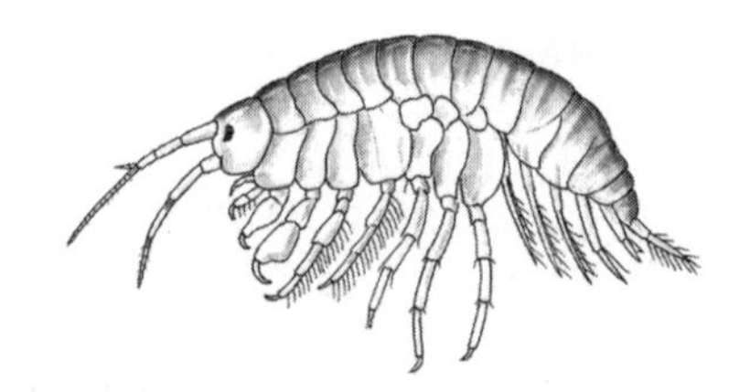

Figure 16. Representative scud or sideswimmer.

- Abdomen consists of the six individual segments to the rear of the thorax.
- Sides of the first four abdomen segments extend down, when viewed from the side, making these segments longer from top to bottom than any of the other body segments.
- Two pairs of antennae on the cephalothorax are about the same length.
- Seven pairs of walking legs present on the thorax, one pair per segment.
- First two pairs of walking legs are enlarged on the ends with a hinged claw for grasping.
- Remaining five pairs of walking legs have a simple pointed claw on the end.
- Six pairs of short appendages are present on the underside of the abdomen, one pair per segment.

Explanation of Names — The common name scud has a Scandinavian origin and refers to the movements of these animals. In Norwegian, *skudda* means to push. This was adopted in English as scud and came to mean to move or run swiftly. Anyone who tries to catch one of these organisms swimming in a pan of water will understand how well this common name applies. Sideswimmer refers to the way that they swim. The scientific name for this order comes from two Greek words: *amphi*, meaning of both kinds and *pod*, meaning foot. This name refers to the two kinds of appendages on the bottom of the body, as seen in side view. Long walking legs protrude down on approximately the front half of the body, while much shorter and simpler swimming appendages protrude down on the rear half of the body.

Habitat — Scuds live in many types of habitats, including lotic-erosional, lotic-depositional, lentic-littoral, and subterranean. They are most common in the shallows of cool streams, springs, seeps, lakes, and ponds. They live in all sizes of habitats, but they are often most abundant in very small habitats that do not have fish populations. Scuds sometimes occur in the backwaters of large rivers. Generally, scuds do not inhabit temporary habitats, but sometimes they live in pools left on the floodplain after high water has receded. A few kinds live in very hot springs. They are rarely in water deeper than 1 m. Scuds are bottom dwellers, primarily in small spaces within tangles of live aquatic plants, roots, coarse detritus, or stones. Sometimes they reside in the upper layer of soft sediment.

Movement — Their most common form of motion is crawling. They accomplish this primarily with the five pairs of legs on the thorax that have simple, pointed claws on the ends. Their crawling is also assisted by pulling themselves forward with the claws on the first two pairs of thorax legs and pushing themselves forward with the short appendages on the end of the abdomen. Scuds also swim just above the bottom by means of the short appendages on the front part of the abdomen. They often roll over on their side when swimming, which explains their other common name, sideswimmers. Scuds are much more active at night, because they have an inherited behavior that makes them avoid bright light (negative phototaxis). Any time they are disturbed, they wiggle down and hide in the loose substrate where they live.

Feeding — Scuds acquire their food by several mechanisms and feed readily on all types of plant and animal matter, thus, they are considered to be omnivores. The functional feeding groups that they can be placed in are collector-gatherers, shredder-detritivores, scrapers, and engulfer-predators. Their most common food is detritus, either as fine or coarse particles. When they feed on coarse particles of detritus, they use the claws on the first two pairs of thorax legs to hold the pieces while they chew on them with their jaws. They also graze on the thin film of algae, fungi, and bacteria that grows on the leaves and stems of submerged plants and other solid objects. They only occasionally attack small living animals, but they promptly eat any recently dead organisms they come upon.

Breathing — They breathe dissolved oxygen by means of gills. Their gills are flat, oval sacs that occur at the bases of most of the thorax legs.

Life History — The reproduction of scuds is similar to that of aquatic sow bugs. Species that live in continuously cool springs reproduce at any time of the year, whereas species that live in waters where temperatures change seasonally usually reproduce in spring. Most kinds produce only one group of offspring. Females brood 15–50 eggs in a protected pouch, called a marsupium, on the bottom of the thorax. The eggs hatch in 1–3 weeks, then the young scuds are retained in the pouch for about another week. The offspring are released the first time the female sheds her skin after mating. Young scuds become mature adults after shedding their skin eight or nine times. Amphipods do not undergo diapause. Most common kinds complete their life cycle and die within 1 year. The exceptions are subterranean species, which may live for 4–6 years.

Significance — Scuds are sometimes amazingly abundant in small habitats without fish. In small, spring-fed streams with thick rooted vegetation for cover and abundant detritus for food, there may be up to 10,000 scuds per square meter of bottom. Because of their abundance and voracious appetite, scuds are often important in the breakdown of organic matter and the pathways that this material moves through food webs. They are important items in the diet of many invertebrate predators. In habitats where fish are present, scuds are usually eaten in large numbers. They are also eaten by amphibians and water birds.

Stress Tolerance — When all common kinds of scuds are considered collectively, they are facultative. However, some individual kinds are typically restricted to permanent bodies of water that are relatively cool, clean, and well oxygenated. Some are sensitive to toxic heavy metals, especially copper, and pesticides.

Crayfishes, Shrimps

Order Decapoda

Crayfish and shrimp form one of the orders in the subphylum Crustacea and class Malacostraca. Another common name for members of this order is decapods. You will also hear crayfish called crawfish or crawdads, and large species of shrimp are sometimes referred to as prawns. Most decapods are marine, but this is a diverse group and about 334 species of crustaceans live in North American freshwater environments. There are many more species of crayfishes than there are shrimps, 315 versus 19, respectively. The appearance and biology of crayfish and shrimp are appreciably different, so this guide presents most of the information separately for these two major groups of decapods. However, much more is known about the biology of crayfishes than shrimps.

Distinguishing Features — The body length of adult decapods ranges from 10 to 150 mm, not including the antennae. The color is different for crayfishes and shrimps, and there is often considerable variation within the same species. Crayfish are usually brown-green, but this ranges from blackish to red or orange. They are often speckled or mottled. Individuals of some species are occasionally blue. Shrimp are usually light gray or cream colored and translucent. Some kinds of shrimp are almost transparent, with the internal organs visible. Both crayfish and shrimp can change their color to resemble the substrate upon which they are living. Decapods can be distinguished from other freshwater crustaceans by the following combination of characteristics (Figure 17):

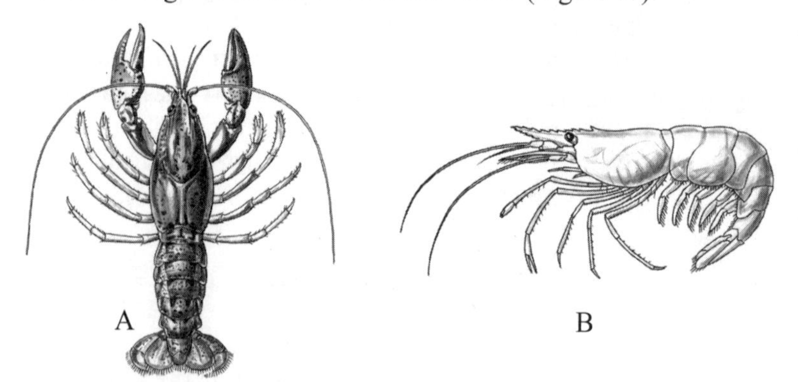

Figure 17. Representative decapods. **(A)** Crayfish. **(B)** Shrimp.

- Body is more or less cylindrical.
- Skin is thick and hard.
- Body segments are grouped into two regions: cephalothorax and abdomen.
- Cephalothorax consists of the head and entire thorax fused into a large region that is enclosed in an additional hard outer covering called a carapace.
- Abdomen is composed of the six individual segments after the cephalothorax.
- Elongate, pointed structure, called the rostrum, extends off the front of the cephalothorax.
- Two large compound eyes, which can be moved around on the ends of stalks, are present on the cephalothorax.
- One pair of antennae on the cephalothorax is much longer than the other.
- Five pairs of walking legs are present on the cephalothorax.
- First two or three pairs of walking legs have a hinged claw on the end.
- Remaining two or three pairs of walking legs have a simple, pointed tip.
- Small appendages are located on the bottom of the first five segments of the abdomen.
- A broad flipper, formed from three flat, paddle-like structures, is situated on the last segment of the abdomen.

Explanation of Names — The first syllable of the common name crayfish has its origins in the Old High German word *krebiz*, meaning crab. This became *crevice* in Middle French and *crevis* in Middle English. The second syllable, fish, has long been used as a combining form to describe a variety of aquatic animals, for example, starfish. The common name for the other family of decapods covered in this guide, shrimp, has its origin in Scandinavia. In Old Norse, *skorpa* meant to shrivel up. This became *schrempen* in

Middle Low German and took on the meaning to contract or wrinkle. In Middle English, this verb kept the same meaning but the spelling became *shrimpe*. Shrimp came to be used as a common name for these animals because of the way they tightly fold their abdomen under their cephalothorax as part of their swimming movement. The scientific name for this order is a combination of two Greek words: *deka*, meaning ten, and *pod*, meaning foot. This name was chosen because of the number of walking legs on the cephalothorax.

Biology of Common Families

Crayfishes

Families Astacidae and Cambaridae; Plate 19

Different Kinds in North America — Many; 12 genera, 315 species.

Distribution in North America — Throughout, except not found in most of the western Great Plains and Rocky Mountain regions. There are two families of crayfish. In general, species in the family Astacidae occur only in the drainage basins on the Pacific slope of the Rocky Mountains. However, this family has been introduced into the eastern slope of the Rocky Mountains in the upper Missouri watershed. Species in the family Cambaridae are much more widespread, occurring throughout the East, from Hudson Bay watersheds in Canada to Atlantic and Gulf drainage basins in the far South.

Habitat — Different kinds of crayfish live in a wide variety of habitat types, including lotic-erosional, lotic-depositional, lentic-littoral, wetlands, and subterranean. They are common inhabitants of all sizes and velocities of running waters from springs to large rivers, lakes, ponds, sloughs, marshes, swamps, ditches, underground waters, and damp meadows. They usually occur only in shallow water (1–2 m), although in lakes they migrate to deeper water (> 30 m) during winter to avoid ice. Most of the commonly encountered kinds of crayfish live in the water, where they spend almost all of their time hidden in spaces among rocks, woody debris, plants, or coarse detritus. Most of the kinds that live in streams have a home range of less than 30 square meters. In clean lakes, the density of crayfish can reach 1–15 organisms per square meter, but they are not as dense in streams. Other kinds that are less commonly seen live on land in damp places near bodies of surface water or in wetlands. Because all crayfish must keep their gills wet in order to breathe, these kinds burrow down to the underground water level. The depth of burrows ranges from 5 cm to 3 m. There is an interesting species on the tall grass prairie of Texas that burrows into the soil where there is no apparent source of underground water. The burrows are on hills, and it appears that rainwater goes in the burrows as it runs downhill, then crayfish cap the burrows.

Movement — Most kinds of crayfish hide during the day and crawl around slowly at night by means of their legs. If disturbed, they swim backward in a quick, darting movement by repeatedly folding the abdomen down and forward. The flipper-like tail on the end of the abdomen helps to propel them through the water. Burrowing crayfish do their digging at night. They excavate pellets of mud, which they bring up and deposit around the opening of the burrow. The pile of mud pellets is called a chimney. Chimneys are usually about 15 cm high, but they can be as tall as 45 cm. Some kinds of burrowing crayfish spend almost all of their lives in the refuge that they build, only leaving occasionally at night to forage for food or to find a mate. Other kinds spend most of their lives in the water and burrow only under special circumstances, such as summer drought or winter freezing.

Feeding — Crayfish are omnivores and most will consume whatever food is available. The various mechanisms that they use to acquire their food place them in the following functional feeding groups: shredder-detritivores, shredder-herbivores, engulfer-predators, and scrapers. Their primary food is usually decaying vegetative material, but they often consume live plants. They greatly reduce beds of aquatic plants because they cut off entire plants at their roots even though they eat only part of the foliage. If vegetation is not abundant, crayfish will prey on live snails, aquatic insects, scuds, small fish, and fish eggs. They also readily eat carrion, such as dead fish, and they are even known to scrape algae and microbes from firm surfaces. The large claws on the first pair of legs are used for crushing or ripping food, while the small claws on the second and third legs are used for handling and chopping food. When feeding on snails, crayfish select species with thin shells that they can easily crush.

Breathing — Crayfish breathe dissolved oxygen by means of gills. Their branched, filamentous gills are located in chambers on each side of the cephalothorax. The gill chambers are spaces between the actual body and the carapace that covers the body in the cephalothorax region. They wave the gills to circulate oxygenated water into the gill chambers.

Life History — When it is time to reproduce, female crayfish secrete a sticky substance that covers the bottom of their abdomen. The eggs are attached among the short appendages on the abdomen by this sticky substance. The abdominal appendages are moved to keep water flowing over the eggs so they have a constant supply of dissolved oxygen. Females brood the eggs until they hatch, then the newly hatched crayfish hold on to their mother until they shed their skin two or three times. At that time, the young crayfish begin to leave the mother for successively longer periods of time until they eventually become free living. Young crayfish grow and develop and become capable of reproducing after they have shed their skin a total of six to ten times. Mature crayfish shed their skin two to four more times after they mate. Most kinds live 2 or 3 years and pass through at least two breeding seasons. Some crayfish have been observed to live as long as 6–8 years. There are some differences in the life history of the two families of crayfish. Species of Cambaridae usually mate in the spring and carry 100–450 eggs (range 25–800) for 2–20 weeks between March and June. Species of Astacidae usually produce 100–250 eggs in autumn. They carry the eggs for 7–8 months until they hatch from April to June.

Crayfish have two major defense mechanisms, in addition to hiding. The large claws on the first pair of legs are used for active defense. In addition, the walking legs have a crease near their base that acts as a breaking point for passive defense. When a crayfish is grabbed or alarmed, muscles in the leg contract suddenly at this crease, and the leg breaks off so the crayfish can escape. Legs may be broken off when predators attack them, or when they fight among themselves. Missing legs are regenerated the next time the skin is shed, but they are not as large as the original legs. They get larger with subsequent molts.

Significance — Because of their large size and omnivorous diet, crayfish play important roles in the ecology of natural freshwater ecosystems. They regulate the distribution and relative abundance of snail species because they prey selectively on the kinds that can be easily crushed. Snails, in turn, regulate the community of algae and microbes growing on firm surfaces. The abundance of crayfish determines the density of aquatic plant beds, or if plants are present at all. Beds of aquatic plants provide habitat that is essential for many kinds of invertebrates and fish. Crayfish break down organic matter and make the nutrients and energy available to other organisms. Many fish, including important sport species such as smallmouth bass, feed on crayfish, especially the juvenile stages. Other wildlife that consume crayfish include water snakes and raccoons. Crayfish also provide direct benefits to humans. They are harvested and sold for fish bait. Crayfish have long been eaten by people in the southern United States, where edible species are commonly caught with traps in natural habitats. The popularity of these tasty and nutritious crustaceans has spread to other areas, and, as a result, they are mass produced in commercial aquaculture facilities, primarily in the warm, southern states.

Stress Tolerance — Crayfish are facultative to most forms of environmental stress and pollutants. They can withstand wide ranges in temperature, pH, and alkalinity. However, they are sensitive to certain toxic substances including metals, insecticides, herbicides, and lampricides. Crayfish bioaccumulate some metals, such as mercury, when the substances occur at levels that are too low to be immediately toxic, so tissue analysis of crayfish can be used to track metal contamination in freshwater environments. Stream species are generally less tolerant of stress than the species that live in lakes and ponds.

Shrimps

Family Palaemonidae; Plate 20

Different Kinds in North America — Few; 2 genera, 15 species. All of the common kinds of freshwater shrimp in North America are in the family Palaemonidae. There is one other family (Atyidae) with four rare species that are seldom found. Two of the rare species live only in caves of the Southeast, and two are known only from coastal California. One of the rare California species is probably extinct.

Distribution in North America — Throughout.

Habitat — Shrimp are still-water organisms, primarily inhabiting the lotic-depositional and lentic-littoral habitat types. They are found mostly in the quiet backwaters of large rivers, but they are also common in shallow water around the margins of lakes and reservoirs. A few kinds live in slow areas of springs or the brackish waters in the upper reaches of estuaries. Shrimp are almost always associated with dense growths of aquatic plants in these habitats.

Movement — They are swimmers that move forward by paddling motions with the short appendages on their abdomen. Shrimp usually stay in protected spaces within beds of aquatic plants, where they also perch for short periods on the plants. When disturbed, they quickly dart backwards, like crayfish, by repeatedly bending their abdomen down and forward.

Feeding — Shrimp feed primarily as scrapers. They remove the algae that grows attached to the leaves and stems of aquatic plants. They also function as engulfer-predators, foraging for immature aquatic insects. Cave-dwelling species feed by straining fine organic matter and microbes from the sediment with their mouthparts. Shrimps in surface waters may also supplement their diet as collector-gatherers in this manner.

Breathing — These organisms breathe dissolved oxygen the same way as crayfish. They wave their branched gills in the space between their body and the carapace that covers the body in the cephalothorax region.

Life History — In northern latitudes of North America, shrimp usually breed between April and August. In most southern regions the breeding period is extended from February to October. Species in the Everglades, Florida, reproduce throughout the year. The number of eggs that females produce per brood varies greatly according to the size of the different kinds of shrimp. The much larger species in the genus *Macrobrachium* produce 6,000–24,000 eggs, while species in the genus *Palaemonetes* produce only 8–160 eggs. Female shrimp hold their eggs on the abdomen in the same manner as crayfish. The eggs incubate for 12–24 days on the female. The larvae that hatch are free-swimming individuals about 4 mm long. They become mature enough to reproduce when they are about 20 mm long, which is usually after they have shed their skin five to eight times. Most shrimp die after reproducing, but some molt shortly after the first brood of young larvae are released and live long enough to produce another brood. Most kinds of shrimp live approximately 1 year. Adults usually disappear by late summer or early fall, and the population consists of immature stages during the winter.

Significance — The ecological roles of shrimp in freshwater ecosystems are probably not as notable as crayfish, because shrimp are smaller and have more restricted feeding habits. However, they are important food for many kinds of fish. Some of the large species of shrimp are consumed by humans, particularly in the southern United States. Freshwater shrimp have become popular enough as food that they are raised for commercial sales in outdoor, aquaculture facilities in warm regions.

Stress Tolerance — Not much is known about the tolerance of shrimps, but they are probably facultative, similar to crayfish.

Aquatic Insects

Subphylum Atelocerata, Class Hexapoda, Subclass Insecta

Some insects are the only members of the arthropod subphylum Atelocerata that live in freshwater environments. The diversity of insects is amazing. There are 751,000 known species, which is 86% of arthropods (875,000 known species), 76% of invertebrates (990,000 known species), and 73% of all animals (1,032,000 known species). To take this a step farther, insects constitute 53% — that is more than one out of two — of all known species of living things on Earth, including plants and microbes. Although most insects are terrestrial, this group of invertebrates is richly represented in freshwater environments. In North America, the recorded number of all insects is 87,000, which belong to 31 orders and 626 families. Of that total, 8,600 species, belonging to 12 orders and 150 families, live in association with freshwater during some stage of their lives.

In addition to there being many different kinds of aquatic insects, they usually occur in high numbers too. Aquatic insects are essential participants in ecological processes that take place in freshwater ecosystems, such as breaking down organic matter and making food available to other organisms. Because of their diversity, abundance, and important ecological roles, it is safe to say that insects are the dominant invertebrates in freshwater environments. However, in contrast to other invertebrates, there are hardly any insects that live submerged in marine environments. Crustaceans take care of the same jobs on the bottoms of marine environments that insects perform in freshwater environments. Most of the very few insects associated with marine environments live on the surface or in damp areas near the margins.

For the reasons explained above, this guide emphasizes aquatic insects. The North American aquatic insects belong to the following orders.

Mayflies (Ephemeroptera)[1]
Dragonflies and damselflies (Odonata)[1]
Stoneflies (Plecoptera)[1]
Dobsonflies and alderflies (Megaloptera)[1]
Caddisflies (Trichoptera)[1]
True Bugs (Hemiptera)[2]
Beetles (Coleoptera)[2]
True Flies (Diptera)[2]
Grasshoppers and crickets (Orthoptera)
Spongillaflies (Neuroptera)
Moths (Lepidoptera)
Wasps (Hymenoptera)

The eight insect orders with superscripts (either [1] or [2]) are covered in this guide. In the five orders marked with [1], almost all members are aquatic during some stage of their lives. In the three orders marked with [2], most of the species are terrestrial, but substantial numbers of species are aquatic. The remaining four orders that are not included in this guide are not likely to be encountered by beginning aquatic biologists. There are many kinds of insects that are associated with the land at the margins of freshwater environments. Scientists refer to these kinds as semi-aquatic. These insects usually depend on certain plant species for their food and habitat, and it is just a coincidence that the plants grow only near water. Some of the insects themselves require damp soil for reasons such as burrowing or laying their eggs, but they have no direct link to water, as do the true aquatic insects. Semi-aquatic kinds are not included in this guide. A total of 73 families of aquatic insects are covered.

Aquatic insects differ markedly from other freshwater invertebrates in regard to their life history. In many kinds of aquatic insects, there are pronounced changes in their appearance and biological features during their metamorphosis from immature larvae (singular larva) to adults. In most of the orders with aquatic species, only the larvae live in the water. The adults are usually winged forms that live out of the water, similar to the other terrestrial insects. For the kinds in which the larvae and adults are both aquatic, almost all of them leave the water for awhile after they have become adults then reenter the water later. In many kinds with aquatic adults, the larvae and adults look different and do not have the same biological features. For the 73 families of aquatic insects included in this guide, all aquatic adults and larvae are shown in color, with whole body illustrations of both stages. The only exception is the true bugs, because the larvae and adults look basically alike. For the aquatic insects that have terrestrial adults, there is a black and white illustration of an adult that is representative of the order.

Studying immature aquatic insects can be a little bit more challenging than studying other freshwater invertebrates. Aquatic insects are very small and hard to see in their early larval stages. Some external features of the body that distinguish a particular family from others in the same order may not even develop until the last part of the larval period. Users of this guide will probably only find and recognize aquatic insect larvae that are fairly mature. Also, there are different types of metamorphosis in the various orders of aquatic insects. It is important to understand the types of metamorphosis because this information is used to identify different kinds of aquatic insects, and the type of metamorphosis is an important part of their life history. To use this guide, it is only necessary to understand two of the major types of metamorphosis, incomplete (hemimetabolous) and complete (holometabolous).

Insects that undergo incomplete metamorphosis emerge from their eggs as larvae that look somewhat like the adults, except that the larvae lack wings (see Figures 19, 20, 23, 24). The larvae have compound eyes and segmented legs. Wings develop externally on the second and usually the third thorax segments of the larvae in pads or buds that are outgrowths of the skin. Developing reproductive structures are also visible at the end of the abdomen. Development proceeds successively in small incremental changes. Each time the larva sheds its skin, it looks a little bit more like the adult. The wing pads and reproductive structures gradually get larger each time the larva sheds its skin to grow. Wing pads are not visible in very young larvae, and you will have to look closely to see them until the insect larvae are at least half way through their development. When larvae have matured to the point that they are ready to transform into winged adults, the wing pads become dark and swollen with some of the wing veins visible through the skin. Organisms are generally active throughout the larval stage. There is no distinct resting stage during which dramatic developmental changes take place. In older books, the immature stages of insects with incomplete metamorphosis were called nymphs, or sometimes naiads for the aquatic kinds. Today, most scientists have adopted the practice of using the term larva for the immature stages of all kinds of insects.

For insects with complete metamorphosis, the appearance of the larvae is usually quite unlike that of the adults (see Figures 29,

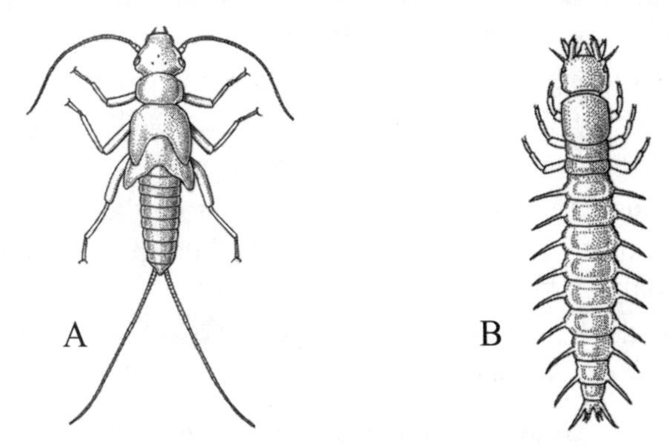

Figure 18. Two types of insect metamorphosis, with representative aquatic insect larvae. **(A)** Incomplete metamorphosis (stonefly). **(B)** Complete metamorphosis (dobsonfly or hellgrammite).

30, 33, 34). These larvae do not have compound eyes and may not have segmented legs. Wing pads and developing reproductive structures are not visible because they develop internally. Thus, larvae of insects with complete metamorphosis do not add any external structures as they shed their skin and develop. Organisms are generally active during most of the immature period, but there is a distinct inactive, resting stage before transforming into adults. The resting stage is called the pupa, and this is when almost all of the developmental changes take place. All of the changes that occur progressively in successive steps in insects with incomplete metamorphosis occur suddenly in one step for insects with complete metamorphosis. External wing pads, segmented legs, and preliminary reproductive structures appear when the larva molts to the pupa. Then, functional wings and reproductive structures develop inside the pupa and appear when the adult emerges from the pupa.

Distinguishing Features of Larvae — The appearance of aquatic insect larvae is different according to the type of metamorphosis.

Incomplete Metamorphosis (Figure 18A):

- Body segments are grouped into three, readily distinguishable regions: head, thorax, and abdomen.
- One pair of compound eyes, which are composed of many tightly packed individual units, are located on the head.
- One pair of antennae, are situated on the head.
- Wing pads (also called wing buds) appear on the second and third thorax segments, except in a few kinds that have wing pads only on the second thorax segment.
- Three pairs of segmented, hardened legs occur on the thorax, one pair per segment.

Complete Metamorphosis (Figure 18B):

- Body segments are grouped into three regions, which are the head, thorax, and abdomen, but these regions may be hard to distinguish.

- One pair of simple eyes, which consist of just a single dark spot or sometimes a cluster of a few dark spots, are found on the head.
- One pair of short antennae are usually present on the head, but these may be so small that they are not visible in some kinds.
- Wing pads are not present on the thorax.
- Three pairs of segmented, hardened legs may be on the thorax, one pair per segment, but these are absent on many common kinds of insects with complete metamorphosis.
- Short, fleshy, unsegmented appendages (prolegs) may be present on the thorax or abdomen.
- Prolegs usually have claws or short, stiff hairs on the ends.

Unfortunately, some aquatic insect larvae with complete metamorphosis look more like aquatic earthworms than insects because they do not have body regions, legs, prolegs, or any other structures. This situation arises mostly with some of the larvae of true flies (Diptera). After some practice, you will learn to recognize these as insects. They do have a head, or at least some parts of a head, which distinguishes them from aquatic earthworms. The head may be withdrawn into the first segment of the thorax or the head may be reduced to just the mouthparts, both of which make the head inconspicuous. The next three regions after the head are still referred to as the thorax, and the remaining segments after the thorax are referred to as the abdomen, even though they all look alike.

Distinguishing Features of Adults — Adult aquatic insects can be readily distinguished from other arthropods by the following combination of characteristics (see Figures 20 and 21):

- Body is usually more or less cylindrical, but may be somewhat flattened from top to bottom.
- Skin varies from somewhat soft to very hard.
- All appendages are composed of a single branch (uniramous).
- Body segments are grouped into three regions: head, thorax, and abdomen.

- Head appears to be one body segment, without any subdivisions.
- Thorax is composed of three body segments, but the individual segments may be difficult to distinguish.
- Abdomen consists of up to ten individual body segments that can be readily distinguished.
- Two jaw-like mouthparts (mandibles) that move laterally, from outside to inside, usually occur on the head; however, in some adult insects, these are modified into a single tube or sponge for feeding on fluids.
- One pair of antennae are found on the head.
- One pair of compound eyes, which are composed of many individual cell-like units, are situated on the sides of the head.
- Three pairs of segmented legs are located on the thorax, with one pair on each segment of the thorax.
- Legs usually have one or two claws on the end.
- One or two pairs of wings are usually present on the thorax, with the front pair of wings on thorax segment two, and the rear pair of wings, if present, on thorax segment three.
- No appendages occur on the abdomen, except on the rear segment.
- Two or three tails and short reproductive structures may protrude from the last abdomen segment.

Explanation of Names — The common name insect and scientific name Insecta come from the Latin word for these creatures, *insectum*. The Romans derived *insectum* from the verb *insecare*, which is a combination of *in*, meaning into, and *secare*, meaning to cut. The origin of this name probably has to do with the fact that so many common insects obtain their food or build protective structures by cutting with their mouthparts.

Mayflies

Order Ephemeroptera; Plates 50 – 59

Mayflies constitute a medium-sized order, in terms of diversity, with 21 families and 676 species in North America. This guide includes 10 families that are common in freshwater habitats. Mayflies have incomplete metamorphosis. All members of the order are aquatic as larvae and terrestrial as adults.

Distinguishing Features of Larvae — Mayfly larvae have elongate bodies. Most kinds are slightly flattened, but some are very flat and others are almost cylindrical. The body length of mature larvae in different species ranges from 2 to 32 mm, not including the antennae or tails. Mayfly larvae can be distinguished from all other aquatic insect larvae by the following combination of characteristics (Figure 19):

- Wing pads are present on the thorax.
- Three pairs of segmented legs extend from the thorax.
- One claw occurs on the end of the segmented legs.
- Gills occur on at least some abdomen segments preceding the last segment.
- Gills are attached to the sides of the abdomen, but they sometimes extend over the top or bottom of the abdomen.
- Gills consist of either flat plates or filaments.
- Three long, thin tails usually occur on the end of the abdomen, but there are only two tails in a few kinds.

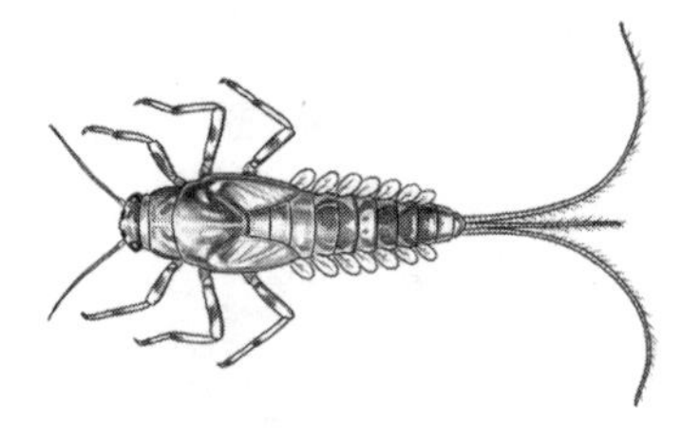

Figure 19. Representative mayfly larva.

Distinguishing Features of Adults — Mayfly adults are not aquatic, but they are often found near the bodies of water where the larvae develop. They have elongate, soft bodies about the same length as the larvae, not including the tails. Mayfly adults can be distinguished from all other aquatic insect adults by the following combination of characteristics (Figure 20):

- All wings are membranous and have many veins.
- Wings are held together extended above the body when not flying.
- Front wings are large and somewhat triangular in shape, while the hind wings are much smaller and usually rounded.
- Hind wings are missing in some kinds.
- Two or three long, thin tails extend from the end of the abdomen.

Explanation of Names — The only common name that is widely recognized for the entire order is mayflies. The explanation of the common name is that many species of Ephemeroptera emerge from the water as adults during the month of May. They may be called a number of different common names in certain localities or regions. Some of the local names for mayflies are shadflies, willowflies, duns, and spinners. Shadflies and willowflies probably refer to the fact that many mayflies emerge in the spring, which is when shad migrate upstream into freshwater rivers to spawn and the willows growing along streams are in bloom. Duns and spinners are common names used by fly-fishing anglers for certain stages in the life history of mayflies. Dun refers to a mayfly in the subimago stage, which is a stage that is unique to mayflies, as explained be-

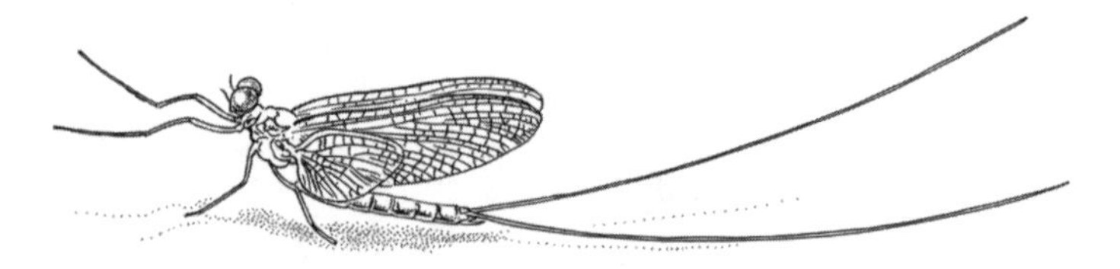

Figure 20. Representative mayfly adult.

low in the section on Life History. Spinner is the name used by anglers for mayfly adults. The scientific name of the order is a combination of two Greek words: *ephemeros*, meaning lasts a day, and *ptera*, meaning wings. The scientific name refers to the short duration of the adult stage.

Habitat — Mayfly larvae live in many different habitats, more than many of the aquatic insect orders. The most different kinds of mayflies live in lotic-erosional habitats. However, there are many kinds that inhabit lotic-depositional and lentic-littoral habitats. Different species of mayflies are adapted to live in parts of aquatic habitats with either firm or soft bottom, but more are found on firm substrate. Preferred places within the habitat are stones, aquatic plants, or coarse detritus, but some kinds have special adaptations to live on top of or down in silt and sand. Different species are found in all sizes of aquatic habitats, ranging from small streams to large rivers and from small ponds to lakes. They are even found in temporary aquatic habitats, both lotic and lentic.

Movement — Because mayfly larvae live successfully in so many different types of habitats, it is logical that larvae would also exhibit many different habits for movement. Common species of mayfly larvae can belong to any of six categories of movement: swimmers, clingers, climbers, crawlers, sprawlers, or burrowers.

Feeding — Most are either collector-gatherers or scrapers. A few kinds of mayfly larvae are collector-filterers. Some rare species are engulfer-predators of small organisms in sand, mostly midge larvae (a family of true flies, Diptera: Chironomidae).

Breathing — Mayfly larvae have a closed breathing system with gills on the abdomen for obtaining dissolved oxygen from the water. The gills are usually flat plates or filaments (Plates 50B, 56B, 59B), but there are many different sizes and shapes of gills in different species. Species that live in warm, still waters tend to have larger gills with more surface area for obtaining dissolved oxygen than species that live in cool, flowing waters. Their gills have muscles attached to their bases, so they can be held out in the current or waved in still water to increase the efficiency of obtaining oxygen. Mayfly gills are delicate, fragile structures. Sand grinds away the delicate gills, whereas silt sticks to the gills and blocks the transfer of dissolved oxygen from the water.

Life History — Most kinds of mayflies have one generation per year. However, some kinds have two or more generations per year, while others require 2 or more years for a generation to develop. As a general rule, ones that have many generations per year tend to be small in size, while those that require 2 or more years for a generation tend to be large. Adults of various mayfly species emerge throughout the warm months of the year, and some will even emerge during the winter if there is a short period of unusually warm weather.

The females always lay the eggs in the water. The time required for the eggs to hatch usually ranges from about a week to several months, but some species diapause in the egg stage for up to 11 months. By temporarily ceasing their development in the egg stage they are able to avoid periods of the year when environmental conditions are naturally unsuitable.

Mayflies usually spend 3–6 months as larvae, but this can range from 10 days to 2 years in different species. They have the distinction of shedding their skin more times than any other aquatic insect. The exact number of molts is not known for most species, but the usual number is thought to be between 12 and 27. This probably varies, even within the same species. Some mayflies may shed their skin as many as 45 times while they are larvae. Mayflies remain active throughout the period that they are larvae. There is no diapause in the larval stage.

When larvae have completed their development, different species of mayflies emerge from the water by one of two ways. In some species, larvae float to the water surface and emerge from the last larval skin by using it as a raft. These kinds usually emerge from the water surface in just a few seconds, because they are vulnerable to being eaten by fish. Species that live in fast streams migrate to areas of still water for emergence. These quiet areas might be in a pool, near the margins of a stream, or in an eddy behind a rock or log. The second way that some mayflies emerge from the water is for the larvae to crawl a short distance out of the water before shedding their skin. These species always migrate to the edge of the water when the mature larvae are ready to emerge. Mayfly larvae exit by means of rocks, vegetation, tree roots, woody debris, or any other protruding objects that provide a suitable surface for pulling themselves out of the water. They usually crawl more horizontally than vertically and usually move no farther than 5 cm away from the water. Gradually slanting surfaces of stones, woody debris, or living plants (aquatic or terrestrial) are preferred places to emerge. Species that crawl out of the water may take a bit longer to emerge, up to a few minutes. All mayflies fly away immediately after emerging from the last larval skin.

In mayflies, the organism that emerges from the last larval skin and flies away is not in the adult stage yet. Mayflies are unique among all insects in having a stage with functional wings in which the organism is not fully developed into an adult. The reproductive system has not finished developing in the first winged stage, so mayflies must shed their skin again when they have completely developed into an adult. The scientific term for an adult insect is imago, so the first winged stage of mayflies is called a subimago. Mayflies are the only insects with a subimago stage. Subimagoes have the same general appearance as adults, but close examination will show subtle differences. The wings of subimagoes look dull and semi-transparent, whereas adult wings are shiny and clear. The wings of subimagoes are covered with fine hairs that cause the dull appearance.

When mayfly subimagoes leave the water, they only fly briefly to nearby vegetation, where they remain still while final adult development takes place. Subimagoes of the same species in a habitat usually emerge together on the same day, or at least within a few days of each other. This sometimes produces large clusters of subimagoes on vegetation near water. The subimago stage does not last long, usually only about 24 hours. The duration of the subimago stage can range from 1 to 2 minutes up to 48 hours, depending on the species. When adult development is complete, the subimago sheds its skin one more time, including the outer covering of the delicate wings.

The adult is the stage that is responsible for the scientific name of mayflies (*ephemero* – lasting a day). The duration of the adult stage is usually only about 24 hours, ranging in different species from as short as 90 minutes to a few days at most. This is much shorter than the adult stage of any other insects, either aquatic or terrestrial. Adult mayflies do not feed. As a result, they have no mouthparts. The only role of the adult stage is to find a mate and reproduce to continue the species. With such a short adult life, mayflies must quickly find a mate of their own species. They accomplish this by using a visual communication behavior called swarming. Mayfly swarms usually take place over the aquatic habitat where the larvae live, or not very far away. Males of the same species congregate in the air and perform a conspicuous aerial dance by all flying together in the same pattern. Females recognize a swarm of males belonging to their species according to the time of day, altitude, number of males, and, most importantly, the motion of the swarm. Each male engages in the same pattern of horizontal and vertical movements in flight. Females in the vicinity enter the swarm, and males quickly grasp them for mating. Females proceed with laying their eggs right away. Both males and females die soon after mating.

Significance — Neither larvae nor adults of mayflies are ever considered serious pests. Large numbers of emerging adults are sometimes a temporary nuisance, but they do not cause any significant economic or health problems. Mayfly larvae are definitely beneficial organisms.

Emergence of dense populations of adult mayflies does cause a short-term nuisance in towns and cities along the upper Mississippi River and some of the Great Lakes. This relatively minor, but interesting, problem is caused by a few species of common burrower mayflies in the family Ephemeridae. People are annoyed by adult mayflies landing on them, or even flying into their mouths. Some persons are allergic to the hairs and fragments from the skins that are shed by the subimagoes. The enormous numbers of adults that accumulate on bridges in these areas create unsafe driving conditions, much like driving on wet, slushy snow. Snowplows have even been brought out in summer to clear away adult mayflies. A huge swarm of these mayflies caused a brownout in northwestern Ohio, where they were attracted to the lights of an electrical substation which caused electricity to short circuit across insulators. However, the occurrence of these large populations of adult mayflies is actually a good sign, because it means that the aquatic environments where the larvae develop are healthy.

Most species of mayflies are very sensitive to pollution, so the number of different kinds of mayflies and their relative abundance compared to other invertebrates provide useful and reliable information about the environmental health of a body of water. When many different kinds of mayflies are found or when mayflies represent a high proportion of the invertebrates at a particular location, that is a reliable indication that the aquatic environment is healthy. Mayfly larvae are probably the most useful aquatic invertebrates for biomonitoring pollution, because various types of mayfly larvae occur naturally in many different habitats and they are usually very abundant and easy to find.

In addition to biomonitoring pollution, mayfly larvae play significant roles in the food webs of aquatic ecosystems, particularly in streams. They are one of the major converters of plant material to animal tissue. Almost all mayfly larvae eat either live plants, usually in the form of algae, or decomposing plant material, called detritus. Mayfly larvae are often very abundant, so the living animal tissue that they produce in their bodies accounts for a significant amount of energy flowing from primary producers (plants) to animal consumers. Many invertebrate and vertebrate predators in the aquatic environment eat mayfly larvae and emerging subimagoes. The role of mayfly larvae in aquatic ecosystems is comparable to rabbits and mice in terrestrial ecosystems. Mayfly larvae account for a high proportion of the diet of many fish that are managed for recreational fishing, especially trout and salmon in cold-water streams. If fisheries biologists need to evaluate the food base for fish populations, mayflies are probably the major invertebrates that should be studied.

Fly fishing is a popular outdoor recreational activity that stimulates a multimillion dollar industry. Mayflies are the most common models for the "flies" that mimic natural aquatic prey in streams. Imitations of all stages of mayflies (larvae, subimagoes, adults) are used by fly-fishing anglers.

Biology of Common Families

Ameletid Minnow Mayflies

Family Ameletidae; Plate 58

Different Kinds in North America — Few; 1 genus, 34 species.

Distribution in North America — More common in the northern part of the continent. In West, south along mountains to California and New Mexico; in Central, south to Illinois; in East, south along mountains to Georgia.

Habitat — Primarily lotic-erosional, also lotic-depositional, sometimes lentic-littoral. The larvae of ameletid minnow mayflies are usually found in small, rapidly flowing streams among pebbles. They can also occur near banks among vegetation and debris or in rocky pools on the sides of boulders. Members of this family can be abundant in very small brooks that are only 20–30 cm wide and less than 2–3 cm deep. They are sometimes in large rivers, lakes, or ponds if there is clean rock or gravel bottom with some current.

Movement — Primarily swimmers, also clingers. The larvae of ameletid minnow mayflies are exceptionally strong swimmers, being able to navigate in water flowing at the rate of 60–90 cm per second (2–3 ft per second). They seek quieter water before coming to rest on solid objects, which is how they spend most of their time.

Feeding — Primarily collector-gatherers, also scrapers. Spines on their maxillae look like they would be effective for removing periphyton, and diatoms have been found in their digestive system.

Other Biology — Some species of ameletid minnow mayflies live in temporary streams by remaining dormant in the egg stage when the streams are dry, then hatching and developing later when water is flowing. Mature larvae crawl out of the water to emerge. They usually climb to the top of a rock or 5–10 cm up a twig and remain quiet for up to 15 minutes before the subimagoes emerge and fly away. There is usually only one generation per year. Adults occur as early as February in the southeast and as late as autumn in other areas. Males have never been collected for some species, indicating that females reproduce by parthenogenesis.

Stress Tolerance — Very sensitive.

Brushlegged Mayflies

Family Isonychiidae; Plate 51

Different Kinds in North America — Few; 1 genus, 17 species.

Distribution in North America — Throughout, but more common in East.

Habitat — Lotic-erosional. Larvae of brushlegged mayflies are found in all sizes of rapidly flowing waters from brooks to large rivers. They occur in shallow riffle areas, mostly in leaf packs, branches, tangles of vegetation, and other debris in swift current. Sometimes they are found on bare stones or rock outcrops, but always in current.

Movement — Primarily clingers, also swimmers. Larvae in this family are very strong swimmers, but they spend most of the time holding on to the substrate and feeding in the current.

Feeding — Collector-filterers. Larvae of brushlegged mayflies face into the current holding on with their middle and hind legs. The front legs are held fairly close together, below the head, in the water. The long hairs on the front legs overlap and filter particles of food from the water (Plate 51B). From time to time they browse on the material that is captured on the hairs. Their diet consists mostly of very fine particles of detritus, diatoms, and other algae. Small insect larvae, such as midges, black flies, or smaller mayflies, may also be consumed.

Other Biology — When it is time to emerge, mature brushlegged mayfly larvae often move to shallow, calm water near the margin of the stream, where they can be observed sitting on the bottom. Larvae crawl a few centimeters out of the water before shedding their last larval skin and changing into the subimago stage. They leave the water by means of rocks, sticks, or other protruding objects. They have either one or two generations per year, depending on the species and environmental factors, especially temperature. Different species emerge as adults from May through August.

Stress Tolerance — Somewhat sensitive.

Common Burrower Mayflies

Family Ephemeridae; Plate 50

Different Kinds in North America — Few; 4 genera, 15 species.

Distribution in North America — Throughout.

Habitat — Lentic and lotic-depositional. Larvae of some common burrower species are usually very abundant in lakes, ponds, large rivers, and reservoirs that have been constructed by damming rivers. These large larvae may represent over half the biomass of bottom-dwelling invertebrates in some lakes. A few species occur in small to medium sized streams, but they are not as numerous as those in larger bodies of water. All live in soft sediment but individual species have characteristic preferences about substrate composition. Some species of common burrower mayflies occur only in very fine particles of silt and clay while others are restricted to areas where the bottom consists of sand and gravel. In lakes and large rivers they are usually most abundant at depths of several m but they can also be found in much deeper water (≥ 18 m). In small to medium sized streams, larvae can be found in very shallow water (< 0.5 m) wherever the appropriate mixture of soft sediment occurs.

Movement — Burrowers. Larvae of common burrower mayflies construct a U-shaped tunnel with holes at both ends. They dig somewhat like moles. The ends of the front legs are flattened and used as shovels. The head is also modified for burrowing. The jaws have two stout pointed tusks that project forward, and there is a shelf-like projection coming off the front of the head. Species that live in lakes and large rivers burrow as deep as 13 cm (5 in). The openings of the burrows can be observed while wading in shallow water of lakes and reservoirs or by diving in deep water with SCUBA or snorkeling gear. The dense populations of common burrowers that often live in lakes and reservoirs make the bottom look like people have been walking around in spiked shoes. Larvae can swim reasonably well by undulating their abdomens and waving the feathery gills, but seldom do so unless disturbed.

Feeding — Primarily collector-gatherers, also collector-filterers. Most species of common burrower mayflies ingest sediment non-selectively. They leave their burrows at night and roam around nearby to feed. Some species may also bring small particles of organic matter into their burrows by establishing a current with movements of their feathery gills.

Other Biology — The feathery gills (Plate 50B) of larvae in this family are waved in the burrows to maintain a current of water carrying dissolved oxygen for breathing. Larvae of some common burrowing mayflies have been observed to swim out of their burrows to shed their skin for growth, then they dig another burrow after molting is completed. Larvae usually float to the surface for emergence of the subimagoes. Some species migrate to very shallow water at the edge of their habitat for emergence. Common burrowers usually have one generation per year, but they often require 2 years to complete a generation in the far North. It is possible that some species produce two generations per year in the far South. Adults of different species have been collected from May through September. Emergence of individual species tends to be highly synchronized in a given location, with all members of the population emerging within a day or two. Mass emergences of some species in this family that inhabit the upper Mississippi River and some of the Great Lakes cause considerable annoyance to people living in the vicinity (described above in the summary of the order).

Stress Tolerance — Mostly facultative, some very sensitive or somewhat sensitive. Populations of common burrowing mayfly larvae increase with moderate additions of organic matter and nutrients because under these conditions more food becomes available on the bottom. However, excessive organic and nutrient pollution will eliminate common burrowers because decomposition causes the dissolved oxygen concentration to become intolerably low. The critical level appears to be about 1 part per million of dissolved oxygen. They are also susceptible to some toxic chemicals that can be chemically bound to the sediments where the larvae reside. Pollution nearly eliminated formerly dense populations of common burrowers in the upper Mississippi River and in some parts of the Great Lakes, especially western Lake Erie, by the late 1950s. They have now recovered to a large extent in these areas because of improved environmental conditions brought about by government regulations that control pollution.

Flatheaded Mayflies

Family Heptageniidae; Plate 55

Different Kinds in North America — Moderate; 14 genera, 126 species.

Distribution in North America — Throughout.

Habitat — Primarily lotic-erosional, some lentic-erosional. Flatheaded mayfly larvae are found in all sizes of flowing waters from brooks to large rivers. Different species occur in currents ranging from slow to exceptionally fast. Many species live only in cool, fast-flowing streams. Fewer species are found in warm slow waters. They occasionally occur on rocks along wave-swept shores of lakes. Larvae are most common on solid, firmly anchored objects that do not have excessive growths of algae or fungi or accumulations of fine sediment. Preferred substrates are large stones (cobble, boulder, bedrock) and water-soaked wood and leaves.

Movement — Clingers. Larvae of flatheaded mayflies hold on to surfaces very tightly by means of their flat bodies and legs (Plate 55B). Some species also have large gills that extend under the abdomen and form suction discs for even greater adhesion. They hold on to the substrate so tenaciously that they are difficult to remove when collecting specimens. Larvae move on the substrate in a crab-like manner and can walk in any direction with equal ease. They occur mostly on the undersides of loose rocks and other debris that have current flowing through the small spaces where they hide. These organisms can also stay on the tops of rocks where the most periphyton grows, because their flat shape allows them to escape the force of the fast-moving water. Larvae of flatheaded mayflies are very poor swimmers and if they become dislodged they can only wiggle their bodies awkwardly until they reach another solid surface.

Feeding — Primarily scrapers, also collector-gatherers. The flat head, especially the large upper lip, shields the mouth underneath the head from the current, allowing the larva to ingest the food dislodged by the mouthparts before it is swept away.

Other Biology — Flatheaded mayfly larvae demonstrate a behavior pattern called positive thigmotaxis. Organisms with this behavior are stimulated to be continuously in contact with some solid surface. If several larvae are placed in a container of water with no other objects, they will hold on to each other in lieu of any other solid surface. Subimagoes emerge from the surface of the water. Different species have one or two generations per year. Adults have been collected from March through October.

Stress Tolerance — Chiefly somewhat sensitive, others very sensitive to facultative.

Little Stout Crawler Mayflies

Family Leptohyphidae; Plate 52

Different Kinds in North America — Few; 2 genera, 24 species.

Distribution in North America — Throughout.

Habitat — Primarily lotic-depositional, some lentic-littoral. Larvae of little stout crawlers are common in flowing waters ranging from small streams to large rivers, but they occur only in areas of slow current. Substrates where they reside include silt, fine sand, gravel, woody debris, moss and other plant growth on stones, exposed roots of terrestrial plants, and at the bases of rooted aquatic vegetation. Sometimes larvae are found along the shores of lakes where they sprawl on the sediments.

Movement — Primarily crawlers, some sprawlers.

Feeding — Collector-gatherers.

Other Biology — The gills of little stout crawlers are specially adapted for existence in silty habitats. The first pair of gills on abdomen segment two is thickened and platelike and does not function for obtaining dissolved oxygen. The platelike gills cover the other functional gills and protect them from being covered with silt, which would prevent the passage of dissolved oxygen from the water to the organism. To obtain dissolved oxygen, little stout crawlers raise the platelike gills slightly and circulate water under them by waving the other gills. Many species diapause in the egg stage over winter, which explains why they may be found in temporary streams. In some species, the larvae transform into subimagoes underwater, which rise to the surface and take flight. Other species of little stout crawlers float to the water surface as nymphs, then the subimagoes emerge. Different species have from one to many generations per year. Adults usually occur from May to late October, but they have been observed as early as February in the southeast.

Stress Tolerance — Facultative. In contrast to most mayflies, larvae of little stout crawlers may be abundant in warm silted streams with reduced dissolved oxygen concentration.

Primitive Minnow Mayflies

Family Siphlonuridae; Plate 59

Different Kinds in North America — Few; 4 genera, 26 species.

Distribution in North America — Throughout most of continent in cooler waters as far south as mountains of California, Arizona, Missouri, and Georgia.

Habitat — Primarily lotic-depositional, also lentic-littoral. The specific substrate where larvae usually dwell is silt, sometimes vegetation. Lotic species of primitive minnow mayflies are found in areas with little or no current, such as in pools or along the margins of riffles. Some lotic species occur in flowing water as young larvae, then migrate to still water as mature larvae. Larvae may become isolated in pools on the floodplain or in the channel as stream flow declines. Beaver ponds are a preferred habitat for lotic species that migrate to still water. Some species of primitive minnow mayflies live in shallow water at the edge of ponds and lakes.

Movement — Primarily swimmers, also climbers. Although larvae of primitive minnow mayflies are strong swimmers, they spend most of their time sitting on silt-covered bottom.

Feeding — Primarily collector-gatherers, also engulfer-predators, sometimes scrapers. Decaying plant material that lies on soft sediment is the most common food of primitive minnow mayflies. Larvae will also eat any soft-bodied insects that they come upon, especially midge larvae (Diptera: Chironomidae).

Other Biology — When sitting on the bottom, larvae constantly wave their gills, probably to obtain as much dissolved oxygen as possible from the still waters where they reside. Primitive minnow mayflies have some of the largest plate-like gills among the mayflies, probably because they need a large amount of gill surface area to acquire enough dissolved oxygen. Mature larvae crawl a short distance out of the water, or at least get about half of their body out of the water, on vegetation or other objects to transform to subimagoes. Most species probably have only one generation per year. Emergence of adults in different species has been reported from March to early May.

Stress Tolerance — Facultative.

Pronggilled Mayflies

Family Leptophlebiidae; Plate 56

Different Kinds in North America — Moderate; 9 genera, 74 species.

Distribution in North America — Throughout.

Habitat — Primarily lotic-erosional, some lentic-littoral. Most species of pronggilled mayflies are common in shallow, fairly rapid streams of small to moderate size, but some species also occur in larger rivers. Although larvae live in all sizes of flowing waters, they occur only in protected areas because they are poorly adapted for fast current. Places where they are found include the undersides of large stones, among pebbles and gravel, in moss or other aquatic vegetation, in tangles of woody debris, and near the bank in accumulations of leaves or exposed roots of terrestrial plants. In spring, some common species of pronggilled mayflies are found in very high numbers in isolated flood plain pools and in the uppermost reaches of small tributary streams because of migratory behavior (see Other Biology).

Movement — Primarily crawlers, some clingers. Pronggilled mayfly larvae are very poor swimmers. Migrations are accomplished by crawling.

Feeding — Primarily collector-gatherers, some scrapers. Larvae tend to be omnivores. Sometimes they eat the cast skins of their own species or other species.

Other Biology — Young larvae of pronggilled mayflies exhibit a behavior pattern known as negative phototaxis, which means they move away from light. During the day, larvae hide within the substrate where they live. At this time they feed on fine detritus that has settled into spaces within the substrate. At night they move to the upper surfaces of their habitat and feed upon the algae that grows there. In spring, mature larvae of some species migrate either upstream or to the stream margin to find still water. This behavior often leaves high numbers of larvae either stranded in pools on the flood plain after high water recedes to normal flows or in temporary seeps and marshes where the smallest tributary streams originate. In Manitoba, one species of pronggilled mayflies migrates up temporary tributaries formed by melting snow, moving as far as 366 m in a day. Migratory species have large gills that make them well adapted for the stagnant conditions that occur in the temporary habitats where they complete their development. When larvae have finished developing and it is time for them to emerge as subimagoes, their phototactic response becomes reversed and they crawl toward light. They emerge from solid surfaces, such as rocks or wood, that are barely covered with water. Some species of pronggilled mayflies diapause in the egg stage for part of the year, either during winter or summer. Pronggilled mayflies usually have one generation per year, but some species may produce two generations per year in the south. There is a wide range of emergence times for adults of different species in this family, from February through November.

Stress Tolerance — Primarily somewhat sensitive, others very sensitive or somewhat sensitive. Species that migrate to still water are facultative.

Small Minnow Mayflies

Family Baetidae; Plate 57

Different Kinds in North America — Moderate; 18 genera, more than 130 species.

Distribution in North America — Throughout.

Habitat — Lotic-erosional, lotic-depositional, lentic littoral. The different species of small minnow mayflies occur commonly in a wide variety of habitats. These mayfly larvae are most commonly found in shallow flowing waters on or under cobbles and pebbles, but they are also found on aquatic plants in flowing waters. Larvae occur in all sizes of flowing waters from small spring-fed brooks to large rivers. Other small minnow mayflies are found around the edges of ponds and lakes, especially on plants. Larvae are common in all types of standing water habitats including: permanent and temporary ponds, roadside ditches, margins of lakes, backwaters of streams, warm desert springs, brackish waters, and even sewage treatment ponds. Some species of small minnow mayflies are often very abundant in areas that are choked with vegetation. Larvae have been reported to develop successfully in water as warm as 32 °C and as cool as 4 °C.

Movement — Swimmers, clingers, climbers. In lotic habitats, larvae of small minnow mayflies are primarily swimmers and clingers. Larvae swim well, darting in short spurts from one spot to another by rapidly bending their abdomen up and down. Their long tails help them swim in this manner. When they stop swimming they always face upstream with their abdomen slightly raised and swinging side to side in the current. In lentic-littoral habitats, they are primarily climbers, sometimes swimmers. Small minnow mayflies that are climbers are very well camouflaged on plants by their greenish translucent color.

Feeding — Collector-gatherers, scrapers. Larvae of the different kinds of small minnow mayflies belong to each of these functional feeding groups in approximately equal proportions, regardless of whether they reside in lotic or lentic habitats.

Other Biology — Most lotic species of small minnow mayflies spend the winter as dormant eggs, but some overwinter as active larvae, as do most lentic species. Females of some lentic species retain the eggs for 5–6 days after fertilization. Embryological development is completed during this time, and when the female deposits the eggs in the water, they hatch immediately. Emergence of subimagoes occurs from the surface of the water, often in very turbulent water for lotic species. After floating to the surface, the subimago bursts free in only 5–10 seconds. Most lotic species of small minnow mayflies have two generations per year, but some have only one. Some lentic species require only 6 weeks for development and have many generations per year. Adults of different species of small minnow mayflies have been found from February through September. Some species are known to reproduce by parthenogenesis in northern areas.

Stress Tolerance — Mainly facultative, others somewhat sensitive to very tolerant. Small minnow mayflies are another family that is an exception to the rule about mayflies being indicators of good water quality. Other exceptions are little stout crawlers (Leptohyphidae) and small squaregills (Caenidae). Small minnow mayflies can be found in degraded waters where high nutrient concentrations cause excessive growths of algae. Sedimentation, elevated water temperature, and reduced dissolved oxygen concentration often accompany this situation.

Small Squaregill Mayflies

Family Caenidae; Plate 53

Different Kinds in North America — Few; 4 genera, 26 species.

Distribution in North America — Throughout.

Habitat — Lentic-littoral and lotic-depositional. Larvae of small squaregills usually inhabit quiet or even stagnant water and are more often found in ponds and lakes than streams. They are probably the most abundant mayflies in many ponds and lakes, but they are often overlooked because they are very small and do not move much. In ponds and lakes, larvae of small squaregills occur mostly on silt bottom in the zone of rooted vegetation in shallow waters. They may also be found among accumulations of leaves and other plant debris and sometimes on living submerged plants. In streams, they are usually found near the bank, dwelling in silt that has accumulated at the bases of plants.

Movement — Sprawlers.

Feeding — Primarily collector-gatherers, also scrapers. Larvae of small squaregills feed chiefly on decaying plant material but sometimes on dead animals.

Other Biology — The gills of small squaregills are adapted for living in silty habitats in the same way as described for little stout crawlers (Leptohyphidae). The number of generations per year varies according to the species, geographic location, and unique temperature conditions. There can be one, two, or many generations per year. Usually there are many generations per year only in the far south (Florida) or in waters that are kept continuously warm by thermal springs. Adults of small squaregills emerge from May through September in temperate regions, but they emerge all year in Florida.

Stress Tolerance — Mostly facultative, others somewhat sensitive to somewhat tolerant. Squaregills, like little stout crawlers (Leptohyphidae), are exceptions to the rule about mayflies being indicators of good water quality. They can be found in degraded conditions — especially those characterized by low dissolved oxygen, sedimentation, and nutrient enrichment — where only a few other facultative and tolerant organisms occur.

Spiny Crawler Mayflies

Family Ephemerellidae; Plate 54

Different Kinds in North America — Moderate; 8 genera, 90 species.

Distribution in North America — Throughout.

Habitat — Primarily lotic-erosional, some lotic-depositional, also lentic-littoral. Larvae of spiny crawlers occur in all sizes of flowing waters, on a variety of substrates, and in currents ranging from slow to fast. In swift current, they usually seek protection in small spaces between rocks, crevices in woody debris, or in various types of vegetation where the flow is reduced. Other preferred locations in streams are in exposed plant roots at the bank and within dense growths of moss on rocks throughout the stream. Larvae of spiny crawlers sometimes occur along the shores of lakes if there is a lot of wave action.

Movement — Primarily crawlers, also clingers. Larvae of most kinds of spiny crawler mayflies crawl about very slowly and stiffly and are very well camouflaged in small, irregular spaces within their habitat. Some cling tightly to surfaces, but they are not exposed to swift current.

Feeding — Primarily collector-gatherers and scrapers, some shredder-detritivores.

Other Biology — Larvae in this family have a very interesting defensive behavior. If threatened by another organism, a spiny crawler larva raises its three tails up in a pose called the "scorpion posture," which may offer protection by making the mayfly look larger. If the intruder does not go away, a spiny crawler larva tips its tails forward over its body and projects the tails in front of its head. The mayfly larva then uses its tails to poke the enemy, much like it is fencing with a sword. To transform into the subimago stage, larvae either float to the water surface or crawl a short distance out of the water. Most species have only one generation per year. Adults of different species have been observed from March through late autumn.

Stress Tolerance — Chiefly somewhat sensitive, others very sensitive to facultative.

Dragonflies, Damselflies

Order Odonata; Plates 34 – 40

This is a medium-sized order, in terms of diversity. There are 9 families and 407 species in North America. This order is divided into two suborders. Both of the suborders are commonly encountered, have some unique aspects of biology, and are easy to tell apart, so this guide includes the information necessary to identify each suborder. Dragonflies are included in the suborder Anisoptera, and damselflies are included in the suborder Zygoptera. This guide covers six families that are common in freshwater habitats, three families of dragonflies and three families of damselflies. Dragonflies and damselflies have incomplete metamorphosis. All members of the order are aquatic as larvae and terrestrial as adults. In the following paragraphs on distinguishing features, the entire order of dragonflies and damselflies are first distinguished from other orders of aquatic insects, then the suborder of dragonflies is distinguished from the suborder of damselflies.

Distinguishing Features of Larvae — Bodies of dragonfly and damselfly larvae have variable shapes. They can be long and slender, long and stout, or oval and somewhat flattened. The body length of mature larvae in different species ranges from 13 to 68 mm, not including any gills that might be present. At the order level, dragonfly and damselfly larvae can be distinguished from all other aquatic insect larvae by the following combination of characteristics (Figure 21 A, B):

- Lower lip (labium) is long and elbowed and is folded back against the head when not feeding, thus concealing the other mouthparts.
- Wing pads are present on the thorax.

- Three pairs of segmented legs extend from the thorax.
- Two claws occur on the end of the segmented legs.
- No gills are found on the sides of the abdomen, but some kinds have three flat, elongate gills on the end of the abdomen.

At the suborder level, dragonfly larvae and damselfly larvae can be distinguished from each other by the following combination of characteristics:

Dragonfly Larvae (Figure 21 A)

- Body is either long and stout or oval and somewhat flattened.
- Head is narrower than the thorax and abdomen.
- No gills are found on the end of the abdomen.
- Three short, stiff, pointed structures occur on the end of the abdomen, forming a pyramid-shaped valve for the opening on the end of the abdomen (rectum).

Damselfly Larvae (Figure 21 B)

- Body is elongate and slender.
- Head is wider than the thorax and abdomen.
- Three flat, elongate gills project from the rear of the abdomen.

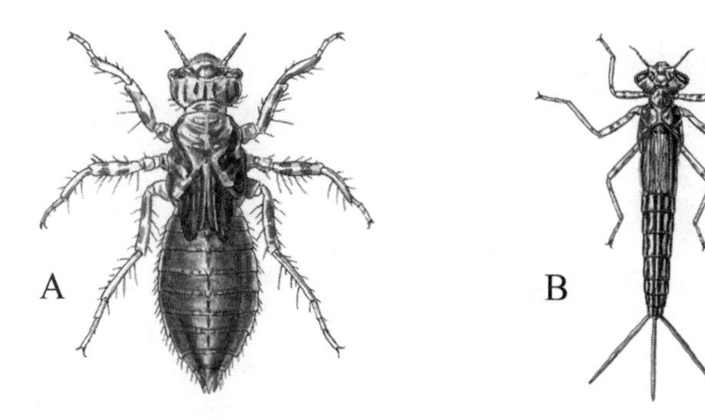

Figure 21. Representative dragonfly larva **(A)** and damselfly larva **(B).**

Distinguishing Features of Adults — Dragonfly and damselfly adults are often brightly colored, which is uncommon among aquatic insects. They are medium to large in size. The body length of different species ranges from 20 to 85 mm. Wingspan (left plus right wings spread) ranges from 20 to 110 mm. Adults in this order are not aquatic, but they are often found near the bodies of water where the larvae develop. At the order level, dragonfly and damselfly adults can be distinguished from other adult aquatic insects by the following combination of characteristics (Figure 22 A, B):

- Head is large in comparison to body.
- Large compound eyes occupy much of the head.
- Both pairs of wings are membranous and have many veins.
- When resting, the wings are either held together extended above the body or outstretched away from the sides of the body.
- All wings are elongate.
- Hind wings are about the same size or larger than front wings.
- Abdomen is very long.

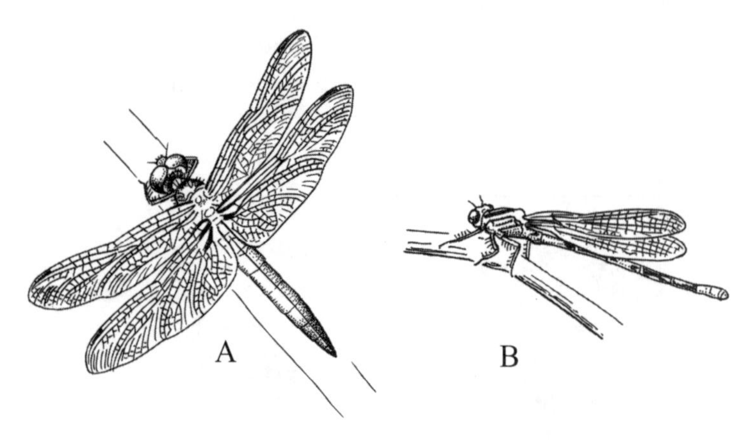

Figure 22. Representative dragonfly adult **(A)** and damselfly adult **(B).**

At the suborder level, dragonfly adults and damselfly adults can be distinguished from each other by the following combination of characteristics:

Dragonfly Adults (Figure 22 A)

- When resting, wings are held outstretched away from the sides of the body.
- Hind wings are wider at the base than the front wings.
- Abdomen is long and stout.

Damselfly Adults (Figure 22 B)

- When resting, the wings are held together extended above the body.
- Front and hind wings are the same shape.
- Abdomen is long and very thin.

Explanation of Names — The only common names that are widely recognized for the order are dragonflies and damselflies. Each of these common names is used for one of the two suborders. Dragonfly is a very old common name that probably arose out of superstitious fear of these insects. They certainly look fierce because of their large compound eyes, prominent sharp mouthparts, and long thin abdomens. In addition, they fly boldly and aggressively. On the other hand, damselflies have a delicate appearance and fly feebly. Hence, they are named for a misconception that was widely believed during medieval times – women are frail, fragile creatures. There are a number of different common names that might be used for dragonflies in certain localities or regions. These local names include mosquito hawks, devil's darning needles, and horse stingers. Mosquito hawk refers to the fact that adult dragonflies capture and consume many adult mosquitoes while flying, and they can often be seen flying back and forth through swarms of mosquitoes. Devil's darning needle refers to the long, thin, pointed abdomen and the strong flying and hovering abilities of the adults. There was a superstition that these insects would use their abdomen to sew up ears, especially those of children who misbehaved. The name horse stinger came about because people often saw dragonflies flying around horses, and the horses had small bleeding wounds on their backs. However, dragonflies help horses rather than harm them. Dragonflies fly around horses to feed upon large horseflies (Diptera: Tabanidae), which cut the skin of horses to obtain blood and leave a nasty open wound. The scientific name of the order comes from the Greek word *odon*, which means tooth. This refers to the mouthparts of the adults. The jaws are prominent and bear many sharp, hard, tooth-like projections.

Habitat — Almost all dragonfly and damselfly larvae live in still waters, either lentic-littoral or lotic-depositional habitats. A few kinds live in lotic-erosional habitats. Different species occur in all sizes of these habitats, from ponds to lakes and small streams to larger rivers. Some dragonflies and damselflies are adapted for life in very small and temporary habitats, such as seeps, bogs, and wet-weather ponds.

Movement — Different kinds of dragonfly and damselfly larvae are adapted to exist in parts of aquatic habitats with either firm or soft substrate. Most larvae fit into one of four categories of habits for movement: climbers, crawlers, sprawlers, or burrowers. Hardly any damselflies are sprawlers or burrowers, but these habits are fairly common among dragonflies. Dragonfly burrowers do not maintain a discrete tunnel. Larvae position themselves in soft sediment so that only their eyes and the tip of their abdomen are above the substrate.

Dragonfly larvae can also propel themselves rapidly forward in short bursts by a special use of their breathing mechanism (see Breathing). They do this by filling the rectum with water, closing the valve at the end of the abdomen, contracting the abdominal muscles to build up pressure, then opening the valve and suddenly expelling the water. This "jet propulsion" mechanism is used mostly for emergency escapes from predators, such as fish, but they sometimes use it just to move to another part of their habitat. Damselfly climbers sometimes swim feebly if they need to move across a short stretch of open water. They accomplish this by waving their long, thin abdomens from side to side. The flat gills on the end of the abdomen (see Breathing) help them with this secondary method of movement.

Feeding — All kinds of dragonflies and damselflies are engulfer-predators. They capture their prey with a unique lower lip, which is not found on any other insects, either terrestrial or aquatic. The lower lip is very long and has an elbow, or hinge, at the middle that allows it to bend (Plates 38B, 39B). When not in use, the lower lip is bent and held tightly against the underside of the head and thorax. The elbow extends back almost to the middle legs. The front half of the lower lip has two movable, finger-like lobes that have sharp teeth and spines. When not in use, the lower lip covers the other mouthparts, and most of the face up to the eyes. For this reason, the lower lip of dragonfly and damselfly larvae is sometimes called a "mask." To feed, larvae rapidly extend the lower lip in front of their head and grab their prey with the sharp, finger-like lobes. They then bring the captured prey back to their mouth. If it is a small organism, it is consumed whole. If the prey is large, larvae use the other sharp, pointed mouthparts to cut and tear the prey into pieces that are small enough to be consumed. This special way of capturing prey with their lower lip is advantageous because it leaves all six of their legs free to maneuver and hold on to the substrate.

Dragonfly and damselfly larvae are visual predators. They employ two strategies to find prey. Climbers and crawlers slowly stalk their prey, while sprawlers and burrowers lie in wait and ambush their prey. All species will eat any type of prey that they can subdue. The size of their prey increases as the larvae grow. Young dragonfly larvae and damselfly larvae often feed on zooplankton, then they switch to larger invertebrates, such as mayflies (Ephemeroptera), as they grow. Mature dragonfly larvae are voracious feeders and sometimes consume small fish.

Breathing — Dragonfly and damselfly larvae have a closed breathing system, but there are some differences between the two suborders. Dragonfly larvae have gills in a chamber inside the rear of their abdomen. They pump water in and out of the gill chamber by expanding and contracting their abdomen. This guarantees that the gills will be supplied continuously with water that is saturated with dissolved oxygen, and the gills are protected from damage. Damselfly larvae have three flat, long gills on the end of the abdomen. These look a lot like narrow leaves (Plates 34D, 35C, 36D). To get dissolved oxygen from the water, they spread the three gills as far as possible. If not enough oxygen is being obtained, larvae move the abdomen from side to side to keep the gills' surfaces in continuous contact with water that is saturated with dissolved oxygen. The gills of damselfly larvae are fragile and are sometimes broken off, but they can survive without them. Apparently, dissolved oxygen also diffuses across the skin of the entire body and enters the air tubes lying underneath.

Dragonfly and damselfly larvae both have emergency mechanisms for breathing if dissolved oxygen becomes too low. One way is to climb headfirst up an object until their body is partially, or completely, out of the water. Oxygen from the atmosphere diffuses into the film of water on their body, then across the body wall into the underlying network of tiny air tubes (tracheae). When they dry out, they submerge themselves to get wet, then repeat the process. Sometimes dragonfly larvae will climb up an object backwards, push the end of their abdomen into the air, and open the end of their abdomen. This allows gaseous oxygen from the atmosphere to come into contact with the moist surfaces of the internal gills. Some damselfly larvae do this too. Even though damselflies do not have internal gills like dragonflies, in emergencies the thin, moist lining inside their abdomen provides an effective site for gas exchange with the atmosphere.

Life History — It is most common for members of this order to have one generation per year. No dragonflies have more than one generation per year in North America, but a very few species of damselflies may have two generations. Some common dragonflies require 2 years to develop. A few kinds of dragonflies require 3–6 years to complete a generation. Different species of dragonflies and damselflies emerge throughout spring, summer, and early fall.

Most dragonflies deposit their eggs directly in the water. However, damselflies and some dragonflies insert their eggs into the tissue of plants growing in the water. The females of those species have a special sword-like egg laying structure that cuts slits in the plants to hold the eggs. Eggs usually hatch in several weeks, but this can be as short as 8 days. Some species undergo a long diapause in the egg stage that can last as long as 11 months.

Some common species are able to inhabit temporary standing-water habitats by means of egg diapause. Females lay the eggs well above the water in the stems of living emergent aquatic plants. The eggs are laid as the habitat is beginning to dry up. The eggs do not hatch but are protected in the plant tissue, which stays moist because the roots grow deep in the sediment to find water. Eggs usually remain dormant for 3–6 months, but dormancy can last as long as 11 months. The plants containing the eggs eventually die and fall over. When the habitat fills with water again and the correct temperature is reached, the eggs are prompted to hatch. This can be anytime from late autumn to the following spring. Larvae then develop rapidly to the adult stage before the habitat dries up again.

Dragonflies and damselflies usually spend about 10 months in the larval stage. This can range from as short as 5 or 6 weeks to as long as 5 or 6 years. Those that spend a short time as larvae are usually inhabitants of temporary waters in warm climates. Lengthy larval development is characteristic of species that live in cold climates (northern latitudes or high elevations), or that obtain food irregularly because of unusual habits and habitat. They usually shed their skin 11 or 12 times as larvae, but this can range from 10 to 15 times. Dragonflies and damselflies remain active throughout the larval stage, without any diapause.

Climbing larvae of damselflies exhibit some interesting territorial behavior. They try to defend locations where prey are plentiful. If an intruder approaches a spot that is already occupied, the defender will usually begin with a movement called a "slash." The defending damselfly bends its abdomen around and strikes the intruder on the head with the end of its abdomen and gills. If a "slash" does not scare away the intruder, the defender resorts to a "strike" with its lower lip. Neither action does any damage, but the intruder will usually let loose and swim away, unless the intruder is larger than the defender. In that case, the defender is likely to vacate the position, and the intruder will assume occupancy.

When dragonfly and damselfly larvae have completed their development and are ready to transform into adults, they do not feed for several days before emergence. All of their energy is going into manufacturing adult tissues inside of the last larval skin. The adult lower lip, which is a normal insect lower lip, forms several days before emergence, leaving the unique grabbing lower lip of the larvae non-functional. Thus, dragonfly and damselfly larvae do not have a way to capture food toward the end of the aquatic part of their life cycle. At this time, larvae reposition themselves in the aquatic habitat to get ready for emergence. Bottom dwellers move close to the edge. Those that inhabit vegetation, move higher on the plants, or, in some cases, change from a submerged species of aquatic plant to an emergent or floating species.

When it is time for the final molt to the adult stage, all dragonflies and damselflies have to climb out of the water before they shed their skin. Common spots where they leave the water are emergent aquatic vegetation, terrestrial vegetation drooping into the water, exposed tree roots, rocks, or woody debris. They usually travel only a short distance (about 5–30 cm) from the water, but larvae have been observed to crawl 2–45 m away from the water and 2–6 m vertically before emerging. They apparently travel longer distances when high populations of larvae emerge at the same time, and the suitable emergence sites become crowded. Dragonflies and damselflies emerge slowly, requiring 1–3 hours from the time the larva first crawls out of the water. During this time they are vulnerable to being eaten by birds, reptiles, amphibians, and spiders. Many species emerge at night and are ready to fly by dawn to minimize predation.

Newly emerged adult dragonflies and damselflies remain in the teneral condition longer than most other aquatic insects. They are soft and pale for several days to a week or more, but they are still able to fly and catch food. Dragonflies and damselflies live a long time as adults, compared to many aquatic insects. On average, they live about 1 month, but adult longevity of different species ranges from 2 weeks to 2 months.

Adults do not reproduce right away, and most disperse away from the larval habitat after emerging. When it is time to find a mate, they return to the vicinity of a suitable larval habitat, but not necessarily the one they emerged from. Adults in this order, especially the dragonflies, are much more active and conspicuous than the adults of other aquatic insects. Some spend most of the daylight hours in flight, either looking for food or seeking a mate, then they rest on vegetation at night. Most of the fliers are dragonflies. All damselflies and some of the dragonflies spend a lot of the time during the day perching on vegetation or other objects, interspersed with short flights to capture food or seek a mate. They too stop flying at night and rest on vegetation.

All adult dragonflies and damselflies are predators that capture their prey while flying. They have excellent vision because their compound eyes contain many individual facets (up to 10,000), and they can rotate their heads almost 360^0. Adults usually capture prey with their legs. The thin legs hang below the body, and long, stout spines on the legs point inward to form a "basket." Because they fly so fast, they can scoop prey out of the air with their basket. After they capture a flying insect, they bring the basket forward to their mouth to consume the prey it contains. Small prey are eaten in flight, whereas adult dragonflies and damselflies usually land to consume large prey that they have captured. Mosquitoes and other flies are some of their most common prey. Large dragonflies sometimes consume butterflies and smaller species of dragonflies and damselflies. When consuming butterflies, they cut off the unpalatable scaly wings and eat only the bodies. Occasionally, adults will hover and pick stationary prey off of a plant or other substrate with their mouthparts. Aphids, beetle larvae, and spiders have been reported to be consumed in this manner. A very large dragonfly species in the family Aeshnidae was observed picking up small frogs from the ground and eating them.

Most male dragonflies defend territories while they are in the reproductive phase, and some damselflies exhibit this behavior also. Males and females usually look different in this order. The sexes may be different colors, and often the wings have different patterns of pigmentation. Because of their excellent vision, males can distinguish females of their own species from all other males and females at considerable distances (at least 20 m, perhaps 40 m). Males patrol an area at a suitable larval habitat and keep other dragonflies and damselflies away, except females of their own species. The size of the territory varies according to the size and population density of the species. Larger species have larger territories, and crowding leads to smaller territories. Territories are usually defended by ritualized displays between two individuals in the air, which do not involve physical contact, until one of the competitors flies away. For example, competing males may hover in front of each other with their abdomens raised. When one gives up, he signals by lowering his abdomen. Physical fighting also occurs in defense of territories. Males will smash into each other and wrestle in the air. They grapple with their legs and noisily strike each other with their wings. These physical encounters are usually very brief. The combatants spiral downward, but seldom reach the ground. Tattered wings and other minor damage results from fights, but they are seldom fatal.

After mating, males of some dragonflies and damselflies remain with the females and participate in egg laying. This is more common among damselflies. The purpose is to prevent other males from mating with the female before she lays her eggs. Male participation ranges from just remaining in the vicinity of the female to holding on to the female while she lays her eggs. Males that hold on to the female, may hover in flight while she deposits her eggs, or they may perch on the same substrate in front of her. In some species, males push the female underwater while holding on to her, while in other species, both individuals go underwater. Lastly, males of some species release the female so that she can go underwater to lay her eggs, and the male flies around in the vicinity. When the female comes out of the water, the male grabs her again, and they fly together to another site to repeat the egg laying process.

Some species of dragonflies are thought to embark on regular migrations in the adult stage. The best known example is the green darner (*Anax junius*) in the family Aeshnidae. This is a common species in the eastern United States and Canada. Larvae occur in most ponds that have vegetation. Adults are conspicuous because of their large size and distinctive coloration (green thorax and bright blue abdomen). There is one generation per year, but there are two distinct populations of larvae that occur together in the same habitat. The resident population spends 11 months as larvae, with adults emerging in late-June and eggs of the next generation hatching in mid-July. Larvae of the migratory population require only 3 months to develop in the same pond. Migratory adults emerge in September and fly south to overwinter, then return in May to lay eggs. Eggs of the migratory population hatch in June, and larvae develop rapidly during summer.

Significance — In spite of some of the scary common names, such as devil's darning needle, dragonflies and damselflies are never pests. Both adults and larvae are beneficial insects.

Adults and larvae eat appreciable numbers of mosquitoes (Culicidae), black flies (Simuliidae), and other biting flies, thereby lessening the likelihood of some diseases being transmitted and allowing people to more thoroughly enjoy outdoor activities that are economically significant.

Dragonfly and damselfly larvae are important food for some migratory shore birds and waterfowl. Many of these birds eat mostly aquatic insects on their breeding grounds. Aquatic insects provide them with protein that is essential nutrition for the production of eggs. Large aquatic insects, such as dragonfly and damselfly larvae, are very useful as food because the birds get a lot of nutrients in each organism they find and eat. Dragonfly and damselfly larvae are usually not an important component of the diet of fish.

Many adult Odonata are conspicuous because of their bright colors and territorial and feeding behaviors. They have long attracted the attention of professional entomologists and amateur natural history enthusiasts. For these reasons, the identification and distribution of species is better known than for many other types of aquatic insects. There are a lot of rare species that only occur in unique habitats, sometimes only in one locality. Some species have been included on lists that designate them with a special status requiring protection (locally rare, threatened, endangered, etc.).

Biology of Common Families

Clubtail Dragonflies

Family Gomphidae; Plate 37

Different Kinds in North America — Moderate; 13 genera, 109 species.

Distribution in North America — Throughout.

Habitat — Primarily lotic-depositional and lentic-littoral, some lotic-erosional. Different species are found in all sizes of habitats, from the very smallest streams and ponds that are permanent to large rivers and lakes. Clubtail larvae are more common in flowing waters, especially large rivers, than other families of dragonflies and damselflies. Larvae are found on soft substrates including silt, sand, gravel, and detritus. A few species prefer accumulations of leaves lying on the bottom. Larvae that live in rocky riffle areas will be found in patches of fine, soft sediment that accumulates along the margins or in eddies behind large rocks in the channel.

Movement — Primarily burrowers, a few sprawlers. Burrowing clubtail larvae do not construct a distinct tunnel in the bottom like the burrowing mayflies (Ephemeridae). Clubtail larvae merely cover themselves with soft sediment. They hold their body with the abdomen slightly curved upward so that only the eyes and the tip of the abdomen project above the sediment.

Feeding — Engulfer-predators. Larvae lie in wait, hidden by being mostly buried in sediment, until prey comes close enough to be grabbed with the hinged lower lip.

Other Biology — Larvae are able to get dissolved oxygen while they are buried because the tip of their abdomen extends above the sediment. When it is time to emerge as adults, many of these burrowers with short, thick legs cannot climb up plant stems like other dragonflies and damselflies. Instead, they crawl onto shore or onto broad, nearly horizontal, surfaces of rocks and logs. Most species in this family require 2 years to develop to the adult stage. Some species produce one generation per year, but others require as long as 3 or 4 years for a generation.

Stress Tolerance — Primarily somewhat sensitive, others very sensitive to somewhat tolerant.

Darner Dragonflies

Family Aeshnidae; Plate 38

Different Kinds in North America — Moderate; 11 genera, 38 species.

Distribution in North America — Throughout.

Habitat — Primarily lentic-littoral, some lotic-depositional and lotic-erosional.

Most darner larvae are found at the edges of ponds and lakes or in marshes and bogs. They usually live on the stems of living or dead aquatic plants, woody debris, or tangled roots. The species that live in running water usually occur under stones or large water-logged pieces of wood.

Movement — Primarily climbers (species in still water), some crawlers (species in running water).

Feeding — Engulfer-predators. Darner larvae stalk their prey, which can include small fishes for large larvae.

Other Biology — Females of most darner species lay their eggs in living or dead plants by means of a sword-like egg laying structure. Many species produce one generation per year, but other common species of darners require 2–4 years for one generation.

Stress Tolerance — Somewhat sensitive.

Skimmer Dragonflies

Family Libellulidae; Plates 39, 40

Different Kinds in North America — Many; 35 genera, 156 species.

Distribution in North America — Throughout.

Habitat — Primarily lentic-littoral and lotic-depositional. Larvae of skimmer dragonflies are usually the most abundant dragonflies in ponds, lakes, swamps, marshes, and ditches. A sweep with an aquatic net at the edge of a pond will almost always yield a few larvae belonging to this family. They are especially abundant in warm, shallow waters. Some species are associated with aquatic plants, but many live on the bottom, particularly around the bases of plants. Some species of skimmer dragonflies are tolerant of saline conditions and are among the few aquatic insects found in brackish, estuarine environments. Other kinds are more common in smaller, cooler waters, such as high altitude bogs (including sphagnum bogs) and pools of shaded woodland streams.

Movement — Primarily sprawlers, some climbers. The kinds of skimmer dragonflies that are sprawlers tend to be rather active, not staying in one place too long. Climbers can be found on all types of vegetation and debris. Some species are sprawlers as young larvae then become climbers as they mature.

Feeding — Engulfer-predators.

Other Biology — Some of the larvae of skimmer dragonflies that are sprawlers are camouflaged by being dull colored and quite hairy. A thick layer of silt adheres to their bodies when they are collected. The kinds of skimmer dragonflies that are climbers often have distinctive patterns of green and brown to help camouflage them. Most members of this family produce a generation per year, but 2 years may be required at northern latitudes or high elevations.

Stress Tolerance — Mainly very tolerant, others facultative to somewhat tolerant. Some species of common skimmers can withstand very low dissolved oxygen concentrations and, as a result, are found in habitats polluted by excessive organic matter and nutrients, such as livestock watering ponds.

Broadwinged Damselflies

Family Calopterygidae; Plate 34

Different Kinds in North America — Few; 2 genera, 8 species.

Distribution in North America — Throughout.

Habitat — Lotic-erosional. Larvae of broadwinged damselflies always live in flowing water, but only in areas of moderate current. They are found on plant materials including aquatic plants, terrestrial grasses that droop into the water, exposed roots of terrestrial plants, and accumulations of small twigs and branches from shrubs and trees. They never occur in fast, splashing water or on rocks. Larvae occur in all sizes of permanent flowing waters.

Movement — Climbers. Larvae of broadwinged damselflies move very slowly in a rather stiff and awkward manner and seldom change locations in their habitat. If they are collected with plant debris, their elongate bodies and lack of movement make them hard to find. This behavior and appearance probably serves them well to avoid detection by prey or predators.

Feeding — Engulfer-predators.

Other Biology — Female broadwinged damselflies lay their eggs singly, just below the surface of the water in plant tissue, either living or dead. Members of this family have one generation per year.

Stress Tolerance — Mostly somewhat tolerant, others facultative.

Narrowwinged Damselflies

Family Coenagrionidae; Plate 36

Different Kinds in North America — Moderate; 16 genera, 115 species.

Distribution in North America — Throughout.

Habitat — Primarily lentic-littoral, some lotic-depositional, also lotic-erosional. This is the family of damselflies that is most likely to be encountered. Collecting in ponds, lakes, swamps, marshes, and ditches will almost always produce narrowwinged damselfly larvae very quickly. The contents from a single sweep with an aquatic net will often contain larvae of this family and those of the most abundant dragonfly family, common skimmers (Libellulidae). The most common habitats are in standing waters on living aquatic plants or tangles of dead vegetation, including terrestrial plants that have fallen in the water. Some species live on soft sediment. Larvae also occur in pools of all sizes in streams and rivers, either on plants, debris, or sediment. A few species live within the rocky substrates of riffle areas in streams and rivers.

Movement — Climbers, some sprawlers, also clingers.

Feeding — Engulfer-predators.

Other Biology — Female broadwinged damselflies lay the eggs in plant tissue, and often this is below the surface of the water. The depth at which narrowwinged damselflies lay their eggs ranges from just a few centimeters underwater, in which case the females just stick their abdomen below the surface, up to 30 cm or more, in which case the females crawl around completely submerged. It is easy to observe the unusual egg-laying behavior in which male broadwinged damselflies hover in the air or hold on to the females while they lay eggs. There is almost always one generation per year, but a few species may have two generations per year.

Stress Tolerance — Chiefly somewhat tolerant, others very tolerant.

Spreadwinged Damselflies

Family Lestidae; Plate 35

Different Kinds in North America — Few; 2 genera, 19 species.

Distribution in North America — Throughout.

Habitat — Primarily lentic-littoral, sometimes in very slow weedy streams. Larvae of spreadwinged damselflies always live on vegetation. They are found in ponds, small lakes, marshes, swamps, and ditches. Some species are very abundant in temporary habitats that are dry throughout the warm months of the year.

Movement — Climbers. If it is necessary to move to another plant, larvae of spreadwinged damselflies can swim slowly, but efficiently, by sideways sweeping movements of their long abdomen and posterior gills.

Feeding — Engulfer-predators. Larvae stalk their prey. Their elongate bodies and coloration make them very well camouflaged on living plants.

Other Biology — Some common species of spreadwinged damselflies are able to inhabit temporary standing-water habitats by having a prolonged egg stage in plant tissue. Members of this family have one generation per year. Species that inhabit temporary habitats emerge as adults only in the spring.

Stress Tolerance — Very tolerant.

Stoneflies

Order Plecoptera; Plates 41–49

This is a medium-sized order, in terms of diversity. There are nine families and 626 species in North America. This guide includes all nine families because they are common in freshwater habitats. Stoneflies have incomplete metamorphosis. All members of the order are aquatic as larvae and terrestrial as adults.

Distinguishing Features of Larvae — Stonefly larvae have elongate bodies that are either somewhat flattened or cylindrical. The body length of mature larvae in different species ranges from 5 to 70 mm, not including the antennae or tails. Stonefly larvae can be distinguished from all other aquatic insect larvae by the following combination of characteristics (Figure 23):

- Long, thin antennae project in front of the head.
- Wing pads are present on the thorax, but these may only be visible in older larvae.
- Three pairs of segmented legs extend from the thorax.
- Two claws are located on the end of the segmented legs.
- Gills usually occur only on the bottom of the thorax, or there are no gills at all, except for one family that has gills on the bottom of the thorax and the bottom of the first two or three abdomen segments.
- Gills are either single or branched filaments.
- Two long, thin tails project from the rear of the abdomen.

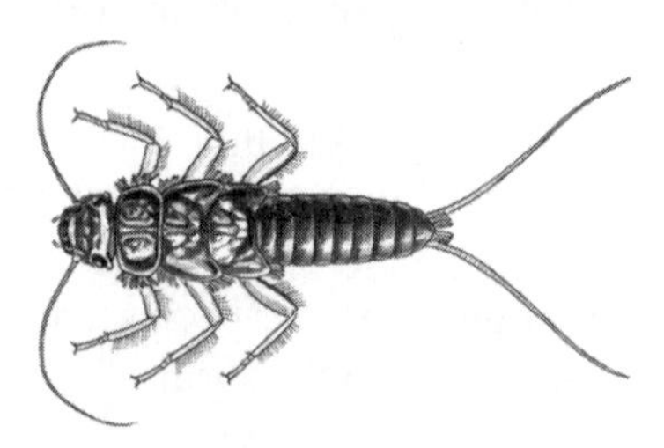

Figure 23. Representative stonefly larva.

Distinguishing Features of Adults — Adult stoneflies have elongate, somewhat flattened, soft bodies. In a few families there are wingless or partially winged forms, in addition to winged forms. They are small to medium in size. The length of different species ranges from 6 to 65 mm, measured from the front of the head to the tip of the wings folded at rest, or to the end of the abdomen for those with reduced wings. Stonefly adults can be distinguished from other adult aquatic insects by the following combination of characteristics (Figure 24):

- Long, thin antennae project in front of the head.
- Both pairs of wings are membranous and have many veins.
- Wings fold, when not in use, so that they lie flat over the abdomen, with the hind wings hidden under the front wings.
- Front wings are straight and about the same width as the body.
- Hind wings, when extended, are much wider than the front wings, thus, the hind wings are pleated so that they can crumple under the front wings when not in use.
- When wings are folded flat, the tips of the front and hind wings are even with each other and extend past the end of the abdomen.
- Two thin tails project from the end of the abdomen, extending at least past the wing tips and sometimes far beyond the wing tips.

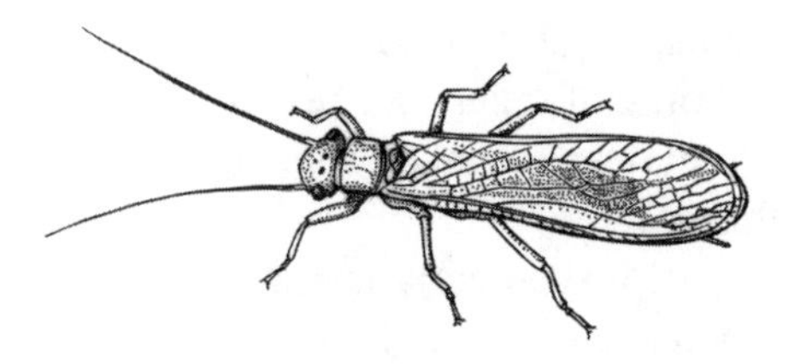

Figure 24. Representative stonefly adult.

Explanation of Names — The only common name that is widely recognized for the entire order is stoneflies. Entomologists assume that this name comes from the fact that larvae are often found among stones in streams or that adults often crawl around on stones on the banks of streams. Members of one family (Pteronarcyidae, giant stoneflies) are called salmonflies by local residents living near western rivers and by persons who fish those waters. This regional common name for the family refers to the importance of larvae and emerging adults in the diet of salmon, and the emphasis that fly fishers put on imitating these insects. The scientific name of the order is a combination of two Greek words: *plekein*, meaning to braid, and *ptera*, meaning wings. The explanation of the name has to do with the large hind wings of the adults that must be folded into pleats in order to fit under the front wings.

Habitat — Almost all stonefly larvae are found exclusively in lotic-erosional habitats. Different kinds of stonefly larvae inhabit all sizes of running waters, including seeps, springs, streams, and rivers. However, the greatest diversity is found in small, cool streams where the water temperature does not exceed 25° C and dissolved oxygen concentrations remain near saturation. Usually these small streams are heavily shaded. Larvae have a strong preference for coarse substrates, such as boulders, cobble, pebbles, pieces of water-soaked wood, and accumulations of coarse detritus, especially leaf packs. It is common to find several families of stoneflies within a single leaf pack in swift current. Detritus-feeding stonefly larvae use them for food and hiding places, while predaceous species invade them in search of prey, including other stoneflies. Hardly any stoneflies inhabit lentic habitats, but a few species live on the wave-washed, rocky shores of lakes. There is one unique species that lives on rocks 70 m deep in Lake Tahoe, but that is very much an exception for this order.

Movement — Most stonefly larvae are crawlers. They move around to find food and places to hide. They usually hide in small spaces within the coarse substrate in riffles where there is ample dissolved oxygen and current velocity. The kinds that are predators are more active than the ones that feed on detritus. The predaceous stonefly larvae crawl around in search of prey, mostly at night. The kinds that feed on detritus tend to stay put if they find a protected space within the substrate where there is plenty of food, such as within a leaf pack.

Feeding — Most stonefly larvae are either shredder-detritivores or engulfer-predators. These feeding habits are usually consistent within a particular family. There are a few species in some families that are scrapers. Some kinds of stonefly larvae may change their diet while they are developing. Some shredder-detritivores and engulfer-predators are collector-gatherers when they are very young larvae, simply because they are not strong enough to handle large, tough pieces of food. It is often easy to determine whether mature larvae are shredders or predators by the appearance of their mouthparts. The mouthparts of shredders are directed downward and are shaped for cutting and grinding, much like terrestrial grasshoppers. Predaceous stonefly mouthparts project forward and are very sharp and pointed. These are adaptations to lunge at and grab live prey, and perhaps tear it into pieces. Predaceous stoneflies usually take small prey, which they consume whole, head first, by gulping movements. However, they will sometimes rip their prey apart and consume only one or a few pieces. Predaceous stoneflies consume a variety of prey, according to what is available at a particular time of year. Studies have shown that their most common prey are midges and black flies in the order of true flies (Diptera: Chironomidae and Simuliidae, respectively) and various mayflies (Ephemeroptera).

Breathing — All stonefly larvae have a closed breathing system, but there are considerable differences among the different kinds of stoneflies regarding the gills. Most have at least some filamentous gills on the thorax. However, the filamentous gills range from a single stout filament (Plate 47A at base of legs) to a profusely branched cluster of fine filaments (Plates 42B, 43B). There is some evidence that species with larger, more branched gills have broader distributions that include larger, warmer streams and rivers. The highly branched gills provide greater surface area to acquire dissolved oxygen. Some species of stonefly larvae do not have any gills and rely on oxygen diffusing across the skin of the entire body and entering the air tubes lying underneath.

Some of the kinds with gills exhibit an interesting behavior when the dissolved oxygen concentration becomes too low or there is too little current to deliver water saturated with dissolved oxygen to the gill surfaces. In some stonefly families (for example, common stoneflies, Perlidae), larvae use their legs to raise and lower their body, in a motion that resembles push ups. In other families (for example, giant stoneflies, Pteronarcyidae), larvae wag their body from side to side. Both types of movement cause water to circulate over the gill surfaces and serve to compensate for low dissolved oxygen concentration or too little current.

Life History — Most kinds of stoneflies have one generation per year, but a considerable number require 2 years to complete a generation. Only a few species require 3 years to complete a generation. These species inhabit northern latitudes or high elevations. No stoneflies in North America are known to have more than one generation per year. Adults of most kinds of stoneflies emerge from early spring to early summer, however, there are also some species in various families that emerge in summer, fall, and even winter. Stoneflies are unique among the aquatic insects in that there are different species that normally emerge in all months of the year.

Stonefly eggs are always produced in a loose egg mass. In most species, the female lands briefly on the water surface, just long enough to dislodge the egg mass, then she flies away. The egg mass quickly falls apart in the water, and the individual eggs stick to solid objects on the bottom. The eggs usually hatch in 3–4 weeks, but there are some kinds of stoneflies that undergo a long diapause in the egg stage to avoid periods of the year when environmental conditions are naturally unsuitable. Egg diapause usually lasts 3–6 months in species that have this adaptation, but in some rare cases stonefly eggs do not hatch for 24 months.

Stoneflies usually spend 10–11 months as larvae, but this can range from 6 months to 3 years in different species. They usually shed their skin 10–22 times, but as few as six molts have been reported for some species. Like most aquatic insects, the larval behavior of stoneflies has not been studied very much. Larvae of some species have been observed to defend themselves from attack by pushing their long, thin tails toward the predator that is attacking them. They also use their tails to strike other stonefly larvae belonging to the same species that get too close. Perhaps this is to defend a location with superior conditions, such as food, current, dissolved oxygen, or a place to hide. Some species, particularly shredder-detritivores, have been observed to act like they are dead when approached by fish that eat insects. When disturbed, these stonefly larvae bend their heads down, curl their bodies, and lie still. Insect-eating fish, such as sculpins, will usually reject a stonefly in this position, whereas active or straightened stoneflies are eaten. Larvae of one family (giant stoneflies, Pteronarcyidae) respond to predators by reflex bleeding, which is also called autohemorrhaging. They squeeze out drops of their own blood through pores located in the soft, thin membrane at one of the joints on the third pair of legs. This fluid is thought to present a bad taste or smell to a predator, or otherwise confuse the attacker.

When stonefly larvae have completed their development and are ready to transform into adults, they often reduce their feeding. Some larvae that are about to emerge spend more time drifting in stream currents, especially at night. All stonefly larvae must crawl out of the water to transform into adults. Mature larvae congregate in the water at the bases of good emergence sites, such as bridge abutments, woody debris, downed trees, exposed tree roots, emergent aquatic plants, and most commonly, rocks. After crawling a short distance out of the water, they secure their tarsal claws, split the skin along lines of weakness on the head and thorax, and a teneral adult climbs out. Most species transform to adults at night to avoid being eaten by birds and other terrestrial insectivores. About 5–10 minutes is usually required for the final molt.

Teneral adults soon move to nearby streamside vegetation, where they wait for their new skin to harden. Adult stoneflies usually live 1–4 weeks. Winter emerging species usually have longer adult lives. Adult stoneflies are secretive, mostly nocturnal organisms that are seldom seen by humans, except at the time of their emergence. After emerging from the water as adults, most stoneflies hide on branches or leaves of streamside vegetation. Other favorite hiding places are in crevices of tree bark, on wood and leaf debris piled up on shore by high water, and in cracks in old fence posts. Winter stonefly adults often seek mineral substrates, such as piles of stones and concrete bridge abutments, probably because of the warmth retained by these objects. Stonefly adults mostly sit still during the day and crawl around at night. They usually only fly to disperse to a new habitat. Most females fly to the water to lay eggs about dusk. Most stonefly adults feed, and all drink water. In many species, the females must feed to complete egg development. Common foods include soft parts of terrestrial plants, such as young leaves, fruit, buds, blossoms, and pollen. In addition, adult stoneflies eat lichens and algae that grow on plants at the edge of the water.

Adult stoneflies have an interesting way of locating each other for mating, which is effective for their secretive life style. In many species, male and female adults of the same species are attracted to each other by "drumming." Stonefly drumming consists of tapping the tip of their elongate abdomen upon the substrate. This makes a vibration that can be felt by other stoneflies through touch receptors in contact with the substrate. Their touch receptors are located on the abdomen, ends of the legs, tails, and possibly antennae. Male and female drumming patterns are specific for each species and for each sex. The drumming pattern of each species and sex is made unique by a combination of amplitude, frequency, or duration. Male stoneflies initiate drumming and the females answer. The male and female begin crawling toward the location from where the other's drumming is coming, each pausing periodically to drum some more. Eventually, the male and female locate each other and mating ensues. Stonefly drumming is a type of auditory communication. It is closely related to the "songs" of terrestrial grasshoppers, crickets, and katydids in the order Orthoptera, which are familiar sounds to most persons during summer. Orthoptera songs are sound waves traveling through air that are loud enough for us to hear, whereas stonefly drumming consists of sound waves traveling through a solid medium that our receptors are not sensitive enough to perceive. Stonefly drumming achieves the same purpose as swarms of male mayflies dancing in the air, except that the former is auditory communciation and the latter is visual communication.

The stoneflies that emerge in winter are a particularly noteworthy group within this order. While other orders have a few species that normally emerge in winter or might emerge in winter because of unusual warm spells, two entire families of stoneflies emerge only during winter. These are among the few insects, terrestrial or aquatic, that have their life history basically reversed so that growth, development, and emergence take place during the coldest part of the year. These two families are appropriately called the winter stoneflies (Taeniopterygidae) and small winter stoneflies (Capniidae). Their eggs hatch in late winter or early spring. When the water begins to warm, young larvae migrate 1 m or more down in the stream bottom, into an area of loose, rocky substrate known as the hyporheic zone. There they go into diapause, and this inactive condition lasts 3–6 months. When the water begins to cool off again, the larvae become active and feed and grow rapidly during fall and winter. They complete their development and emerge during winter and sometimes into early spring. Adult winter stoneflies are often seen crawling around on the snow and ice. They are darker than most adult stoneflies, which possibly is an adaptation to absorb more heat from the sun. Adult winter stoneflies avoid severe cold by remaining within tiny caverns in the snow and ice. In northern latitudes with particularly harsh winter weather, adults may spend their entire lives under the ice that covers the streams. Wing polymorphism occurs in small winter stoneflies. Adults of the same species can have fully developed wings, half-developed wings, or no wings. Wing polymorphism probably reflects the fact that some species live where it is too cold for insect flight muscles to work, so they do not need functional wings.

Significance — Stonefly adults are not pests, or even nuisances. They do not have large populations of larvae in a particular body of water, so mass emergence of adults does not occur. The adults are secretive and do not eat our crops or possessions, so most people do not even know they exist. There have been some isolated reports in the Pacific Northwest of winter stonefly adults nibbling on the blossoms of fruit trees in early spring, but those events are inconsequential in comparison to the harmless activities of all the kinds of stoneflies throughout North America.

Stonefly larvae are clearly beneficial organisms. One reason is that they play important roles in stream communities and ecosystems. Shredder-detritivore stonefly larvae are important in breaking down the leaves, flowers, and twigs that fall into streams from land and making this material available to other organisms. Some stonefly larvae are voracious predators that roam around a lot in riffles and eat many other invertebrates as prey. While this may seem to be a bad thing, it is actually a beneficial activity that keeps populations of other species in check, thereby increasing diversity (see Fundamentals of Freshwater Ecology, Biological Factors).

All kinds of stoneflies serve as food for other invertebrate predators, fish, salamanders, and birds. Both the larvae and adults of stoneflies are important in this regard. Some nesting ducks have been observed to feed on large numbers of emerging stonefly adults before the insects could take flight. It might be important for ducks to have a readily available source of protein such as this for successful reproduction. Adults and larvae of stoneflies are imitated by fly-fishing anglers, although not to the same extent as mayflies

(Ephemeroptera). A species called the salmonfly in the family of giant stoneflies (Pteronarcyidae) is very important in this regard to anglers who fish in western streams. In addition to serving as a model for fly tying, this species is harvested and used as live bait by some anglers. Sometimes they are sold alive in fishing shops near streams and rivers in the West.

Almost all species of stoneflies are very sensitive to pollution, so the number of different kinds and their relative abundance compared to other invertebrates provide useful and reliable information about the environmental condition of a body of water. The EPT Index has become a widely used measure of environmental condition. This simple index is calculated as the number of different kinds of Ephemeroptera (mayflies), Plecoptera (stoneflies), and Trichoptera (caddisflies). The EPT Index is useful because most of the species in these three orders of aquatic insects are sensitive to pollution and environmental stress. The order Plecoptera is probably the most pollution sensitive order of aquatic insects. However, by itself, this may not be the most useful order for biomonitoring pollution. Most species of stonefly larvae do not occur naturally in very high densities, and their natural habitat requirements usually restrict them to small, cool, swift streams. Thus, the absence of stoneflies in large, warm rivers would not tell us anything about pollution.

Biology of Common Families

Common Stoneflies

Family Perlidae; Plate 43

Different Kinds in North America — Moderate; 15 genera, 72 species. This family may also be referred to simply as the stones.

Distribution in North America — Throughout, but more kinds in the East.

Habitat — Primarily lotic-erosional, but occasionally lentic-erosional (rocky, wave-swept shores). Most kinds of common stoneflies are found in cool, clear streams of small to medium size, but a few kinds inhabit large, warm rivers that carry a considerable amount of silt. Larvae are most commonly found on or under large stones, such as cobbles and boulders, but they may also live on logs and within accumulations of debris. Common stoneflies are often found in leaf packs because of the abundance of prey found there. A few kinds live in areas of streams with gravel bottom. Most kinds prefer the faster currents in riffles, but some live near the edges of streams where the current is slower.

Movement — Crawlers. Larvae of common stoneflies are very agile and can move quickly. If they are transferred to still water they cannot obtain enough dissolved oxygen from the water, so they create currents over their gills by using their legs to raise and lower the body in a manner resembling "push-ups."

Feeding — Engulfer-predators. Studies of their diets have shown that common stonefly larvae consume midges, black flies, mayflies, caddisflies, other stoneflies, beetles, moths, and crustaceans. They are opportunistic and will consume whatever prey is available. Very young larvae may be collector-gatherers, until they are strong enough to subdue prey.

Other Biology — The length of the common stonefly life cycle varies from 1 to 3 years, according to species and geography. Many kinds require 2 years to complete their life cycle. Adults emerge primarily during the summer, May through August. They are more active than most adult stoneflies and are strongly attracted to lights.

Stress Tolerance — Chiefly very sensitive, others facultative.

Giant Stoneflies

Family Pteronarcyidae; Plate 42

Different Kinds in North America — Few; 2 genera, 10 species. Sometimes salmonflies is used as a common name for the whole family, but it originally pertained to some of the western species.

Distribution in North America — 1 genus (*Pteronarcys*) occurs throughout; the other (*Pteronarcella*) only in the West.

Habitat — Lotic-erosional. Larvae of giant stoneflies are usually found in cool streams of small to medium size. Larvae live in leaf packs and accumulations of small woody debris in fast current. They are always in the swiftest portions of riffles, but in the spaces under and between loose, large rocks (boulders, cobbles), where they are not exposed to the full force of the current.

Movement — Crawlers. Larvae of giant stoneflies move very sluggishly. They will curl their bodies and pretend to be dead when disturbed. During respiratory stress they wag their abdomen from side to side.

Feeding — Primarily shredders, some scrapers, especially as young larvae. There is also some evidence that mature larvae of giant stoneflies are occasionally engulfer-predators.

Other Biology — Local residents around western rivers and persons who fish those waters call members of this family salmonflies. This regional common name demonstrates the importance of larvae and emerging adults in the diet of fish and the emphasis that fly fishers put on imitating these insects. Giant stoneflies use reflex bleeding to discourage would-be predators. The life cycle requires from 1 to 3 years, according to species and geographic location. Adults emerge during late spring and early summer, April through June.

Stress Tolerance — Very sensitive.

Green Stoneflies

Family Chloroperlidae; Plate 45

Different Kinds in North America — Moderate; 13 genera, 103 species. Members of this family are also called sallflies.

Distribution in North America — Throughout.

Habitat — Lotic-erosional. Most kinds of green stonefly larvae inhabit only the clean, cool, swift waters that are usually found in mountainous areas. The size of their habitat is usually spring-fed brooks or small streams, but some occur in medium-sized streams. Many kinds are found primarily on gravel bottom in current, but they also occur on larger stones and in leaf packs and other accumulations of detritus.

Movement — Crawlers.

Feeding — Mostly engulfer-predators, some scrapers and collector-gatherers. Green stonefly larvae of a few western species are scavengers of recently hatched salmon that die while still attached to their yolk sacs. This is probably an opportunistic source of food, and not one that they regularly depend upon for their successful development.

Other Biology — There is usually one generation per year, but some kinds of green stoneflies require 2 years to complete their life cycle. Adults emerge from April to June.

Stress Tolerance —Very sensitive.

Nemourid Stoneflies

Family Nemouridae; Plate 46

Different Kinds in North America — Moderate; 12 genera, 71 species. Forestflies is another common name that is used for members of this family.

Distribution in North America — Throughout, from the Arctic into Mexico.

Habitat — Primarily lotic-erosional, but occasionally lentic-erosional (rocky, wave-swept shores). Larvae of nemourid stoneflies mostly inhabit spring-fed brooks and small streams in the mountains, but they may also be found in small rivers at lower elevations. Larvae are usually found on coarse substrates including leaf packs, roots, woody debris, and stones.

Movement — Crawlers.

Feeding — Primarily shredders, some collector-gatherers.

Other Biology — Eastern species of nemourid stoneflies have one generation per year, but some western species may require 2 years to complete their life cycle. These have sometimes been called the spring stoneflies, but this is a misleading common name. Most emergence does not take place until late spring and early summer, April to June, and different species in this family may emerge throughout the year. In addition, adult stoneflies in other families emerge in the spring. Larvae are usually not found in the summer and early fall because many kinds halt their development in the egg stage until later in the fall.

Stress Tolerance — Primarily somewhat sensitive, others very sensitive to facultative.

Perlodid Stoneflies

Family Perlodidae; Plate 44

Different Kinds in North America — Many; 30 genera, at least 122 species. Stripetails and springflies are other common names that may be used for members of this family.

Distribution in North America — Throughout, from Alaska to northern Mexico.

Habitat — Primarily lotic-erosional, but occasionally lentic-erosional. Most kinds of perlodid stonefly larvae are found in small to medium streams that are cool, swift, and clear, with rocky bottoms. Other kinds live in all sizes of flowing waters, ranging from small, temporary, trickles that feed into headwater streams to large rivers that carry a considerable amount of silt. A few kinds of perlodid stonefly larvae can live in lakes that remain cold because of high altitude or northern latitude, as long as the shoreline is stony. In streams, they are most commonly found on or under cobbles and boulders, but these larvae may also live among accumulations of woody debris and coarse detritus. Perlodid stoneflies are often found in leaf packs because these microhabitats attract a lot of prey organisms and provide good hiding places for the predaceous stoneflies. Many of the prevalent kinds of perlodid stonefly larvae prefer the faster currents in riffles. However, a few kinds that live in small seeps can exist under stones that are merely kept wet by trickling, splashing water.

Movement — Crawlers. Larvae of perlodid stoneflies are very agile and can scurry quickly to catch prey or avoid danger.

Feeding — Mostly engulfer-predators, a few scrapers and collector-gatherers. Perlodid stonefly larvae are opportunistic and will consume whatever small to medium-sized invertebrate prey is available, but non-biting midges are the most common item in their diet. Other organisms that are frequently eaten include mayflies, caddisflies, and black flies. Although larvae of perlodid stoneflies have mouthparts that are obviously adapted for catching other organisms and tearing them into pieces, scientific studies of their stomach contents have consistently reported a lot of plant material in their diet, including diatoms, other algae, and detritus. In some kinds of perlodid stonefly larvae, this appears to depend upon their age. Very young larvae eat mostly plant material, then, when they become older and larger, they switch to eating small animals. A few kinds of perlodid stoneflies are omnivorous throughout their larval development and always eat considerable plant material in combination with animal prey.

Other Biology — Perlodid stoneflies always have one generation per year. Adults of the most prevalent species emerge from April to June, but other species may emerge earlier in the spring, as well as later during summer and fall. Many kinds of perlodid stoneflies undergo diapause in the egg stage to avoid high water temperature during the warm months, some for as long as 7 months. Egg diapause enables some kinds to inhabit very small seeps and headwater streams that only flow temporarily during the seasons with ample precipitation.

Stress Tolerance — Mainly somewhat sensitive, others very sensitive to facultative.

Roachlike Stoneflies

Family Peltoperlidae; Plate 41

Different Kinds in North America — Few; 6 genera, 20 species. Another variation of the common name is roachflies.

Distribution in North America — Occur in East and West, but not in Central.

Habitat — Lotic-erosional. Roachlike stoneflies are restricted to cool springs and small streams in the mountains, including seeps and intermittent streams. Larvae are almost always found in leaf packs or on other plant debris.

Movement — Crawlers. Larvae of roachlike stoneflies can move quickly.

Feeding — Shredder-detritivores.

Other Biology — Roachlike stonefly larvae are often very abundant in accumulations of leaves in undisturbed streams flowing through deciduous forests. Studies have reported densities of roachlike stonefly larvae greater than 500 organisms per square meter of stream bottom. Their significance in breaking down the leaves that enter streams in autumn is easy to observe. If you pick up a leaf pack in early spring from a stream inhabited by roachlike stoneflies, dozens of these organisms will scurry out and the leaves will be mostly "skeletonized" by their feeding. The life cycle requires 1 or 2 years in different species. Adults usually emerge during late spring and early summer, April through June.

Stress Tolerance — Very sensitive.

Rolledwinged Stoneflies

Family Leuctridae; Plate 49

Different Kinds in North America — Moderate; 7 genera, 55 species. Members of this family are sometimes called needleflies.

Distribution in North America — Throughout.

Habitat — Lotic-erosional. Larvae of rolledwinged stoneflies occur in a wide range of sizes of flowing waters, ranging from spring-fed brooks to small rivers. They live primarily in leaf packs but also in other coarse components of the bottom, such as rocks and woody debris.

Movement — Crawlers.

Feeding — Shredder-detritivores.

Other Biology — The life cycle of rolledwinged stoneflies requires 1 or 2 years according to species and geographic location. Adults emerge primarily in fall, winter, and spring, but there are some species in this family that emerge all year. Adult rolledwinged stoneflies emerge over an extended period, even within a single species. In some species, the adults emerge over a period of 4 months, even though they all belong to the same generation. Part of the reason for the spreading out of adult emergence is that the eggs do not all hatch at the same time. Egg hatching is spread out over 9 months in some species of rolledwinged stoneflies. The net result is very asynchronous life cycles in this family.

Stress Tolerance — Mostly very sensitive, except one important exception. Some species of rolledwinged stoneflies in the genus *Leuctra* are very tolerant of low pH. Acid deposition from the atmosphere (acid rain) often leads to an increase in the numbers of these species in affected streams.

Small Winter Stoneflies

Family Capniidae; Plate 48

Different Kinds in North America — Moderate; 10 genera, 152 species.

Distribution in North America — Throughout, from the Arctic into Mexico.

Habitat — Lotic-erosional. Larvae of small winter stoneflies live in small streams to medium rivers. They are most common in small streams, including temporary spring seeps. Larvae are mostly found in leaf packs but also within other coarse substrate where there is a current, such as accumulations of small woody debris or loose stones. Some kinds of small winter stoneflies spend much of the year as young larvae deep in the bottom of the stream, in the area known as the hyporheic zone, as explained in the overview of the order.

Movement — Crawlers.

Feeding — Shredder-detritivores. Larvae of small winter stoneflies play a significant role in the breakdown of terrestrial leaves that enter streams. Studies have shown that larvae consume an amount of leaves equivalent to 30% of their body weight each day, even during the coldest water temperatures of winter.

Other Biology — This is one of the two families of stoneflies that have their life history reversed from most aquatic insects, with the adults of small winter stoneflies emerging during the cold months rather than during the warm months. There is one generation per year. In most kinds of small winter stoneflies, adult emergence takes place from December to April, depending on species, latitude, and altitude. One unique species in Lake Tahoe completes its entire life cycle underwater.

Stress Tolerance — Very sensitive.

Winter Stoneflies

Family Taeniopterygidae; Plate 47

Different Kinds in North America — Few; 6 genera, 35 species. Another common name that is used for members of this family is early stones.

Distribution in North America — Throughout.

Habitat — Almost all lotic-erosional, but a wide range of ecological conditions. Larvae of winter stoneflies occur in permanent streams of all sizes and temperatures, from small, cool, mountain streams to large, warm rivers on plains. One unusual species occurs in the bayous in the plains of the Gulf of Mexico. Larvae live in a variety of coarse substrates, such as leaf packs, debris jams, tree roots, submerged plants, and on large rocks with algae growing on them. Winter stonefly larvae are often found near the stream margins where the current is reduced. Some members of this family spend much of the year as young larvae deep in the bottom of the stream, in the area known as the hyporheic zone.

Movement — Crawlers.

Feeding — Mostly shredder-detritivores, some collector-gatherers and scrapers.

Other Biology — This is another of the two families of stoneflies that have their life history reversed from most aquatic insects, with the adults of winter stoneflies emerging during the cold months rather than during the warm months. There is one generation per year. In most kinds of winter stoneflies, adult emergence takes place from December to April, depending on species, latitude, and altitude.

Stress Tolerance — Chiefly somewhat sensitive, others facultative.

True Bugs

Order Hemiptera, Suborder Heteroptera; Plates 21– 27

This is a large order, in terms of diversity, but most members of the order are not aquatic in any part of their life cycle. In the entire suborder, there are 40 families and 3,600 species in North America. However, of that total, only 15 families and 400 species are associated with freshwater environments. This guide includes seven families that are common in freshwater habitats. True bugs have incomplete metamorphosis. All of the aquatic true bugs live in the same habitat as larvae and adults.

Distinguishing Features of Larvae — True bug larvae are almost identical to the adults, except larvae lack fully developed wings. The body shape is variable, ranging from oval and somewhat flattened to elongate and slender. The body length of mature larvae ranges from about 2 to 60 mm, excluding breathing tubes, in different species. True bug larvae can be distinguished from other immature aquatic insects by the following combination of characteristics (Figure 25):

- Mouthparts are highly modified into an elongate beak or a short cone, which comes off the front of the head and projects back under the head when not feeding.
- Wing pads are present on the thorax.
- Three pairs of segmented legs extend from the thorax.
- Two claws are located on the end of at least some of the segmented legs.
- No gills are present.

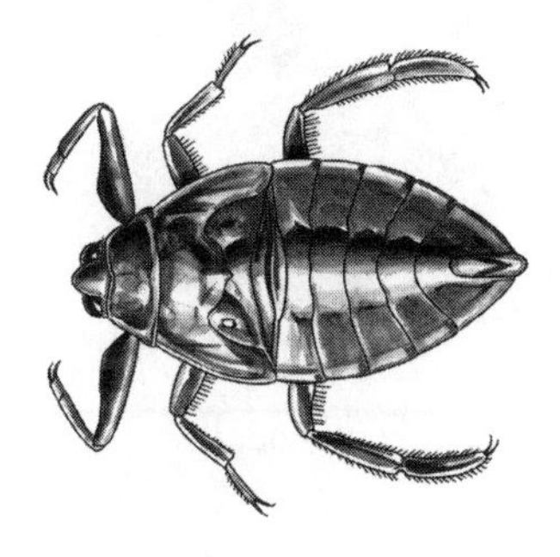

Figure 25. Representative true bug larva.

Distinguishing Features of Adults — The body shape of adult true bugs is always the same as their respective larvae, ranging from oval and somewhat flattened to elongate and slender. In several families there are wingless or partially winged forms, in addition to winged forms. They are small to medium in size. The body length of different species ranges from 3 to 65 mm, measured from the front of the head to the tip of the wings folded at rest, or to the end of the abdomen, excluding breathing tubes, for those with reduced wings. True bug adults can be distinguished from other adult aquatic insects by the following combination of characteristics (Figure 26):

- Mouthparts are highly modified into an elongate beak or a short cone, which comes off the front of the head and projects back under the head when not feeding.

- Bases of the front wings are leathery and not transparent for about one-half to two-thirds of the wing length, while the ends of the front wings are membranous and semi-transparent for the remaining one-third to one-half of the wing length.
- Hind wings are completely membranous and slightly shorter than the front wings.
- Wings fold, when not in use, so that they lie flat over the abdomen, with the membranous, semi-transparent ends of the front wings overlapping.

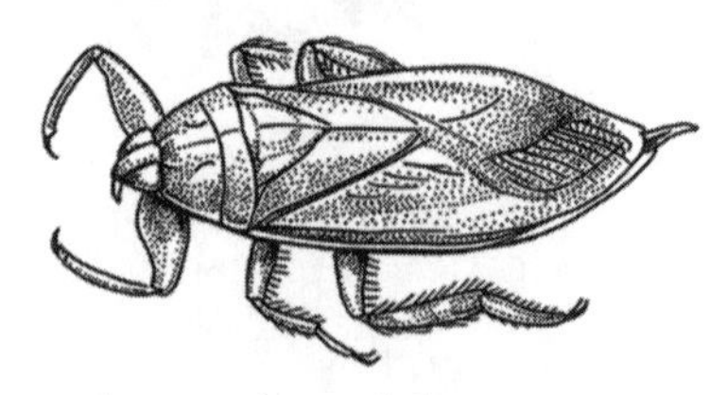

Figure 26. Representative true bug adult.

Explanation of Names — The only common name that is widely recognized for the entire order is true bugs. To an entomologist, the name "bug" means a member of the order Hemiptera. However, this name has come to be used by the public for all insects and even some other arthropods. Therefore the common name that is used by entomologists for members of the Hemiptera is "true" bugs. There are no local common names that are used for the whole order, but all of the individual families have their own accepted common names. The scientific name of the order is a combination of two Greek words: *heteros*, meaning different, and *ptera*, meaning wings. This name refers to the structure of the front wings — the part of the wing close to the body is leathery, while the part away from the body is membranous.

Habitat — Almost all larvae and adults of true bugs inhabit only standing waters or the quiet portions of running waters. These are lentic-littoral, lentic-limnetic, or lotic-depositional habitats. The habitats preferred by true bugs are the same as most dragonflies and damselflies (Odonata) and many water beetles (Coleoptera). Members of all three orders are often collected at the same time. The habitats where true bugs live represent a wide range of sizes, including springs, streams, rivers, ponds, and lakes. Some of the places where they live have unusual features of water quality, including situations with high temperature, high salinity, or pollution. Examples of these unusual habitats include marshes, hot springs, estuaries, tidal pools on beaches, and sewage ponds. One kind of true bug (water striders, Gerridae) is the only insect to live on the open ocean. True bugs also inhabit temporary habitats, such as ditches and puddles, because the adults can fly to another location when the habitat dries up. They also disperse into artificial habitats constructed by humans, such as swimming pools, livestock watering tanks, and structures for culturing fish.

Movement — Almost all of the aquatic true bugs are either skaters, swimmers, or climbers. Adults and larvae of the same kind always exhibit the same habits.

Feeding — All aquatic true bugs are piercer-predators, except for one family (water boatmen, Corixidae). The mouthparts of true bugs are highly modified into a long, thin, sharp beak (Figure 26). They use this beak to stab their prey. There are salivary glands in the head near the beak that produce enzymes. At first, these enzymes are pumped into the prey to poison and immobilize it. Then, the enzymes dissolve most of the internal structures of the prey. After the insides of the prey have become a liquid, the contents are withdrawn through the beak into the true bug's mouth. Adults and larvae feed the same way.

Breathing — All aquatic true bugs have an open breathing system and depend on oxygen from the air for their respiration, just like the terrestrial members of the order. The surface dwellers spend very little time underwater, so they require no special modifications for breathing. They simply open their spiracles when they need to acquire oxygen. Many of the true bugs that live underwater use transportable air stores for breathing. These are bubbles or thin layers of air that are held under the wings or in body hairs. Sometimes the air stores function as physical gills, allowing the organism to stay underwater longer before needing to replenish the air store. Other underwater-dwelling true bugs have breathing tubes on the end of their bodies that push above the water to breathe from the air, similar to a person snorkeling (Plates 23, 24).

Life History — Most aquatic true bugs have two or more generations per year. Some kinds have only one generation per year. Almost all of them follow the same seasonal pattern. Eggs hatch in late spring or early summer and the growth and development of larvae takes place during the following warm months. They spend cool months as either active or hibernating adults. There is no diapause in the true bugs. When adults become inactive during winter, that is hibernation not diapause. The distinction is that diapause involves a state of arrested development. The adults that become inactive during winter have already completed their development, so the correct term for the condition is hibernation.

True bugs lay their eggs individually or in loose clusters. Eggs are always attached to solid objects, usually underwater, but sometimes just above the water surface. The substrates used for laying eggs vary widely, but many types of true bugs lay their eggs on aquatic plants. Moist soil at the margin of the aquatic habitat is another site commonly used for laying eggs. The time required to hatch is about 1–2 weeks. Hardly any aquatic true bugs undergo diapause in the egg stage.

Aquatic true bugs usually spend about 2 months as larvae, but this can be as short as 1 month or as long as 8 months in different species. They almost always shed their skin five times, but a few kinds molt only four times. True bug larvae proceed through the larval stage without any interruptions in their development. When the larvae have completed their development and are ready to transform into adults, they do not leave the water. This is not especially challenging for true bugs because the larvae and adults both breathe oxygen from the air.

With very few exceptions, aquatic true bugs become adults in the fall, but they do not mate until the following spring. Thus, they are long lived as adults compared to most other aquatic insects. Adults may be 6 months old before they mate, and they may live several weeks to several months after mating. Only some of the water beetles (Coleoptera) have an adult life span as long as that of true bugs. Adults of some true bugs remain active during the winter, but many kinds overwinter in a state of hibernation. When organisms hibernate, their metabolic rate is dramatically reduced and they cease to move. Hibernating organisms may appear to be dead. Hibernation may take place in shallow water at the edge of the habitat or a short distance out of the water on shore. In either location, they hide themselves in mud or plant debris during hibernation. When they mate in the spring, most true bugs do not leave the water. Most can fly as adults but do not do so unless they run out of food in their habitat, they become too crowded, or their habitat dries up.

Various types of aquatic true bugs display diverse and interesting behaviors for detecting prey, defense, communication, and parental care. However, these behaviors are not exhibited in all families within the order, so they will be explained in the parts on the biology of individual families. One behavior that is common to several families is stridulation. This behavior is performed by many insects, terrestrial and aquatic, and consists of them making a noise by rubbing two body parts together. The best known examples of stridulation are the "songs" of crickets and katydids in the summer. Some aquatic insects do this underwater too. Human ears cannot hear the sound waves as they pass through the water, but aquatic insects belonging to the same species can perceive them.

Significance — None of the aquatic true bugs are serious pests, but some can be minor nuisances. Some are significantly beneficial. Although humans occasionally receive painful bites from some kinds of aquatic true bugs, the bites are not serious. Bites usually only happen when humans deliberately attempt to handle these insects, but do so carelessly. Bites from true bugs can always be avoided by simply being careful. More often, they are merely a nuisance by being attracted to lights near where people have gathered outdoors at night, especially at swimming pools. Some of the large species of giant water bugs (Belostomatidae) sometimes become residents in fish culturing facilities, where they kill appreciable numbers of fish fry.

On the other side of the coin, some true bugs are consumed by humans. Like many insects, true bugs provide a good source of protein and are considered to be tasty delicacies, mostly outside of North America. For this reason, some aquatic true bugs have been sold commercially for many centuries, and they still have a minor role in commerce today. Edible species are sold in local markets in Central America and the Orient. Today, the greatest economic use of true bugs is as pet food for fish, turtles, and birds. Water boatmen (Corixidae) are the primary true bugs that are used for this purpose. They are either harvested from natural environments or grown in culturing facilities, then they are dried before being packaged as pet food.

Many kinds of aquatic true bugs eat mosquito larvae and, therefore, help suppress the numbers of biting insects that emerge from aquatic habitats. The extent of their role in controlling mosquitoes is not known, and they have never been used formally in biocontrol programs. Although most true bugs have scent glands that make them unpalatable to fish, there are exceptions in the aquatic families. The most notable exception is water boatmen (Corixidae), which are consumed in large quantities by many species of fish.

Because aquatic true bugs do not depend on dissolved oxygen for their respiration, some kinds can live in very polluted waters, such as where sewage is discharged. If many true bugs but few other kinds of aquatic invertebrates are found in a habitat, this is usually an indication that some type of pollution is adversely affecting that habitat.

Biology of Common Families

Backswimmers

Family Notonectidae; Plate 22

Different Kinds in North America — Few; 3 genera, 32 species.

Distribution in North America — Throughout.

Habitat — Lentic-littoral and lotic-depositional. Backswimmers are common in all types of quiet waters. They are often found together with water boatmen (Corixidae). Small ponds and littoral zones of large lakes with abundant vegetation are preferred habitats. Other habitats where backswimmers live include sewage lagoons, spring-fed pools, roadside ditches, stagnant pools, thermal springs, and brackish waters.

Movement — Swimmers. All backswimmers swim powerfully for short distances by rapid oar-like strokes of the hind legs. They spend much of the time hanging upside down from the water surface. The head of backswimmers is angled downward, and the tip of the abdomen is in contact with the surface film. The hind legs are held in a forward position, poised for a sudden rowing stroke. Some kinds float freely, while others hold on to vegetation at the water surface.

Feeding — Piercer-predators. Backswimmers feed on any prey they can subdue. Their prey is mostly a variety of invertebrates, such as insects, crustaceans, and snails, but they also feed on some vertebrates, such as fish fry and tadpoles. The front and middle pairs of legs of backswimmers have strong spines that are used to hold prey while their beak is being used to kill the prey and suck out its body fluids.

Other Biology — Backswimmers can inflict a painful bite with their beak if handled carelessly. They breathe by taking an air store underwater. The air store is held in two troughs with fringes of hair on the bottom side of the abdomen. Additional air is carried under the wings and between the head and thorax. Backswimmers renew their air store by breaking the surface with the tip of their abdomen. The color pattern is reversed in accordance with the upside down position (light on top of body and dark on underside), hence the common name backswimmers. They remain upside down while swimming. Backswimmers overwinter as adults, either remaining active or hibernating within the substrate. Eggs are deposited in spring or summer in or on plant tissue or on rock surfaces. It takes 6–11 weeks to develop from the egg to the adult stage. Most backswimmers have at least two generations per year. Males in this family stridulate by rubbing rough patches on the front legs against the innermost segment of the beak. Backswimmers are strong fliers capable of dispersing considerable distances.

Stress Tolerance — Very tolerant.

Broad-Shouldered Water Striders

Family Veliidae; Plate 27

Different Kinds in North America — Moderate; 6 genera, 34 species. Other common names that are used for this family are small water striders or little water striders.

Distribution in North America — Throughout.

Habitat — Most lentic-littoral and lotic-depositional (surface), some lotic-erosional (surface). Broad-shouldered water striders are surface inhabitants and share the same general habitats with water striders (Gerridae). However, most kinds of veliids inhabit more protected areas within the habitat than gerrids. Broad-shouldered water striders usually live on the water surface among vegetation near the margins of the quieter sections of streams or ponds. They are often difficult to observe among the vegetation. Sometimes broad-shouldered water striders also occur on shore, immediately adjacent to the water. Favored hiding places near the water include on boulders, in rock crevices, in leaf accumulations, on wet sandy soil, and beneath undercut banks. The species that are seen most often belong to the genus *Rhagovelia*. Unlike other veliids, those in the genus *Rhagovelia* occur almost exclusively in lotic-erosional habitats and, thus, are sometimes called riffle bugs. Riffle bugs are found on the surface of moving water or in eddies below rocks that are in the current. They are gregarious, which makes them conspicuous on the surface of the water, even though individual organisms are quite small. A single sweep of a dip net may catch 50 or more riffle bugs. A few other kinds of broad-shouldered water striders live in salt marshes and near-shore marine environments.

Movement — Skaters. Broad-shouldered water striders are very swift and agile on the water surface, just like water striders (Gerridae). They too have their claws located a short distance up the outermost section of the legs, rather than at the ends of the legs, to avoid puncturing the surface film. The riffle bugs in the genus *Rhagovelia* are especially well adapted for skating on the surface of moving water, because they have a thick plume of non-wettable hairs inserted in a deep notch in the segment on the end of the middle legs.

Feeding — Piercer-predators. Veliids feed the same way as gerrids, except their prey is smaller. Broad-shouldered water striders prey on almost any small organisms that they encounter on the surface, such as water fleas, ostracods, and some insects. They will feed on living, crippled, or dead insects. Some species of broad-shouldered water striders are known to feed heavily on mosquito eggs and larvae. Like gerrids, veliids locate potential prey by detecting surface waves with vibration sensors in their tarsi.

Other Biology — Because they are surface inhabitants, broad-shouldered water striders breathe from the atmosphere by means of spiracles, like terrestrial insects. However, they can remain underwater and breathe from a thin layer of air trapped in velvety, non-wettable hairs that cover their body. They take on a silvery appearance while underwater because of the layer of trapped air. Broad-shouldered water striders overwinter as hibernating adults, aggregating in debris at the edge of the water or beneath undercut banks. Eggs are laid on floating or stationary objects at the edge of the water. The larval development period of broad-shouldered water striders lasts 2–5 weeks. There may be as many as six generations per year, according to species and climate. All types of wing polymorphism occur, wingless (apterous), small wings (brachypterous), and normal wings (macropterous), but most broad-shouldered water striders are wingless as adults.

Stress Tolerance — Somewhat tolerant.

Creeping Water Bugs

Family Naucoridae; Plate 25

Different Kinds in North America — Few; 4 genera, 23 species.

Distribution in North America — Throughout, but the greatest number of species occurs in the South and West.

Habitat — Lentic-littoral, lotic-erosional. Different species of creeping water bugs are common in a wide variety of standing and running waters, including pools, ponds, lakes, saline waters, small swift streams, rivers, and hot springs. Living vegetation and accumulations of plant debris in ponds and pools are common habitats. Some kinds are found in streams on and under pebbles and larger stones. A few kinds of creeping water bugs live in hot springs with temperatures up to 36° C.

Movement — Crawlers, climbers, swimmers. As their common name implies, many kinds of creeping water bugs crawl about on the bottom in search of prey. The bases of aquatic plants and plant debris are favorite places for them to search for prey. Some creeping water bugs climb about on aquatic vegetation. Others spend a lot of time moving freely in the water, much like water boatmen (Corixidae).

Feeding — Piercer-predators. Creeping water bugs feed on any tasty aquatic invertebrates they can overpower. They have been observed in captivity to feed on mosquito larvae (Culicidae), midge larvae (Chironomidae), water boatmen (Corixidae), and various types of small crustaceans.

Other Biology — Creeping water bugs can inflict a painful bite with their beak if handled carelessly. As larvae, they acquire dissolved oxygen through their general body surface. When they become adults, creeping water bugs breathe from an air bubble held under their wings. They periodically break the surface with the tip of their abdomen to replenish the oxygen in the bubble. Some air is also held in short, fine hairs on their underside, giving them a silvery luster while underwater. Creeping water bugs extend the time before they have to go back to the surface by extracting some dissolved oxygen from the water and placing it into their air store (physical gill). When they are sitting still underwater, they can be seen moving their hind legs to keep water that is saturated with oxygen flowing over the bubble and thin layer of air on their underside. Creeping water bugs spend longer times in the egg (3–8 weeks) and larval (7–10 weeks) stages than most other true bugs. Adults spend the winter in soft mud among aquatic vegetation. Eggs are attached to solid objects in shallow water, usually plants in still water and stones in running water. Even though creeping water bugs are aggressive predators, they are quite defenseless when molting. At that time they are often consumed by other predators, such as damselfly larvae, that normally would be their victims. Creeping water bugs have one generation per year.

Stress Tolerance — Probably somewhat tolerant, but little information.

Giant Water Bugs

Family Belostomatidae; Plate 24

Different Kinds in North America — Few; 3 genera, 22 species.

Distribution in North America — Throughout.

Habitat — Primarily lentic-littoral, some lotic-depositional. Giant water bugs mostly inhabit ponds, edges of lakes, marshes, ditches, and pools of streams. They are found among living aquatic vegetation, terrestrial plants that have drooped over into the water, and all types of plant debris.

Movement — Combination of climbers and swimmers. Giant water bugs spend most of their time climbing or perching on vegetation or debris. They position themselves approximately vertical, just under the water surface with the head down. The tip of their abdomen is close enough to the water surface that the retractile breathing straps can reach the air. Giant water bugs sometimes hang down from the water surface by the tip of their abdomen. They can also swim rapidly, usually in pursuit of prey, by propelling themselves with strokes of their powerful middle and hind legs. These two pairs of legs are adapted for moving in water by being flattened and having a fringe of long hairs.

Feeding — Piercer-predators. Giant water bugs have powerful raptorial front legs to capture prey and hold it while it is being subdued and consumed. They are regarded as the masters of their immediate environment, feeding upon a variety of aquatic insects, other arthropods, tadpoles, and even fish. If an unfortunate organism happens to come close to a resting giant water bug, it will merely grab the prey with its front legs. If the prey is out of easy reach, they swim and chase it down. If the prey is large, they may have to ride it around while the poisonous salivary enzymes take effect. Some voracious feeding has been recorded for giant water bugs kept in aquariums. One individual ate an 8-cm trout, several young frogs, tadpoles, snails, and various fish fry. Another ate dozens of tadpoles in 24 hours. A member of the genus with the largest individuals (*Lethocerus*) captured a 9-cm chain pickerel. Obviously, giant water bugs can become a nuisance in fish hatcheries, and care has to be taken to keep them out. The most bizarre story of feeding by giant waterbugs also involved a large species in the genus *Lethocerus*. A naturalist hiking in the woods heard a bird making distress noises. Upon investigation, he found a woodpecker thrashing around on the ground. There was a giant water bug clutching the bird's beak with its legs, and its own beak was inserted into the bird's head. There have also been reports of giant water bugs attacking small ducklings.

Other Biology — Giant water bugs will inflict a painful bite with their beak in defense if they are handled carelessly. Another common name for them is "toe biters," but they do not seek human appendages for food in spite of the feeding habits described above. Giant water bugs breathe oxygen from the atmosphere by means of two elongate, flat structures at the end of the abdomen. These structures are usually referred to as air straps or breathing straps. Unlike the breathing tubes of water scorpions (Nepidae), the straps of giant water bugs can be retracted into the abdomen. Air is also stored under the wings for prolonged dives. Some common kinds of giant water bugs have curious egg-laying behavior. Females deposit their eggs on the backs of males, which carry them until they hatch. Males may carry over 100 eggs. The males provide protection from predators and maintain a flow of fresh water over the eggs by stroking them with their hind legs. Some kinds of giant water bugs lay their eggs on the stems of emergent vegetation above the water or on and under logs and boards on shore. All giant water bugs are strong and frequent fliers. They are strongly attracted to lights, and for this reason they are sometimes called "electric light bugs." They overwinter as adults buried in the mud. Larval development takes 5–9 weeks. All giant water bugs are thought to have only one generation per year.

Stress Tolerance — Very tolerant.

Water Boatmen

Family Corixidae; Plate 21

Different Kinds in North America — Moderate; 17 genera, 125 species.

Distribution in North America — Throughout.

Habitat — Lentic-littoral and lotic-depositional. Water boatmen are often found on or around aquatic plants. Sometimes they occur on bare soft substrate. Water boatmen are very common in shallow water around the edges of all sizes of ponds and lakes and in pools of streams and rivers. Some kinds also live in the relatively calm water along the margins of riffles. They are found in some rather harsh environments such as ditches, brackish waters, temporary pools, and intertidal ocean waters.

Movement — Swimmers. Water boatmen spend most of their time swimming head down along the bottom in search of food. They swim in a quick, darting manner by oar-like movements of their hind legs.

Feeding — Collector-gatherers. Water boatmen feed differently from all other families of aquatic true bugs. The head ends in a broad cone rather than a sharp beak. There is a small round mouth opening on the front side of the cone, near the bottom. Water boatmen stir up fine material on the bottom with the scoop-like tarsi (palae) on their front legs. Then they ingest whatever living material they encounter in the sediment, including diatoms, filamentous algae, other algae, protozoa, rotifers, nematodes, and small insects such as midges and mosquitoes. Inside the mouth there are tooth-like structures in a muscular area, where the food is ground up before it moves along the digestive system. Some water boatmen feed on plant fluids, especially from filamentous algae, by puncturing plant cells with slender stylets from inside their mouth, then sucking up the juices.

Other Biology — Water boatmen breathe from an air bubble that is held under the wings and extended around the abdomen. To renew the air bubble, they break the surface of the water very quickly with the head and top of the thorax. Water boatmen are able to stay underwater longer than most other true bugs, because their air bubble functions as a physical gill. As the oxygen in the bubble is used up, some of it is replaced by dissolved oxygen from the surrounding water. The movements of the hind legs during swimming circulate water over the air bubble, which increases the efficiency of the air bubble for obtaining dissolved oxygen. Eventually the air bubble must be restored from the atmosphere. Stridulation is common in the different kinds of water boatmen, usually only among males. There are pegs on the front legs, which they rub over the side margin of the head to produce a chirping sound underwater. Some species of water boatmen hibernate as adults during the winter in the mud, while others remain active, even under ice. Some pond species fly to larger lakes for overwintering, which is probably an adaptation to avoid habitats that might freeze solid. In the spring, eggs are usually deposited on plants, but sometimes they will be deposited on snails or crayfish. Larval development requires 5–7 weeks. Water boatmen can pass through several generations per year, according to species and climate. Adults are strong fliers that are capable of dispersing to new habitats. They are often attracted to lights in large numbers.

Stress Tolerance — Very tolerant.

Water Scorpions

Family Nepidae; Plate 23

Different Kinds in North America — Few; 3 genera, 13 species.

Distribution in North America — Throughout.

Habitat — Lentic-littoral and lotic-depositional. Water scorpions occur in a variety of ponds, swamps, and streams, usually among debris or vegetation in shallow areas. They are often found with giant water bugs (Belostomatidae).

Movement — Climbers. Water scorpions spend most of their time on vegetation and debris, where they lie in wait or slowly stalk their prey. They are very poor swimmers, but they are capable of clumsily paddling through the water. Water scorpions move so seldom and so slowly that microorganisms, such as algae and protozoans, sometimes colonize them. Other aquatic insects, such as water boatmen (Corixidae), backswimmers (Notonectidae), and caddisflies (Trichoptera) sometimes lay their eggs on them.

Feeding — Piercer-predators. Water scorpions feed chiefly on small swimming crustaceans and insects, which they capture with their raptorial front legs. The method of feeding is very similar to praying mantids on land. Their most common prey is probably water boatmen (Corixidae). They have also been reported to capture fish and tadpoles.

Other Biology — Water scorpions breathe from the atmosphere by a pair of elongate tubes on the end of their abdomen. They push the tubes up so that they just touch the surface film. Therefore, they spend most of their time hanging head downward. They transfer air to the space under their wings when they venture deeper in the water. Water scorpions stridulate by rubbing the innermost leg segment (coxa) against the first thorax segment (prothorax). Eggs are attached to plants or woody materials at the water surface or just above. Larval development takes about 2 months. They have well developed wings but seldom fly. Water scorpions produce up to three generations per year according to species and climate.

Stress Tolerance — Somewhat tolerant.

Water Striders

Family Gerridae; Plate 26

Different Kinds in North America — Moderate; 9 genera, 47 species. These are also called pond skaters.

Distribution in North America — Throughout.

Habitat — Most lentic-limnetic or lotic-depositional (surface); some lotic-erosional (surface). Specific habitats of water striders include streams, lakes, ponds, marshes, ditches, estuaries, and brackish waters. One genus (*Halobates*) is strictly marine, living on the surface of coastal areas and the open ocean.

Movement — Skaters. Water striders have special adaptations that enable them to remain suspended on the surface film, or surface tension, created by the attraction of water molecules. The ends of their legs have non-wettable hairs that do not disrupt the surface film. Their claws are located a short distance up the outermost section of the legs (tarsus), rather than at the end of the legs. This keeps the sharp points of the claws from breaking through the surface film. Water striders move swiftly and agilely on the surface of the water, hence the common name skaters. Only the middle and hind legs touch the water. The short front legs are held up. The middle legs are used for the rowing motion. In clear, shallow water, large, round shadows are cast on the bottom for the end of each slender leg that touches the water. The shadows result from the depressed, but unbroken, area of surface film upon which they rest. When disturbed, water striders can dive without becoming wet by virtue of water-repellant scales and hairs covering their bodies. When it rains, they leave the water surface and hide under leaves on shore.

Feeding — Piercer-predators. Water striders use their short, raptorial front legs to grasp their prey on the water surface. The prey includes terrestrial and aquatic insects, both living and dead. Cannibalism is common, especially under crowded conditions. Water striders locate their prey by means of vibration sensors located in the flexible membranes between the tarsal segments of the middle and hind legs. When very small surface waves strike the vibration sensors, they turn toward the direction the waves are coming from by moving one leg forward and one backward, just like turning a rowboat. Then they move forward toward the source of the surface waves. They also locate prey with their eyes.

Other Biology — While skating on the water surface, water striders breathe from the atmosphere by means of openings in the sides of their bodies (spiracles), like terrestrial insects. When they occasionally dive or become submerged, water striders continue to obtain oxygen from a thin layer of air trapped in the fine, non-wettable hairs that cover their body. They take on a silvery appearance while underwater. They overwinter as adults, hiding in protected places on shore, not far from the water. Eggs of water striders are glued to floating objects at the edge of the water, just beneath the surface. Larval development requires 3–6 weeks. All types of wing polymorphism occur, wingless (apterous), small wings (brachypterous), and normal wings (macropterous), even among the same population (polymorphic). Water striders have from one to three generations per year, according to species and climate.

Stress Tolerance — Somewhat tolerant.

Dobsonflies, Fishflies, Hellgrammites, Alderflies

Order Megaloptera; Plates 60 – 61

This is a small order, in terms of diversity, with only 2 families and 46 species in North America. This guide includes both families because they are common in freshwater habitats. Members of this order have complete metamorphosis, and all are aquatic as larvae and terrestrial as pupae and adults.

Distinguishing Features of Larvae — Hellgrammites and other larvae in this order have elongate bodies that are somewhat flattened. The body length of mature larvae ranges from 10 to 90 mm in different species. Larvae in this order can be distinguished from other immature aquatic insects by the following combination of characteristics (Figure 27 A, B):

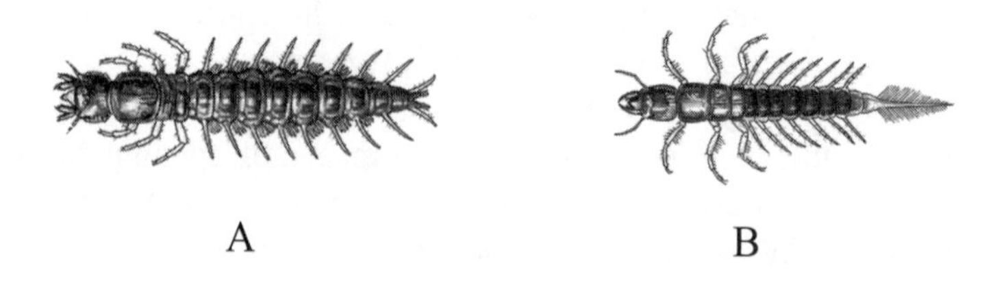

Figure 27. Representative dobsonfly or hellgrammite larva **(A)** and alderfly larva **(B).**

- Head and thorax have thick, hardened skin, while the abdomen has thin, soft skin.
- Prominent chewing mouthparts project in front of the head.
- No wing pads occur on the thorax.
- Three pairs of segmented legs extend from the thorax.
- Seven or eight pairs of stout, tapering filaments stick out from the sides of the abdomen.
- End of the abdomen has either a pair of prolegs with two claws on each proleg, or a single long, tapering filament with no claws.

Distinguishing Features of Adults — Adult dobsonflies, fishflies, and alderflies have bodies that are elongate, stout, and more or less cylindrical. The abdomen is very soft. They are medium to large in size. The body length of different species ranges from 10 to 70 mm, measured from the front of the head to the tip of the wings folded at rest. The wingspan of the largest species can be more than 100 mm. Adult dobsonflies, fishflies, and alderflies can be distinguished from other adult aquatic insects by the following combination of characteristics (Figure 28 A, B):

- Wings are very large and elongate.
- Wings are membranous with many veins.
- Wings are held slanted and roof-like over the abdomen, when the organism is not flying.

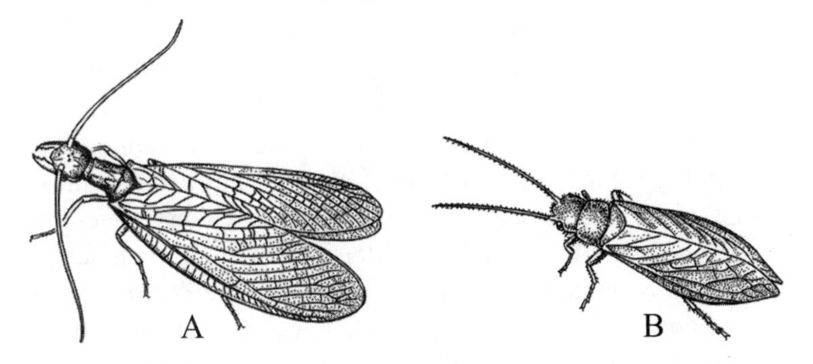

Figure 28. Representative dobsonfly adult **(A)** and alderfly adult **(B).**

- Wings have many short veins (cross veins) that extend between the first two long veins that run from the base of the wing to the tip (longitudinal veins).
- Numerous cross veins create a fairly straight row of small cells along the front of the wings.
- Hind wings, when extended, are much wider than the front wings, thus, the hind wings are pleated so that they can crumple under the front wings when not in use.
- Tips of both pairs of wings extend far past the end of the abdomen when the wings are folded.
- No tails are present on the end of the abdomen.

Explanation of Names — There are no common names that are recognized for the entire order. Several acceptable common names are used for the adults and larvae, but they really pertain only to certain families and genera. For the adults these are dobsonflies (genus *Corydalus* in the family Corydalidae), fishflies (all other genera in the family Corydalidae), and alderflies (all species in the family Sialidae). The larvae of dobsonflies and fishflies (all species in the family Corydalidae) are called hellgrammites. Larvae in the other family (Sialidae) are just referred to by the common name of the adults, alderfly larvae. The scientific name of the order is a combination of two Greek words: *megal*, meaning large or of giant size, and *ptera*, meaning wings. The scientific name refers to how large the wings are in comparison to the body.

Habitat — Even though Megaloptera is a small order, there are some distinct differences among the families and genera in regard to the habitat of the larvae. These will be covered in the parts on the biology of individual families. Larvae of the different kinds of dobsonflies, fishflies, and alderflies live in lotic-erosional, lotic-depositional, and lentic-littoral habitats. Members of this order inhabit all sizes of these habitats, as well as intermittent and damp semi-terrestrial habitats. Flowing-water habitats include spring seeps, intermittent streams, small woodland streams, and large rivers. Still-water habitats include wetlands, bogs, swamps, marshes, ponds, and small lakes. The specific locations within the habitats where different kinds of larvae prefer to reside are very diverse in this order. Some kinds of dobsonflies, fishflies, and alderflies only occur in microhabitats composed of coarse substrate, such as boulders, cobbles, and logs, in fast current and with no accumulations of organic matter. Other kinds of larvae live only in microhabitats with fine substrate, such as mud and silt, in little or no current and with large accumulations of organic matter, such as dead leaves, other decaying plant material, and fine detritus.

Movement — The different kinds of dobsonflies, fishflies, and alderflies are either crawlers, sprawlers, or burrowers as larvae.

Feeding — All members of the order are engulfer-predators that actively roam around in search of prey. Larvae of dobsonflies, fishflies, and alderflies are opportunistic in their feeding, taking whatever is available at a particular time and place. They usually do not attempt to capture very large prey.

Breathing — Larvae of dobsonflies, fishflies, and alderflies have a variety of ways they can breathe, even within the same species. All kinds in this order have an open breathing system, like terrestrial insects, but of course they do not open their spiracles while they are underwater. While underwater, larvae of dobsonflies, fishflies, and alderflies rely mostly on dissolved oxygen diffusing into their body across all of the soft, fleshy tissue. However, the stout filaments on the sides of the abdomen are gills that are specialized to acquire dissolved oxygen, and these structures assist with breathing underwater. Some kinds of larvae in this order also have dense tufts of finely branched gills beneath the stout filaments, which creates more gill surface area for obtaining dissolved oxygen. Other kinds that live in shallow water have a pair of long breathing tubes at the end of the abdomen for obtaining oxygen from the air. When larvae of dobsonflies, fishflies, and alderflies leave the water to prepare for the pupal stage (see Life History), they use the spiracles of their open breathing system to get oxygen from the air.

Life History — The different kinds of dobsonflies, fishflies, and alderflies either have one generation per year or they require 2 or 3 years to complete a generation. The length of time even varies within the same species according to natural temperature differences. One of the most common and widespread species in eastern and central North America, *Corydalus cornutus*, requires either 1, 2, or 3 years to complete its life cycle at different latitudes and elevations. Adults of different species of dobsonflies, fishflies, and alderflies usually emerge from late spring to early summer, but records of adult emergence extend from March through September.

All kinds of dobsonflies, fishflies, and alderflies deposit their eggs out of the water on overhanging objects. The usual sites for laying eggs are rocks, log jams, and the undersides of leaves and twigs on trees and other terrestrial vegetation. This is done at night. The eggs are glued together in a mass. Egg masses are characteristic for individual kinds of dobsonflies, fishflies, and alderflies, according to color, number of eggs, number of layers, and whether there is a covering or not. The number of eggs in a mass ranges from 300 to 3,000 for the various species. Dobsonfly egg masses are most likely to be found. They are circular, about 2 cm in diameter, and consist of three layers of eggs covered by a white secretion. Dobsonfly egg masses are especially conspicuous on large boulders, concrete bridges, and log jams. Remnants of old egg masses persist as white rings until the next year. Dobsonflies tend to use the same sites for laying eggs, so there are often dense patches of egg masses or remnants of ones that have hatched. The eggs of all kinds hatch in 1–2 weeks. There is no diapause in the egg stage in this group. The eggs always hatch at night, presumably to prevent the young larvae from being eaten by birds.

The young larvae of dobsonflies, fishflies, and alderflies immediately drop into the water below their egg mass. Newly hatched larvae often have gas bubbles in their gut that keep them afloat. They float and swim by flexing their bodies for a short time until they come to a suitable location. The different kinds of dobsonflies, fishflies, and alderflies spend from 1 to 3 years as larvae, and during this period they shed their skin 10–12 times. There is no diapause in the larval stage. When larvae have completed their development, they crawl out of the water for the pupal stage. This event is fairly synchronous. All larvae in a particular population of the same species may leave the water within a few days. Larvae of hellgrammites have been observed leaving the water simultaneously when strong thunderstorms occur. Apparently, the vibration caused by the thunder acts as a triggering mechanism for mature larvae that are ready to pupate. This behavior is conspicuous, and local residents refer to it as "hellgrammite crawling." The distance that larvae move away from the water varies with the different species. The smallest members of the order (alderfly larvae) proceed only about 2–3 m from the edge of the water, whereas the largest species (hellgrammites) will go 10 m or more before they stop to pupate. Hellgrammites have been found crawling across bridges, and one larva was found on top of the stone chimney of a riverside cabin.

After coming out on shore, mature larvae of dobsonflies, fishflies, and alderflies prepare a spot where they will be safe during the helpless pupal stage. This usually involves digging a cavity in moist soil with the jaws and legs. The space that is excavated is referred to as the "cell." In most instances, larvae prepare the cell underneath a rock or log lying on the ground. Some will use a space within a rotten log. Larvae in this order do not have silk glands, so there is no lining or cocoon inside the cell.

After completing the cell, the mature larvae lie inactive inside the cell for 1–14 days. During this time, the internal tissues are undergoing many physiological changes. Externally, the body of the larva begins to look more squat, and the soft parts of the skin begin to have a leathery texture. The organism is referred to as a prepupa during this period. However, this prepupal period does not involve a molt. Eventually, the organism does molt, and major morphological changes are revealed in the form of a pupa. Another 7–14 days is spent in the cell as a pupa. Adult dobsonflies, fishflies, and alderflies emerge from the pupal skin while still inside the cell.

Newly emerged adults of dobsonflies, fishflies, and alderflies dig and crawl their way out of the cell. The duration of the adult stage is about 3 days for males and 8–10 days for females. Most adults in this order usually reside in trees, shrubs, or tall grasses close to the aquatic habitat of the larvae. They sit still or crawl around a little in vegetation, except for their inherent flight periods to find mates. Most dobsonflies fly at night, while all alderflies and some dobsonflies fly during the mid-hours of sunny days. Most kinds do not venture far from the aquatic habitat. The exception is that dobsonflies are strongly attracted to lights, which often lead them great distances away from their larval habitat. In spite of having large, sharp mouthparts, adults of dobsonflies, fishflies, and alderflies probably do not eat any solid food. They have been observed to drink liquids in captivity, especially substances with sugar. Adult dobsonflies have sexual dimorphism. All males have exceptionally long and exaggerated jaws. These structures have no function for feeding. They are not even useful for defense, because they do not have enough muscular strength to exert a firm pinch. However, they are apparently used as part of the mating behavior in this species. Male dobsonflies have been observed to place their jaws on top of the wings of the female as part of a courtship ritual. The female becomes receptive to mating with the male after this behavior takes place.

Significance — Dobsonflies, fishflies, and alderflies are never considered to be pests or nuisances, either as adults or larvae. The larvae are beneficial organisms. The hellgrammites that live in lotic-erosional habitats are very active and aggressive predators and probably improve diversity in the community, as discussed for some of the stoneflies (Plecoptera). Some hellgrammite larvae are used quite a bit as live bait for fishing, especially for smallmouth bass. Some persons harvest their own hellgrammites, but they are also sold in stores along rivers. This provides a small economic enterprise in some areas. The extent that hellgrammites are used as live bait has become a concern. In some states, the agencies that are responsible for managing fishing and conservation have imposed regulations for harvesting hellgrammites, and permits are usually required. In spite of their popularity as bait for fishing, larvae in this order do not appear to be a significant part of the natural diet of fish. Analyses of stomachs from fish that are known to consume invertebrates have shown that very few hellgrammites are consumed in comparison to other invertebrate prey. Data on the kinds and relative abundance of the different kinds of larvae in this order can be useful for biomonitoring studies. Larvae of most dobsonflies and fishflies are found only in clean to moderately clean environmental conditions and are considered to be facultative to pollution. Alderfly larvae can be found in a wide range of environmental conditions, ranging from clean to poor. Some species of alderflies live in environments that are seriously impaired (for example, organic wastes, low dissolved oxygen, high metal concentrations), so this family is generally considered to be tolerant of pollution.

Biology of Common Families

Alderflies

Family Sialidae; Plate 61

Different Kinds in North America — Few; 1 genus, 24 species.

Distribution in North America — Throughout the United States and well into Canada.

Habitat — Lentic-littoral, lotic-depositional. Alderfly larvae occur in all types and sizes of freshwater habitats. Standing water habitats include bogs, marshes, ponds, and lakes. Running water habitats include springs, streams, and large rivers. Most species live where the bottom is soft. Alderfly larvae are most common in mud and silt where a lot of organic matter, such as dead leaves, other decaying plant material, and fine detritus, has accumulated on the bottom. A few species live on sandy bottom. Larvae of alderflies seldom occur in flowing water. If they are in a stream or river, they will be in pools or back eddies.

Movement — Burrowers. Larvae of alderflies usually dig in the upper 10 cm of the substrate, but they have been found as deep as 30 cm.

Feeding — Engulfer-predators. Alderfly larvae search for prey in the mud and on the bottom.

Other Biology — Eggs of alderflies are dark brown. The egg mass consists of one layer with no protective covering. Females lay the eggs on the undersides of leaves and twigs of terrestrial vegetation that hangs over the appropriate habitat for the larvae. Alderflies usually produce one generation per year, but sometimes 2 years are required for a generation, depending on the species and climate. Adults emerge primarily in April and May.

Stress Tolerance — Chiefly facultative, others very tolerant. Some kinds of alderflies can endure adverse conditions such as organic wastes, low dissolved oxygen, and high concentrations of metals.

Dobsonflies, Fishflies, Hellgrammites

Family Corydalidae; Plate 60

Different Kinds in North America — Few; 7 genera, 22 species.

Distribution in North America — Throughout, but not as common in the West.

Habitat — Primarily lotic-erosional, occasionally lotic-depositional and lentic-littoral. Larvae of dobsonflies and fishflies, which are known as hellgrammites, usually live in running waters, ranging from small woodland streams to large rivers. Hellgrammites usually live in the areas of swiftest current within the habitat. They reside in small sheltered spaces underneath boulders and cobbles or large pieces of woody debris. However, there are some less common kinds of hellgrammites that live in the very shallow standing waters of ponds, bogs, marshes, and swamps or in damp, semi-terrestrial habitats, such as spring seeps, intermittent streams, and wetlands. The latter kinds of hellgrammites live in or on soft sediment that is rich in organic matter, such as dead leaves or fine detritus.

Movement — Primarily crawlers, some burrowers, sprawlers. The most common kinds of hellgrammites that live in running waters crawl around in spaces within the substrate. They hold on to rocks and wood with the strong claws on their legs and on the short, fleshy prolegs on the end of their abdomen. If hellgrammites become dislodged, they swim backwards for a short distance by bending their abdomen down and toward the head, then reversing the motion, like a crayfish. The less common species of hellgrammites that live in standing water or damp environments burrow in or sprawl on soft sediment during the day. Then at night they come out of the water and crawl around on damp mud and decaying vegetation.

Feeding — Engulfer-predators. All kinds of hellgrammites actively search for prey within their habitat, especially at night. They will consume any small invertebrates, but larvae of mayflies, black flies, net-spinning caddisflies, and midges are some of their most common prey.

Other Biology — All kinds of hellgrammites obtain dissolved oxygen through the soft portions of their skin and by the filaments on the sides of the abdomen, which are gills. Larvae in one genus (*Corydalus*) also have dense tufts of filamentous gills at the bases of each side filament. These filamentous gills have muscles attached to their bases, so the tufts can be spread open and contracted to increase breathing efficiency. The additional filamentous gills probably account for *Corydalus* being able to inhabit larger, warmer rivers that have naturally lower concentrations of dissolved oxygen. Larvae in one genus that live in standing water (*Chauliodes*) have their last spiracles on a pair of long breathing tubes at the end of the abdomen. These hellgrammites live in shallow, muddy, organically rich habitats that have very low concentrations of dissolved oxygen. Dobsonflies and fishflies complete their life cycle in 1, 2, or 3 years, according to species and climate. Adults can emerge from April to August, but the peak emergence of common species is from May to early July.

Stress Tolerance — Primarily facultative, others very tolerant.

Water Beetles

Order Coleoptera; Plates 28 – 33, 78 – 83

This is the largest order of insects, in terms of diversity. In the entire order, there are 113 families and about 24,000 species in North America. More than one-third of all insect species in the world are in this single order. There are 18 times more species of beetles than all species of vertebrate animals. There are more species of beetles than there are species of flowering plants. Most of this tremendous number of beetles are entirely terrestrial and will not be encountered in freshwater habitats. However, because some beetles have managed to be successful in almost all habitats on the planet, there are 20 families with about 1,000 species in North America that are aquatic in at least one stage of their life cycle. Even though only 4% of beetles are aquatic, that still accounts for a lot of different kinds of water beetles. This guide includes seven families that are common in freshwater habitats. Water beetles have complete metamorphosis. In most of the families of water beetles, the larvae and adults are both aquatic. There are a few exceptional families in which only the larvae or the adults are aquatic. All of the water beetles spend their pupal stage out of the water.

Distinguishing Features of Larvae — Bodies of water beetle larvae are variable. Most are elongate and cylindrical. Some are oval and vary from somewhat flattened to very flat. The body length of mature larvae in different species ranges from about 2–70 mm, excluding tails. Water beetle larvae can be distinguished from other immature aquatic insects by the following combination of characteristics (Figure 29):

- Head has thick, hardened skin.
- Thorax and abdomen of most kinds have moderately hardened skin, but the abdomen has thin, soft skin in some kinds.
- No wing pads occur on the thorax.
- Three pairs of segmented legs extend from the thorax in most kinds, but some kinds have no segmented legs.
- No structures project from the sides of the abdomen in most kinds, but some kinds have flat plates or stout filaments sticking out from the sides of the abdomen.
- No prolegs or long, tapering filament occurs on the end of the abdomen.

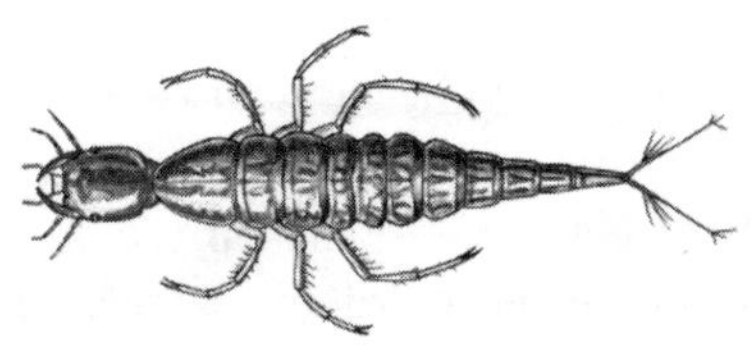

Figure 29. Representative water beetle larva.

Distinguishing Features of Adults — Water beetle adults vary considerably in shape, from short, stocky, and cylindrical to oval and somewhat flattened. The body is always very hard. They are small to medium in size. The body length of different species ranges from 1 to 40 mm, measured from the front of the head to the tip of the wings folded at rest. Water beetle adults can be distinguished from other adult aquatic insects by the following combination of characteristics (Figure 30):

- Front wings are greatly modified into protective structures (elytra) that are thick and hard, without veins.
- Front wings, when folded, meet in a straight line down the middle of the back, covering the hind wings and most of the abdomen.

- Hind wings are membranous.
- Hind wings are usually longer than the front wings, so the hind wings have a crease along their length that allows the ends of the hind wings to fold toward the body, thus fitting under the front wings.

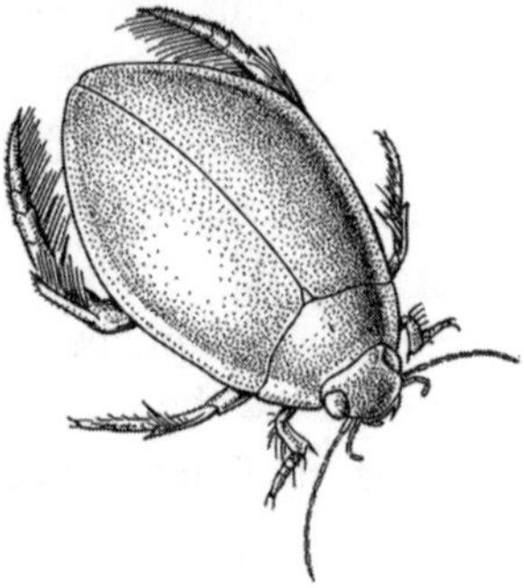

Figure 30. Representative water beetle adult.

Explanation of Names — The only common name that is widely recognized for the entire order is beetles. The common name is derived from the Old English word *bitula* (to bite) and refers to the prominent, strong jaws found on the adults in this order. The aquatic members of the order are simply called water beetles. All of the families have individual common names that are widely recognized, including the aquatic families. These are given in the parts on biology of common families. The scientific name of the order is a combination of two Greek words: *koleon*, meaning sheath, and *ptera*, meaning wings. This name refers to the hardened front wings that provide a protective cover for the hind wings and abdomen.

Habitat — Larvae and adults of water beetles generally share the same habitat and are often collected together. Almost all freshwater habitats are inhabited by some kind of water beetle, but many more species are found in still water than in flowing water. Aquatic plants in shallow lentic-littoral habitats are the prime location for members of this order. However, they are also found in other places within the lentic-littoral habitat, such as on muddy bottom or in accumulations of plant debris, as well as similar places within lotic-depositional habitats. The fewest kinds of water beetles are found in lotic-erosional habitats, but there are some kinds that are quite well adapted for life in currents. They live in all sizes of standing and running waters, including temporary habitats. Even though the adults of most species are strong fliers, they usually only fly when it is necessary to disperse to another location. Dispersal can be brought on by a habitat drying up, becoming too crowded, or not having enough food. Some species disperse from small to large lentic habitats at the onset of winter to be in a location where the water will not freeze all the way to the bottom.

Movement — Larvae of water beetles exhibit a wide range of habits for moving. Most kinds of water beetle larvae are categorized as climbers or crawlers, but it is also fairly common for larvae to be clingers or swimmers. Many adult water beetles are very effective swimmers. They move by powerful strokes of their hind legs (Figure 30), which are fringed with long hairs to help propel them through the water. Other kinds of adults are climbers, clingers, or crawlers.

Feeding — Water beetles are so diverse in their feeding habits that no trends can be summarized for the entire order, either for larvae or adults. All of the feeding categories that are mentioned in the introductory chapter, except shredder-detritivores, are common among the different kinds of water beetles, either for larvae or adults.

Breathing — Most water beetles have a closed breathing system during the larval stage. The primary mechanism for respiration is for dissolved oxygen to diffuse in over the entire body surface. In some families, there are also gills to enhance breathing efficiency. If gills are present, they are usually the branched filamentous type. Some kinds of water beetle larvae have an open breathing system with a pair of functional spiracles at the end of the abdomen. These larvae position themselves with the end of their abdomen touching the water surface to breathe air. Water beetle larvae do not have extended breathing tubes.

Adult beetles that are aquatic have an open breathing system just like terrestrial species. Almost all adult water beetles use transportable air stores for breathing below the water surface. The hardened front wings (elytra) provide an ideal means for holding a bubble of air. The abdomen of adult water beetles is constructed so that there is a hollow space between the wings and the top of the abdomen. The spiracles are modified so that they open on the top of the abdomen, rather than on the sides. Thus, a bubble of air can be retained in the space under the wings, and the adult beetle can breathe out of it through the spiracles. Some adult water beetles also have patches of non-wettable hairs on the sides and bottom of the body that hold a very thin film of additional air. This film of air gives them a silvery appearance while they are underwater. In a few families of water beetles, most notably riffle beetles (Elmidae) and long-toed water beetles (Dryopidae), this film of air functions as a plastron. A plastron is so thin and has so much surface area that the gaseous oxygen it contains is constantly in equilibrium with the dissolved oxygen in the water, and these adults never have to leave the water to breathe.

Life History — Almost all water beetles produce one generation per year. Adults live a long time, more than a year for some kinds. All kinds pass the winter as adults, hence, adult water beetles can be found throughout the year.

Females lay the eggs below the surface of the water, either singly or in small masses. They usually place the eggs on solid objects, with living plants being the site that is used most often. Some water beetles lay their eggs on the bottom in soft mud. Usually there is no covering for the eggs, but some species insert the eggs into plant tissue. In a few kinds of water beetles, the females have silk glands that they use to place a thin layer of silk over individual eggs or to place groups of eggs in a silk cocoon. In most of the families, egg laying takes place in the spring, and the eggs hatch in 1–2 weeks. Hardly any water beetles undergo diapause in the egg stage.

Water beetles usually complete their larval development in the next 6–8 months of warm weather. Almost all kinds shed their skin just three times as larvae, but a few kinds molt more often, up to as many as eight times. There is no diapause in the larval stage. When larval development is finished in the late summer or fall, the mature larvae leave the water for pupation. Most species crawl a short distance on the shore and dig a cell in damp soil. The cells are usually less than 5 m from the water. Sometimes the water beetle larvae crawl under solid objects, such as rocks or logs, to make their cells. The cells are discrete cavities with fairly smooth surfaces on the inside, but no silk is involved in the construction. The larvae choose soil or mud that is damp enough to be made smooth, but not so damp that it will collapse. Some kinds of water beetles make their pupal cells on top of the ground or up on the stalks of plants. Those kinds that construct their pupal cells above ground pick up small pellets of damp mud with their mouthparts and stick them together to form a dome-like structure.

Water beetle larvae molt into pupae within the cells. Many of the pupae have long hairs or spines on their bodies. These function to hold the pupae off of the damp soil, presumably to prevent damage that would occur if the body became soaked with water. After a few weeks as pupae, they transform into adult water beetles within the pupal cell. After molting to the adult stage, they remain in the cell for another 1–2 weeks in a teneral condition. During this time the adult skin hardens and acquires its final colors. After this process is complete, the adults dig and crawl through the soil to the surface. The total amount of time that water beetles spend in the pupal cell, from the time it is completed by the mature larva to the time that the adult digs out, is about 4–6 weeks.

Adult water beetles emerge from their pupal cells in the fall, but they do not mate and lay eggs until the following spring. This life cycle pattern is very similar to the aquatic true bugs (Hemiptera). Most of the water beetles hibernate as adults in the mud at the bottom of their habitat over the winter. Some remain active during the winter, even under ice. Most of the water beetles live a long time as adults, usually about 1 year. Thus, in late summer and fall, populations of adults from the previous year often live together with populations of adults that have recently emerged. New-generation adults are often shinier than the old-generation adults that have overwintered. Some species live as long as 3 years as adults. The exceptionally long-lived water beetles are usually the very large species in the families of predaceous diving beetles (Dytiscidae) and water scavenger beetles (Hydrophilidae).

Some kinds of water beetles exhibit very interesting behavior, however, behavior is quite varied and not the same in all families throughout this large and diverse order. Any unusual behaviors will be explained in the parts on the biology of individual families.

Significance — Although many terrestrial beetles are serious pests of agricultural crops, only a few of the water beetles cause such problems. Most water beetles are considered to have little significance. A few are deemed to be beneficial.

A few species of water beetles that feed as shredder-herbivores attack cultivated rice and cause significant economic losses. Chemical insecticides are often necessary to protect rice crops from these pests. Beneficial uses have been found for other water beetle species that are shredder-herbivores. Alien species of aquatic plants have invaded North America, particularly in the southern United States. These imported weeds become serious pests by clogging lakes and canals, thereby eliminating native species, preventing navigation, and disrupting significant economic gains from recreation. Species of water beetles that feed on these aquatic weeds have been located in other parts of the world and imported into North America for use as biological control agents. These programs are slow and expensive to get started because the biological control agents have to be carefully studied before they can be released, and it takes years for their populations to become high enough to reduce the pest weeds. However, once established, the biological control agents are self-perpetuating and it is not necessary to use chemical herbicides. The water beetles that are pests of rice or useful biological control agents are mostly in two families, leaf beetles (Chrysomelidae) and weevils (Curculionidae). These two families are not included in the guide because most beginning aquatic biologists are not likely to come upon them.

Although water beetles seem to be in almost all habitats, some individual species have very narrow environmental requirements and are found only in a particular type of natural, undisturbed habitat. Thus, an inventory of water beetle species can provide useful information on the uniqueness of aquatic habitats and help decide if habitats are worthy of protection by special conservation measures. In Great Britain, aquatic biologists have developed a system for using species of water beetles to categorize the conservation status of special lentic habitats, including ponds, swamps, marshes, and bogs. Water beetles are also useful for biomonitoring pollution because various kinds range from being facultative to sensitive to substances introduced into the water or to other types of environmental stress caused by human activities.

Biology of Common Families

Crawling Water Beetles

Family Haliplidae; Plates 28, 79

Different Kinds in North America — Few; 4 genera, 67 species.

Distribution in North America — Throughout.

Habitat — Larvae and adults - primarily lentic-littoral, also lotic-depositional. Crawling water beetles are common among dense growths of aquatic vegetation in ponds, shallow regions of lakes, and pools of flowing waters. They are frequently associated with filamentous algae such as *Spirogyra*, macroalgae such as *Chara* or *Nitella*, and aquatic vascular plants such as *Ceratophyllum* or *Elodea*.

Movement — Larvae – crawlers, climbers; adults – crawlers, climbers, swimmers. Larvae of crawling water beetles move slowly and feign death when disturbed. The elongate body, often with long filaments, and slow movements make the larvae very well camouflaged in their habitat. Novice aquatic biologists will have a hard time distinguishing these larvae from the plants and debris where they reside. Adult crawling water beetles are not particularly good swimmers, and they move slowly and clumsily through the water. Adults spend most of their time climbing in submerged aquatic vegetation and crawling around the bases of the plants, hence the common name of the family.

Feeding — Larvae and adults – primarily shredder-herbivores, also piercer-herbivores and engulfer-predators. Many kinds of crawling water beetles feed on filamentous or macroalgae, either biting and chewing or piercing cells and sucking juices. Some species seem to be able to exist only on one particular type of algae. Young larvae may not be able to chew through the strong cell walls of these algae, in which case they eat the material that grows on the outside, such as single-celled algae, fungi, and bacteria. Some larvae and adults have been observed to feed on small animals, but this is probably not a regular part of their diet.

Other Biology — Larvae of crawling water beetles breathe over the whole body surface, but they develop functional spiracles toward the end of the larval period of their lives. The spiracles are probably for the larvae to breathe air when they leave the water for the terrestrial pupal stage. Like many water beetles, the adults of crawling water beetles breathe from an air bubble carried in the space under their front wings. However, adult crawling water beetles are unique in that they also carry an air bubble underneath a segment of the hind legs (coxa) that is flat and greatly expanded (Plate 28A). The two bubbles held under the wings and legs are connected. Most of their underwater breathing comes from the air held under the wings. The air held by the legs of adult crawling water beetles is used as a reserve supply and for maintaining positive buoyancy. This extra air allows them to stay submerged longer and to float to the surface when more air is needed, rather than swimming to the surface. Some kinds of crawling water beetles spend the winter as larvae, rather than adults. Overwintering larvae move out of the water into damp soil near the edge of their aquatic habitat.

Stress Tolerance — Somewhat tolerant.

Long-Toed Water Beetles

Family Dryopidae; Plate 31

Different Kinds in North America — Few; 5 genera, 13 species.

Distribution in North America — Throughout.

Habitat — Adults – most lotic-erosional, a few lentic-littoral; larvae – terrestrial (not included in this guide). Long-toed water beetles are unique among insects in that they have aquatic adults and terrestrial larvae. Adults of the most common species occur exclusively in riffles of shallow streams on objects such as woody debris or rocks. Other less commonly encountered kinds of adults are in swamps, ponds, or depressions in wet woodlands, where there are accumulations of decomposing leaves. Larvae of long-toed water beetles live on land in damp soil or decaying wood or vegetation.

Movement — Adults – clingers. Long-toed water beetles most often use the long, sharp claws at the ends of their legs to adhere tightly to woody debris and sometimes to rocks in flowing waters. Riffle beetles (Elmidae) are often found clinging to the same objects. Sometimes long-toed water beetles crawl about on the bottom, or along the shore.

Feeding — Adults – scrapers and collector-gatherers.

Other Biology — Long-toed water beetles have not been studied a great deal. They use plastron breathing, so they need not come to the surface for air. Adults may leave the stream and fly about, especially at night. Females have a sword-like structure for laying eggs (ovipositor) and put the eggs in soil or plant tissue. The life cycle of long-toed water beetles is completed in a few months in warm climates but may take up to 2 years in cooler climates. Adults are long lived.

Stress Tolerance — Facultative.

Predaceous Diving Beetles

Family Dytiscidae; Plates 32, 82

Different Kinds in North America — Many; 44 genera, 509 species.

Distribution in North America — Throughout.

Habitat — Larvae and adults – primarily lentic-littoral, also lotic-depositional. Different kinds of predaceous diving beetles can be found in every type of aquatic habitat, but most are found in shallow, weedy areas in small bodies of still water, such as the margins of ponds and slow streams. Other habitats include tiny seeps, temporary puddles, potholes in rock outcrops, water-filled holes of land crabs, mineral springs, ditches, bogs, marshes, swamps, and brackish waters.

Movement — Larvae – climbers, swimmers; adults – swimmers. Larvae of predaceous diving beetles spend most of their time climbing about on plants or debris, but they can swim by flexing their body from top to bottom if they need to. Adults are fast and agile in the water because they move both hind legs in unison, like rowing a boat. The strokes are made more powerful by a fringe of long hairs on the legs, like an oar or paddle. Adult predaceous diving beetles often rest with the tip of their abdomen touching the water surface and their head hanging down in the water. Adults may leave the water at night and are often attracted to lights.

Feeding — Larvae – piercer-predators; adults – engulfer-predators. Larvae of predaceous diving beetles are also called water tigers, and they feed in a unique way compared to other water beetles. They have sharp, sickle-shaped jaws with channels on the inner margins (Plate 82B). Water tigers stick their mouthparts into a prey organism and inject a brown fluid from their mouth through the channels into the body of their prey. The fluid quickly kills the prey then slowly dissolves its insides. When the internal parts of the prey have been converted to a liquid state, the water tigers withdraw the fluid with suction through the channels in their jaws, eventually leaving only the empty skin of their unfortunate prey. Adult predaceous diving beetles have normal chewing mouthparts and feed by catching and holding their prey with their legs while they tear it into pieces with their mouthparts. Larvae and adults are voracious eaters and will consume anything that they can catch and subdue. The size of prey ranges from small organisms, such as worms, leeches, crustaceans, and true fly larvae, up to large invertebrate predators, such as dragonfly larvae, and even vertebrates, such as tadpoles, salamanders, and fish. Some of the large species of predaceous diving beetles are occasionally destructive to fingerlings in fish hatcheries. Adults also scavenge dead animals, including vertebrates such as fish.

Other Biology — Predaceous diving beetles, accounting for about one-half of all water beetle species, are the most diverse family and the one best adapted for aquatic life. In standing-water habitats, the vast majority of water beetles will belong to this family or the water scavenger beetles (Hydrophilidae). Most predaceous diving beetle larvae breathe air by two spiracles on the end of their abdomen. They must periodically make contact with the water surface, although they often rest with the end of their abdomen continuously touching the water surface. Some larvae in this family obtain dissolved oxygen by gills or special areas of their body surface and do not come to the surface to breathe. Adult predaceous diving beetles breathe from an air bubble carried under their front wings. They renew the bubble by breaking the surface of the water with the end of their abdomen. If you stop and observe the surface of standing-water habitats, you can distinguish predaceous diving beetle adults from the adults in the other ubiquitous family, water scavenger beetles (Hydrophilidae), by the way they renew their air bubble. Adult water scavenger beetles come head first to the water surface. Adult predaceous diving beetles live for 2 or 3 years.

Stress Tolerance — Mainly facultative, others somewhat tolerant.

Riffle Beetles

Family Elmidae; Plates 30, 81

Different Kinds in North America — Many; 24 genera, 97 species.

Distribution in North America — Throughout.

Habitat — Larvae and adults – primarily lotic-erosional, also lentic-littoral. Riffle beetles are common inhabitants of the swifter portions of streams and small rivers. They are most abundant and diverse in clear, cool waters. Stones, such as cobble, pebbles, and gravel, are the most common substrate for most species, but some species live in crevices or under the bark of decaying woody debris, in water moss, or on exposed roots. A few species of riffle beetles live in ponds and lakes on aquatic vegetation, and a few inhabit the sandy bottom of slow sections of streams.

Movement — Larvae and adults – primarily clingers, also climbers (lentic species). Riffle beetles are efficient clingers by virtue of their long, sharp claws at the end of the legs and their small, compact, hard bodies. When they do move, it is very slowly.

Feeding — Larvae and adults – scrapers, collector-gatherers. Most species of riffle beetles feed on periphyton, but some depend more on detritus.

Other Biology — Riffle beetle larvae breathe dissolved oxygen with gills that are on the end of their abdomen in a pocket with a door (Plate 81B). They protrude the gills out in the water and wave them to obtain dissolved oxygen. They withdraw the gills into the pocket in their abdomen and close the door to protect them from abrasion by sediment carried in the moving water. Adult riffle beetles breathe by means of a highly developed plastron, with microscopic-length hairs as dense as several million per square millimeter of body surface. This plastron is so efficient that most riffle beetle adults never have to come to the surface for air again after they enter the water. Most riffle beetles require a lot of oxygen and are only found in waters with dissolved oxygen at or near the saturation point. Larvae are different from most other kinds of water beetles because riffle beetle larvae shed their skin six to eight times, instead of the usual three times. Most riffle beetles spend 1 or 2 years as larvae, but some species take up to 3 years to complete the larval stage. Newly emerged adult riffle beetles undergo a short flight period, but after they enter the water they lose the ability to fly. The unneeded hind wings progressively waste away by some unknown process. Adult life spans are not known, but riffle beetle adults are thought to be long lived. It is speculated that some species do not reach sexual maturity until their second year of adult life, and some may live on into a third year.

Stress Tolerance — Mostly facultative, others somewhat sensitive.

Water Pennies

Family Psephenidae; Plate 78

Different Kinds in North America — Few; 6 genera, 16 species.

Distribution in North America — The single most common and abundant species, *Psephenus herricki*, occurs only in the East (Georgia to Oklahoma and north into Canada), but other less common species of water pennies occur throughout the United States (especially the mountainous regions of the Southwest) and adjoining provinces of Canada.

Habitat — Larvae – lotic-erosional; adults – terrestrial (not included in this guide). Water penny larvae occur on stones in areas of riffles with moderate to fast current. Occasionally they are found on rocks along the wave-washed shores of lakes.

Movement — Larvae – clingers. Water pennies are very effective at holding on to rocks, because the thin, flat plates extending away from their body are flexible and collectively assume the shape of whatever surface they are on. In addition, their grip on rock surfaces is made tighter by a dense fringe of short, fine hairs around the outer edge of the extended plates. Water pennies are seldom dislodged into a net merely by moving rocks. They have to be picked from the rocks with forceps or fingernails, and even then they are sometimes hard to remove.

Feeding — Larvae – scrapers. Water pennies are highly adapted for removing the thin layer of algae, especially diatoms, that occurs on stones in swift current. Their jaws have a thin, sharp inner edge, much like a paint scraper. The cupped shape of the jaws, along with hairs at the bases, help push the dislodged material into their mouths. Water pennies feed under the protection of the extended body plates, so the current does not wash their food away.

Other Biology — Water pennies obtain dissolved oxygen through gills on the underside of the abdomen (Plate 78B), as well as through the general body surface. During the day they reside underneath stones, then at night they move around to the top, where the most nutritious algae is located. Water pennies take 1 or 2 years to complete their life cycle. Mature larvae crawl out of the water a short distance to pupate in protected locations on rocks. The pupa is further protected under the last larval skin, which is tightly sealed to the rock surface. Little is known about the adults. They are thought to be short lived and probably do not feed. Adults of water pennies are usually observed on the sides and bottoms of rocks and logs just above the water surface in riffles, where they congregate for mating in summer. They appear to be attracted to protruding rocks in splashing water. Females enter the water to deposit their eggs in small patches (about 5 x 7 mm) on stones. Each patch contains 400–600 bright yellow eggs in a single layer.

Stress Tolerance — Facultative. Like most clingers, water pennies cannot persist in habitats where the rocks acquire a thick layer of algae, fungi, or inorganic sediment. However, they are somewhat tolerant of metal pollution.

Water Scavenger Beetles

Family Hydrophilidae; Plates 33, 83

Different Kinds in North America — Many; 20 genera, 192 species.

Distribution in North America — Throughout.

Habitat — Adults and larvae – primarily lentic-littoral, also lotic-depositional. Water scavenger beetles may be found in every still, shallow habitat, including ponds, margins of lakes, pools and other slow-moving sections of streams, puddles, and ditches. They are more common close to the shoreline where there is a lot of aquatic vegetation. Water scavenger beetles occur in the same habitats as predaceous diving beetles (Dytiscidae), and are often collected simultaneously. A few species in this family are terrestrial and occur in dung or decaying vegetation.

Movement — Larvae – climbers; adults – swimmers. Neither the larvae nor the adults of water scavenger beetles move as effectively as the predaceous diving beetles. Larvae tend to spend most of their time lying in wait for prey or crawling slowly. Adult water scavenger beetles swim by moving their legs alternately.

Feeding — Larvae – engulfer-predators; adults – collector-gatherers, engulfer-predators. Larvae of water scavenger beetles are almost exclusively predaceous. They are voracious and consume many kinds of aquatic organisms. Some kinds of larvae even crush and consume snails, shells and all (note their large jaws with teeth in Plate 83B). The common name for this family was probably conceived because the adults are often observed feeding on the bodies of decomposing vertebrates, especially fish. However, adult water scavenger beetles, like their larvae, also eat a considerable amount of live invertebrates.

Other Biology — Water scavenger beetles are the second largest family of water beetles, only exceeded by predaceous diving beetles (Dytiscidae) in the number of species. In standing-water habitats, the vast majority of water beetles collected will belong to one of these two families. Almost all larvae of water scavenger beetles breathe directly from the air by means of spiracles on the end of their abdomen. Adults take an air store underwater. The air store consists of a bubble under the wings as well as a thin layer held in non-wettable hairs on the bottom of the body. In well-oxygenated water, water scavenger beetle adults can remain submerged for long periods because the thin air store acts as a physical gill. This gives them a silvery appearance while underwater. When renewing their air store, water scavenger beetle adults break the water surface headfirst. They project only one antenna through the water surface into the air. Fresh air with oxygen rushes down the antenna and around the body to refill the space under the front wings. If you stop and observe the surface of standing-water habitats, you can distinguish water scavenger beetle adults from the adults in the other ubiquitous family, predaceous diving beetles (Dytiscidae), by the way they renew their air bubble. Predaceous diving beetles bring the end of their abdomen to the water surface. In most species of water scavenger beetles, females enclose their eggs in a case of hardened silk, which is produced by glands in the abdomen. The egg case is either carried by the female, wrapped in a leaf, fastened to plants or debris, or left floating free. In some species of water scavenger beetles, the egg cases are elaborate structures with tubes that protrude above the water or long ribbons that extend down into the water.

Stress Tolerance — Chiefly somewhat tolerant, others very tolerant.

Whirligig Beetles

Family Gyrinidae; Plates 29, 80

Different Kinds in North America — Few; 4 genera, 56 species.

Distribution in North America — Throughout.

Habitat — Larvae and adults – lentic-littoral, lotic-depositonal. Whirligig beetles occur commonly in streams, rivers, ponds, and lakes, especially in the quieter waters near the margins. Adults are often conspicuous on the water surface. Larvae live underwater, usually on aquatic plants, in the same habitat. The larvae of whirligig beetles are difficult to find, even though many adults may be observed.

Movement — Larvae – climbers, swimmers; adults – surface swimmers. Larvae of whirligig beetles usually climb about on submerged vegetation or debris. Adults are found on the water surface congregated in dense groups, sometimes called "schools." They get their common name from the behavior of swimming rapidly in circular patterns. Up to 13 species of whirligig beetles have been observed swimming together in the same group. They are the only beetles that spend most of their time on the water surface. Whirligig beetles swim by means of short, flat middle and hind legs (Plate 29B). They swim exceptionally fast, up to 1 m per second. Adults are equally adept at underwater movements and dive quickly when disturbed.

Feeding — Larvae and adults – engulfer-predators. Larvae of whirligig beetles are predators on small aquatic invertebrates, especially worms and midge larvae. They are also cannibalistic. Adults feed upon small animals trapped on the water surface, or they scavenge any organic material found there. The waves that adult whirligig beetles generate while swimming may function for the echolocation of food, much like the sonar used by ships at sea.

Other Biology — Larvae of whirligig beetles breathe dissolved oxygen in the water. The conspicuous filaments on the sides of the abdomen function as gills, but larvae also obtain dissolved oxygen over the entire surface of their soft, fleshy bodies. Adult whirligig beetles breathe from the atmosphere when they are swimming around on the water surface. When they dive underwater, they take a bubble of air under their front wings. Adult whirligig beetles have eyes that are specially adapted for living on the water surface (Plate 29B). Each eye is divided on the side into two halves, one looking up and one looking down. The bottom pair of eyes serve for underwater vision, while the top pair allows them to see in the air. Even though adult whirligig beetles spend most of their lives in plain view on the water surface, fish and other predators do not eat them. They emit defensive secretions that somehow repel potential predators, probably just by making them taste bad. In some common species of whirligig beetles the defensive secretion smells like ripe apples and is easily detected when handling these organisms. The defensive secretion is not harmful to humans. Mature larvae of whirligig beetles construct a pupal chamber rather than dig a cell in the soil. They use a sticky substance to glue together pieces of vegetation, mud, or sand into a dome-shaped structure. The pupal chamber of whirligig beetles is constructed on top of the ground near the water or on vertical surfaces of plant stems or rocks that emerge from the water. Wasps often parasitize the pupae.

Stress Tolerance — Facultative.

Caddisflies

Order Trichoptera; Plates 62 –77

Caddisflies is a medium-sized order, in terms of diversity, when compared to all of the insect orders. However, it is the largest order of insects that is entirely aquatic. There are 21 families and about 1,400 species of caddisflies in North America. This guide includes 14 families that are common in freshwater habitats. Caddisflies have complete metamorphosis. All members of the order are aquatic as larvae and pupae and terrestrial as adults. One of the major reasons for the success of caddisflies is that the larvae can make a silk thread. Details are provided throughout this chapter on various ways that silk enables caddisflies to thrive in different ecological conditions.

Distinguishing Features of Larvae — Caddisfly larvae have elongate, mostly cylindrical bodies. They resemble caterpillars, which are the larvae of moths and butterflies in the order Lepidoptera. The body length of mature larvae ranges from 2 to 43 mm in different species. Caddisfly larvae can be distinguished from other immature aquatic insects by the following combination of characteristics (Figure 31A, B, C):

- Head has thick, hardened skin.
- Antennae are so short that they are usually not visible.
- No wing pads occur on the thorax.
- Top of the first thorax segment always has a hardened plate, and in some kinds the top of the second thorax segment, or all three thorax segments may have hardened plates.
- Three pairs of segmented legs extend from the thorax.
- Abdomen has thin, soft skin.
- Single or branched gills are located on the abdomen in many kinds, but some kinds have no gills.
- Pair of prolegs with one claw on each is situated on the end of the abdomen.
- Larvae live in various types of portable cases or retreats attached to the substrate, except for the larvae in one family that live uncovered (Freeliving caddisflies, Rhyacophilidae).

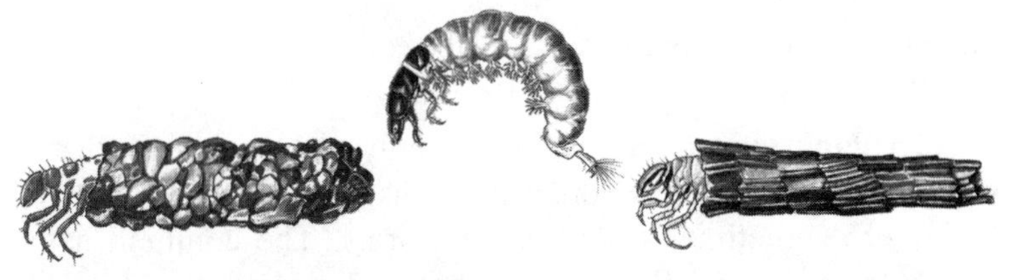

Figure 31. Representative caddisfly larvae.

Distinguishing Features of Adults — Adult caddisflies have bodies that are elongate, cylindrical, hairy, and soft. Their hairiness and manner of holding their wings make them look like moths in the order Lepidoptera. They are small to medium in size. The body length of different species ranges from 2 to 25 mm, measured from the front of the head (not including the antennae) to the tip of the wings folded at rest. Caddisfly adults can be distinguished from other adult aquatic insects by the following combination of characteristics (Figure 32):

- Long, thin antennae project in front of the head.
- Two long, hairy, finger-like structures (palps) extend down from head.
- Front and hind wings are elongate and similar in shape, with the hind wings slightly shorter.
- Wings are covered with very short hairs.
- Wings are held slanted and roof-like over the abdomen, when the organism is not flying.
- Tips of both pairs of wings, when folded, extend past the end of the abdomen.

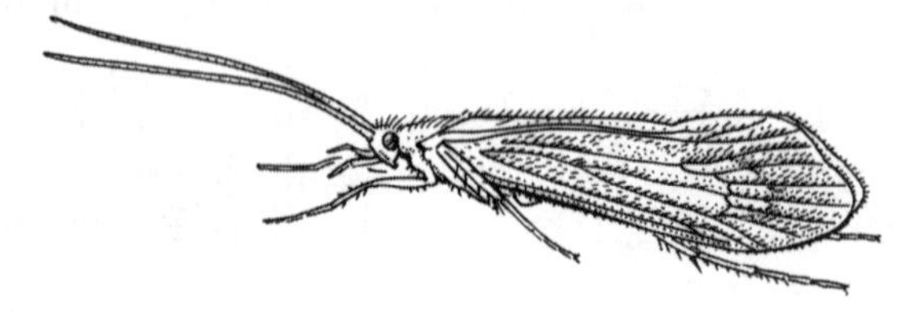

Figure 32. Representative caddisfly adult.

Explanation of Names — The only common name that is widely recognized for the order is caddisflies, or just caddis. The larvae are sometimes called caddisworms. The common name comes from the Middle English word *cadaz*, dating back to the early fifteenth century. Originally the word referred to cotton or silk used as padding. Other spellings that sprang up in England during the fifteenth to seventeenth centuries included *cadace*, *caddis*, *cadice*, and *cados*. During this period the word changed to mean worsted yarn, which is a smooth compact yarn made from long wool fibers, and finally it meant tape or ribbon made out of worsted yarn. In those days, various types of peddlers roamed the countryside in England, calling on small villages and isolated cottages. Those who sold cloth goods, including ribbon (*cadice*), routinely attached many small samples of their inventory onto their coats to entice customers. This early type of advertising earned them the name *cadice men*. The use of caddis as a common name for this order refers to the habit of almost all larvae of making a dwelling by gluing small pieces of various substances together, akin to the coats of *cadice men*. Common names that might be used in certain localities for adult caddisflies are sedges, sandflies, and shadflies. The latter common name is also used for mayflies (Ephemeroptera). Sedges is a term used by fly anglers. Some local names for larvae in particular families and genera are stick bait (Limnephilidae: *Pycnopsyche*) and rock rollers (Rhyacophilidae: *Rhyacophila*). The scientific name of the order is a combination of two Greek words: *trich*, meaning hair, and *ptera*, meaning wings. The scientific name refers to the very small hairs that are abundant on the wings.

Habitat — Different kinds of caddisfly larvae live in all types of lotic and lentic habitats. They also live in all sizes of these habitats, including temporary ones. Every family has some species that inhabit small to medium sized, cool streams. There are progressively fewer families of caddisflies represented in large, warm rivers, standing waters, and temporary habitats.

Movement — Caddisfly larvae are equally as diverse in their habits as they are in their habitats. Their successful adaptations to move about or remain in place are generally associated with their use of silk. Many kinds of caddisfly larvae build elongate tubes out of pieces of live plants or detritus. These tubes are portable cases that the larvae reside in and drag with them as they climb about on aquatic plants or in accumulations of plant debris in lentic-littoral and lotic-depositional habitats. They cut the pieces of vegetative material for their cases with their mouthparts. These cases protect the larvae by providing a physical barrier and camouflage, but they are not too difficult to drag around because they are elongate and have neutral buoyancy.

Some kinds of caddisfly larvae are very effective clingers in lotic-erosional habitats by either of two approaches that use silk. Some kinds make portable cases out of coarse sand or pebbles, and the weight of the cases helps to keep them from being displaced by the current (Plates 67, 68, 70). Some of these heavy cases are also streamlined to reduce the force that the current exerts on them. The caddisfly larvae in the cases assist by holding on with their sharp claws. Other kinds of caddisfly larvae are very effective clingers by constructing attached retreats out of silk (Plate 74). These structures are glued firmly to immovable objects, such as large rocks or woody debris, in lotic-erosional habitats. Some can withstand the force of very swift current, but many are built in small crevices that escape the brunt of the force from the current. Attached retreats usually have to be observed in their natural position because they collapse and come apart if you attempt to dislodge them from their attachment site or if you remove the object they are attached to.

A few kinds of caddisfly larvae are able to sprawl on top of soft sediment by virtue of lightweight portable cases that have a large surface area. Some of the sprawler cases are constructed of pieces of dead leaves and twigs. A few kinds use fine sand to construct a tubular case with a flat, thin extension on the front and sides that keeps them from sinking in the soft sediment, much the same as snowshoes keep people from sinking in deep snow. Lastly, a very few caddisfly larvae are able to burrow in soft sediment. They use their silk to construct long tubes that they live in without their bodies coming in contact with the sediment.

Feeding — Almost all of the ways of acquiring food are used by some type of caddisfly larvae. The types of feeding that are common among the different kinds of caddisfly larvae include shredder-detritivores, shredder-herbivores, collector-gatherers, collector-filterers, and scrapers. A few kinds are engulfer-predators, either stalking their prey or lying in wait to ambush it. A very few kinds of caddisfly larvae are piercer-herbivores. These are some small larvae in the family of microcaddisflies (Hydroptilidae) that bite through the cell walls of filamentous algae and consume the liquid contents.

Silk is not always involved in larval feeding, but there are some examples of its use here too. The most common collector-filterers obtain their food by constructing fine-mesh silk nets that strain edible particles from the water as it passes through. Other caddisfly larvae use hairs on their legs to filter water, just like brushlegged mayflies (Ephemeroptera: Isonychiidae), but these caddisflies have the additional advantage of being able to attach with their silk. Caddisfly scrapers are able to maintain their position over the best periphyton in swift current by means of the heavy stone cases that they construct with silk. Some of the caddisfly larvae that are ambush-type engulfer-predators lie in wait hidden in soft, flat silk structures that blend in with the bottom.

Breathing — Caddisfly larvae have a closed breathing system. The primary way they breathe is by dissolved oxygen diffusing into their body across all of the soft, fleshy tissue. In some families, there are also gills to improve the efficiency of acquiring dissolved oxygen from the water. If gills are present, they are either individual or branched filaments. The ability to produce silk also gives caddisfly larvae an advantage over other organisms for acquiring dissolved oxygen. Almost all larvae reside in cases or attached retreats that are essentially a tube with a diameter not much wider than their body. By wiggling their body inside the case or retreat, they create a current of water from front to back. Thus, larvae are able to generate a constant, one-way current of water inside their home, even though they may live in a habitat where there is little or no current. This guarantees they always have a fresh supply of dissolved oxygen. Caddisfly larvae would not be able to improve their ability to breathe in this manner if they did not produce silk.

Life History — Most kinds of caddisflies have one generation per year. Some produce two generations per year, and some require 2 years for a generation to develop. Adults of most caddisfly species emerge from late spring to early fall. A very few species emerge in winter, some of which are wingless.

Females of different caddisfly species lay their eggs either in a soft, gelatinous mass or bare, without any covering. The number of eggs varies from just a few to about 800. Egg masses are sometimes a bright color, such as green or yellow-orange, and may be shaped like a sphere, an ellipse, a flat spiral, or a doughnut. Almost all caddisflies lay their eggs in the water. Some fly down to the water surface and scatter individual eggs or an egg mass. Individual eggs and egg masses become sticky when wet and attach to immovable objects, such as stones, sticks, or plants. In other species of caddisflies, the females crawl under water and put the eggs directly on solid objects, usually stones. The dense hairs that are characteristic of adult caddisflies are non-wettable and keep the females from becoming water logged while crawling around laying eggs. The females have a silvery sheen while underwater because they are covered with a thin layer of air. A very few kinds of caddisflies, mostly in the family of northern case makers (Limnephilidae), deposit an egg mass above the water on vegetation. This is similar to the egg laying habits of dobsonflies, fishflies, and alderflies (Megaloptera). These kinds of caddisfly larvae must drop into the water as soon as they hatch from the egg. Eggs usually hatch within a few weeks, but some kinds of caddisflies undergo diapause in the egg stage, which can last as long as 10 months. Most caddisflies that diapause do so to avoid adverse conditions presented by hot, dry weather in summer.

The duration of the larval stage can be as short as 2 or 3 months or as long as 2 years, according to the species of caddisfly. Most kinds spend the winter as active larvae. In a few kinds, the larvae diapause during the summer months when the water is warm and stream flows are low. Almost all caddisflies shed their skin five times as larvae, but a few species molt six or seven times.

The most conspicuous and interesting behavior of caddisflies is the building of individual homes by larvae. Caddisflies are closely related to butterflies and moths (Lepidoptera), and one of the features that they have in common is that the larvae have silk glands in their lower lip. Caddisfly larvae in 20 of the 21 families use strands of silk to make portable cases or attached retreats. The larvae use these structures for their abode, and they gain advantages for survival, such as camouflage, physical protection, food acquisition, and respiratory efficiency. Each species of caddisfly larvae always constructs the same type of abode. They use the same materials and proceed through the same sequence of steps during construction. Cases and retreats are sometimes characteristic for an entire genus, or in some instances an entire family. In order for caddisfly larvae to build these dwellings they must choose the correct material and size of the components, or, for vegetative material, they must chew the components into the correct size and shape.

When larval development is complete, mature larvae in all caddisfly families prepare a cocoon in the water for the protection of the pupa. Larvae that live in swift current migrate to areas where there is only moderate current, such as near shore or underneath loose rocks. For the kinds of caddisfly larvae that live in a portable case or an attached retreat, the cocoon is a modification of their dwelling. Case-making caddisfly larvae use silk to glue their case to a solid, immovable object, usually a large rock. They often shorten the case before attaching it. Caddisfly larvae construct a porous sieve plate out of silk in the rear opening of the case, so that water can circulate but predators cannot enter. Finally, they plug the front end with stones or plant material and glue it in place with silk. Retreat-making caddisfly larvae use their mouthparts to cut away any feeding nets or excess sections of their retreat, then they solidify the remaining part of the retreat with sand or stones. This is sealed to a solid, immovable object, usually a large rock. Even the one caddisfly family that does not make a portable case or attached retreat as larvae (Rhyacophilidae) makes a cocoon out of silk and stones for the pupal stage.

After sealing themselves in cocoons, mature caddisfly larvae lie still for several weeks as prepupae. This is very similar to what happens with mature larvae of dobsonflies, fishflies, and alderflies (Megaloptera). The body becomes shorter and stouter. The soft skin acquires a leathery texture, and the legs are held at various odd angles. Eventually, the mature larva sheds its skin and the pupa emerges. These aquatic caddisfly pupae breathe the same way as the larvae, even though pupae are sealed in a cocoon. The pupae also wiggle their bodies and set up a current of water through the cocoon. Water enters through the walls of the cocoon and exits through the sieve plate at the end. The pupal stage of caddisflies lasts 2–3 weeks. Pupae emerge from their cocoon before molting to the adult stage. Most caddisfly pupae have sharp jaws for this purpose. After cutting a hole in their cocoons, caddisfly pupae float to the water surface assisted by wiggling their bodies. Many kinds emerge from the water surface by using the pupal skin as a raft for support, while other kinds squirm slightly out of the water onto a solid object.

After adult caddisflies emerge from the pupae, they fly to nearby vegetation where they spend most of their time. They live about 30 days. Most adult caddisflies fly only at night (nocturnal), or perhaps at dawn or dusk. During the day, they remain still in vegetation near the water. They often rest in an upside down position. Adult caddisflies consume only liquid food, such as nectar, with their sponge-like mouthparts. Adult caddisflies have no tooth-like jaws, but they do have long, hairy, finger-like structures on some of their other mouthparts. These are effective for sponging up liquids. Adult males form mating swarms for visual communication with females. These swarms are similar to those in mayflies (Ephemeroptera), but caddisfly swarms are not as conspicuous.

Significance — Caddisflies are never considered serious pests, either as adults or larvae. Adults are sometimes a nuisance in localities near streams or rivers, because some species emerge in very high numbers. Caddisflies are beneficial as larvae.

Adult caddisflies are strongly attracted to lights after dark. Most caddisflies do not occur in high numbers, so they are hardly noticed. The main exception is some of the common netspinners (Hydropsychidae). Larvae of common netspinners are often very abundant in medium to large streams and rivers, so adults emerge during summer in numbers that are high enough to be troublesome. The attraction of adult caddisflies to lights can be a nuisance for outdoor recreation on warm evenings, such as at ball fields, picnic shelters, or swimming pools. They also ruin fresh paint when their hairy bodies get stuck in the paint before it dries. Some people are allergic to the hairs that break off of their bodies and become air-borne. Such nuisance problems have occurred in the United States in the Midwest along the upper sections of the Mississippi River and in the Northeast along the Niagara River. The decision on where to hold the 1901 Pan-American Exposition within the state of New York was supposedly influenced by the annoyance of adult caddisflies at one of the prospective sites.

Many caddisfly larvae are thought to play important roles in the function of freshwater ecosystems. Some kinds that are shred-ders are abundant in the small streams of deciduous forests. These caddisfly larvae are an important part of the invertebrate assem-blage that eat intact leaves and other large parts of plants that fall from the forest into the water. When caddisfly larvae shred this coarse material, they create very small pieces of organic matter that thereafter serve as food for other organisms. Conversely, some of the filter-feeding caddisfly larvae capture very fine particles of detritus suspended in the water and change it into larger particles of detritus in the form of their feces, which become food for other organisms. Caddisfly larvae are important food items for many types of fish in all habitats. Some birds, such as swallows, consume swarm-ing adult caddisflies.

Most species of caddisflies are sensitive to pollution, so the number of different kinds and their relative abundance compared to other invertebrates provide useful and reliable information about the environmental condition of a body of water. The EPT Index has become a widely used measure of environmental condition. This simple index is calculated as the number of different kinds of Ephemeroptera (mayflies), Plecoptera (stoneflies), and Trichoptera (caddisflies). The EPT Index is useful because most of the species in these three orders of aquatic insects are sensitive to pollution and environmental stress. Of these three sensitive orders, Trichoptera larvae are probably the second most useful for biomonitoring pollution. They are useful because different kinds of larvae occur naturally in many different habitats. However, some caution must be used in interpreting data on caddisfly larvae for biomonitoring purposes. One of the most widespread and abundant families, the common netspinners (Hydropsychidae), contains many species that are facultative towards pollution and environmental stress.

Caddisflies are commonly mimicked in various types of "flies" that are made or purchased by fly-fishing anglers. Fly fishing is a multimillion-dollar industry. Although they are not imitated as much as mayflies (Ephemeroptera) for this purpose, caddisflies are important enough for fly anglers to have given the adults a special name, sedges.

Biology of Common Families

Common Netspinner Caddisflies

Family Hydropsychidae; Plates 73, 74

Different Kinds in North America — Moderate; 12 genera, 149 species.

Distribution in North America — Throughout.

Habitat — Lotic-erosional. Common netspinners are restricted to flowing waters, but different kinds within the family occur in a wide variety of habitats from very fast, cold, mountain streams to large, warm, lowland rivers. The most common location for larvae of common netspinners is on rocks, or within spaces between rocks, that are large enough to be stable in the current, such as outcrops, cobble, and pebble. Water-soaked logs and other woody debris are also inhabited by larvae of common netspinners. They are often very abundant in any type of submerged plants that grow in flowing water.

Movement — Clingers. Common netspinner larvae live in tubular silk retreats that are attached to solid objects in current (Plate 74).

Feeding — Collector-filterers. Larvae of common netspinner caddisflies construct a mesh net exposed to the current near the opening of their tubular retreat (Plate 74). The net acts as a filter to remove particles of food from the water as it passes through the mesh. The filter net is either stretched across parts of the substrate, such as a rock crevice or plant stems, or is suspended on arms constructed by the larvae. The larvae of common netspinners periodically stick their head out of their retreat to eat whatever has been caught on the net. The mesh size and thread diameter of the filter net vary according to species and are determined by the physical structure of the lower lip where the silk comes from. Species of common netspinners living in fast current make coarse mesh nets composed of thick strands of silk. Species living in slow current make fine mesh nets composed of thin silk threads. Larvae of common netspinner caddisflies consume whatever is caught on their nets. Those with coarse mesh nets (250–500 microns) tend to feed more on small invertebrates and pieces of plants, while those with fine mesh nets (5–250 microns) tend to feed more on algae and fine detritus. Many different species can coexist in the same riffle by positioning themselves in the current velocity that is suitable for their nets.

Other Biology — Some of the common netspinners are the most abundant of all caddisfly larvae. Exceptionally high densities of these larvae usually occur in large rivers that carry a lot of fine detritus suspended in the water. Another likely site for high densities of common netspinner larvae is below dams where a lot of algae produced in the reservoir above the dam has been released into the river downstream. The density of one species of common netspinner has been measured at 47,000 larvae per square meter of stream bottom below a dam. Because of the high numbers of common netspinner caddisflies that occur close together, the larvae establish a territory and defend it against intruders. The first line of defense against an intruder is stridulation. Common netspinner larvae stridulate by rubbing sharp, pointed structures at the base of their front legs against the sides of their head, one side at a time. If stridulation fails to repel the intruder, physical aggression ensues. Larvae face each other, rear up the front end of their body, then nip with their mouthparts and grapple with their legs. Usually the larger of the two common netspinner larvae wins and takes possession of the retreat, while the loser slips away without any major damage. When mature larvae are ready to pupate, they construct a protective cocoon similar to their retreat. Some kinds cut away the filter net and fortify their existing retreat with sand and additional plant debris, then seal the opening. Other kinds of common netspinner larvae move to areas of slower current on the underside of a rock or other object where they reside. There, they make a strong, sealed retreat, usually of sand, to protect the pupa.

Stress Tolerance — Primarily facultative, others very sensitive to somewhat sensitive. The facultative kinds of common netspinners become very abundant in streams that are subjected to moderate levels of pollution from organic wastes or nutrients. If common netspinners account for a majority of the community, that is a reliable indicator of organic or nutrient pollution. These types of pollution put more particles of food in suspension, where it is readily captured on the mesh nets spun by the larvae in this family of caddisflies. Black fly larvae (Diptera: Simuliidae) are often the other dominant organisms under these circumstances.

Fingernet Caddisflies

Family Philopotamidae; Plate 75

Different Kinds in North America — Few; 3 genera, 42 species.

Distribution in North America — Throughout.

Habitat — Lotic-erosional. Fingernet caddisflies are most common in small, cool, clear streams, but some kinds also live in large, warm rivers. Larvae inhabit the swift sections of streams, but they are in protected places under rocks or logs, where the current is greatly reduced. Their silk retreats are very fragile and can hardly withstand any current.

Movement — Clingers. Larvae live in silk retreats that are attached to solid objects.

Feeding — Collector-filterers. Larvae of fingernet caddisflies construct filter nets with the finest mesh (<1–15 microns) of any of the caddisflies. This exceptionally fine-mesh net filters out the smallest particles of detritus that are suspended in the water. A slime consisting of fungi and bacteria grows on the material after it is filtered out. The fleshy, T-shaped upper lip has a row of delicate, hair-like spines on the front edge. Larvae of fingernet caddisflies gently rake the inside of their filter net with the front of their upper lip to remove the mixture of fine detritus and the microbes that have grown on it. Small animals that are caught in the net are also eaten, but that is not a regular part of their diet.

Other Biology — There are one or two generations per year, and adults of most fingernet caddisflies emerge throughout the warm months. There is one species that emerges in the winter, often as wingless adults.

Stress Tolerance — Mainly somewhat sensitive, others very sensitive.

Freeliving Caddisflies

Family Rhyacophilidae; Plate 77

Different Kinds in North America — Moderate; 2 genera, 127 species.

Distribution in North America — Throughout, wherever the terrain is steep enough for flowing waters.

Habitat — Lotic-erosional. Most kinds of freeliving caddisflies live only in clear, rapid, mountain streams, but some inhabit warm, slightly silty streams. They occur mostly in the fastest sections of riffles under loose, course mineral substrates (boulder, cobble, pebble). Sometimes they are in slower sections of streams, but always in current. Other substrates where they reside are within clumps of moss or algae.

Movement — Crawlers. Freeliving caddisflies are the only caddisflies in which the larvae do not live in a case or retreat. Larvae roam around freely under rocks and within spaces in the substrate. As they crawl, they spin a silk thread and attach it to the substrate, which helps them hold on to rocks in fast current.

Feeding — Mostly engulfer-predators; a few shredder-herbivores, shredder-detritivores, scrapers. The common kinds of freeliving caddisflies consume a wide variety of small invertebrates, probably anything they encounter and can subdue. Some of these have been observed feeding on fish eggs. A few kinds feed on plant material, including live submerged plants, detritus, and algae.

Other Biology — Freeliving caddisflies do construct a protective structure for the pupa. Mature larvae assemble small stones in a pile and glue them together with silk. Some of the stones may be larger than the larvae, and this habit has earned them the name "rock rollers." Freeliving caddisflies often attach this structure to the bottom of a large rock. If the large rock is lifted from the water, the pupal structure partially breaks up. After completing the rough dome of stones, the mature larva then spins a silk cocoon inside. The inner cocoon is a tough, brown, oval capsule. There is one generation per year, and adults of different freeliving caddisflies emerge throughout the warm months.

Stress Tolerance — Very sensitive. The lack of a portable case or attached retreat in this family probably lessens the ability of freeliving caddisflies to cope with pollution or other kinds of environmental stress.

Giant Case Maker Caddisflies

Family Phryganeidae; Plate 72

Different Kinds in North America — Moderate; 10 genera, 27 species.

Distribution in North America — Some occur throughout, but there are more kinds at northern latitudes and higher elevations.

Habitat — Lentic-littoral, lotic-depositional. Larvae of giant case makers are most common in marshes, ditches, ponds, lakes, and temporary pools. They are also in streams and rivers, but only in areas with little or no current. They are usually in submerged aquatic plants, terrestrial grasses that have bent over into the water, or accumulations of plant debris, especially fallen leaves.

Movement — Climbers. Larvae live in a portable case.

Feeding — Shredder-herbivores, engulfer-predators, shredder-detritivores. Most larvae of giant case makers have two feeding habits. They chew pieces of live aquatic plants and filamentous algae, and they also capture small invertebrates. Some kinds consume plants when they are young larvae, then switch to a diet of animals when the time for the pupal stage approaches. Animal matter contains more protein, which might be necessary for the tissues that develop during metamorphosis. Larvae of some kinds of giant case makers have been observed feeding on zooplankton. These caddisfly larvae whirl their front legs, which sets up a current that brings the small animals within their grasp. A few kinds of giant case makers feed on decomposing plant material, especially leaves from trees.

Other Biology — Giant case makers produce one generation per year, and adults of different kinds emerge throughout the warm months.

Stress Tolerance — Mostly facultative, others very sensitive to somewhat sensitive.

Humpless Case Maker Caddisflies

Family Brachycentridae; Plate 66

Different Kinds in North America — Few; 5 genera, 36 species.

Distribution in North America — Throughout.

Habitat — Lotic-erosional. Larvae of humpless case makers live in a variety of flowing waters, but usually in cool habitats ranging from spring-fed streams to small rivers. The different common kinds within this family generally live in either of two specific places within these habitats. Those in small, spring-fed streams reside in submerged moss growing on rocks. The common kinds in larger streams and small rivers attach their cases to the upper surfaces of stones, water-soaked logs, or tree branches trailing in the water.

Movement — Climbers, clingers. The larvae of humpless case makers that live in moss are climbers. Those kinds that attach their cases to the surfaces of various substrates are clingers. The latter orient the case facing into the current and attach only the front edge to the substrate with silk.

Feeding — Shredder-herbivores, collector-gatherers, collector-filterers, scrapers. The kinds of humpless case makers that live in moss eat living parts of that plant or browse on bits of organic matter that accumulate within the tangled clumps of moss. Those kinds that attach their cases to the surfaces of various substrates feed primarily by filtering small particles of organic matter from the water. Because the case is glued to the substrate with silk, these kinds of humpless case makers are free to hold their middle and hind legs up in the current. A fringe of long, fine hairs on these legs removes particles of food from the water. Periodically, larvae bend one leg at a time to their mouth and remove the accumulated material from the hairs. These kinds of humpless case makers also feed by scraping algae off of the substrate where they reside.

Other Biology — Humpless case makers produce one generation per year, and most common kinds emerge as adults in the spring.

Stress Tolerance — Very sensitive.

Lepidostomatid Case Maker Caddisflies

Family Lepidostomatidae; Plate 65

Different Kinds in North America — Moderate; 2 genera, more than 80 species.

Distribution in North America — Throughout.

Habitat — Most lotic-erosional and lotic-depositional, some lentic-littoral. Larvae of lepidostomatid case makers are most common in small, cool springs and streams, in areas where the current is slow. Some kinds live in the backwaters of large rivers and around the edges of ponds and lakes. They are usually found in places where plant debris, especially dead leaves, has accumulated on the bottom.

Movement — Crawlers, climbers. Larvae live in a portable case.

Feeding — Shredder-detritivores. Larvae of lepidostomatid case makers usually feed on fallen leaves that have softened and begun to decompose.

Other Biology — Lepidostomatid case makers produce one generation per year, and adults of different kinds emerge throughout the warm months.

Stress Tolerance — Very sensitive.

Longhorned Case Maker Caddisflies

Family Leptoceridae; Plate 64

Different Kinds in North America — Moderate; 8 genera, 113 species.

Distribution in North America — Throughout.

Habitat — Most common in lentic-littoral, but also common in lotic-depositional and lotic-erosional. The longhorned case makers are second only to the northern case makers (Limnephilidae) in diversity and widespread occurrence. They inhabit all types of freshwater habitats, but many common kinds tend to be more prevalent in warm, still waters. Their lentic habitats include marshes, ponds, small glacial lakes, large lakes, and reservoirs. Larvae of longhorned case makers are also found in small, cold, clear streams and large, warm, silty rivers. Many common kinds live among all types of submerged vegetation, but others occur on mineral substrates ranging from outcrops to sand and exposed roots of terrestrial plants.

Movement — Climbers, crawlers, sprawlers. Larvae of longhorned case makers live in a portable case. Most common kinds climb about in plant beds or crawl on the bottom. Some species use their long hind legs to swim about with their lightweight vegetative cases. Usually longhorned case makers only swim for a short distance to get from one plant to another. Some species are able to sprawl on soft sediment by virtue of flat extensions on their tubular cases.

Feeding — Engulfer-predators, shredder-herbivores, collector-gatherers. All of these feeding habits are common among the larvae of different kinds of longhorned case makers. Quite a few kinds consume small invertebrates, which is not a widespread feeding habit among the families of caddisflies. There are even a few species of longhorned case makers that are predacious on freshwater sponges. Many of the kinds that live in plant beds and make cases out of pieces of vegetation, also consume pieces of the plants as food.

Other Biology — Longhorned case makers produce one generation per year, and adults emerge throughout the warm months.

Stress Tolerance — Chiefly facultative, others sensitive to somewhat sensitive.

Micro Caddisflies

Family Hydroptilidae; Plate 63

Different Kinds in North America — Moderate; 15 genera, more than 220 species.

Distribution in North America — Throughout.

Habitat — Many lentic-littoral, some lotic-depositional, few lotic-erosional. Micro caddisflies are diverse in their habitats, occurring in all types of permanent waters including springs, streams, rivers, ponds, and lakes. They most frequently live among submerged aquatic plants or filamentous algae. Many kinds are common, but larvae of micro caddisflies are seldom noticed because they are small and camouflaged by their cases.

Movement — Mostly climbers, some clingers. Almost all mature larvae of micro caddisflies live in a portable case. Some kinds attach the case to the substrate.

Feeding — Piercer-herbivores, scrapers. The primary food of micro caddisfly larvae is algae. Most of the common kinds pierce individual cells and consume the fluid contents from strands of multicellular filamentous algae. Others scrape off and consume the thin layer of periphyton.

Other Biology — Micro caddisflies have an unusual life history. Like most caddisflies, the larvae shed their skin five times. During the first four of the larval stages, the larvae roam around freely without a case, and they have narrow bodies. In the fifth larval stage, the abdomen of micro caddisflies becomes very enlarged, the larvae take up residence in a portable case, and they do not move around nearly as much. The free-living stages are seldom collected because they are exceptionally small, they probably pass through these stages rather quickly, and they may live in different locations within the habitat during this time. Most of the time that micro caddisflies spend as larvae takes place in the fifth stage. During this time, the abdomen continuously increases in size. Micro caddisflies enlarge their same case to accommodate the growing abdomen. Those that make purse cases cut the bottom seam, freeing the two halves of the case, add more material to both sides, then glue the two halves back together again. Micro caddisflies produce one or two generations per year, and adults of different kinds emerge throughout the warm months.

Stress Tolerance — Primarily facultative, others very sensitive to somewhat tolerant. One facultative kind of micro caddisflies becomes very abundant in streams when nutrient pollution causes excessive growths of algae on rocks. These larvae are scrapers that can take advantage of the abundant food provided by the algae. They live in conspicuous flat cases that are glued very tightly to rocks, and they often occur in dense patches in streams moderately polluted by nutrients.

Northern Case Maker Caddisflies

Family Limnephilidae; Plates 69, 70

Different Kinds in North America — Many; 51 genera, more than 300 species.

Distribution in North America — Throughout, but there are more different kinds at northern latitudes and higher elevations.

Habitat — Lentic-littoral, lotic-depositional, lotic-erosional. Northern case makers are the most diverse and widespread of the caddisfly families, and they occur in all types of lentic and lotic habitats. These habitats range from small springs to large rivers and from ponds to marshes and lakes. Some northern case makers are adapted to live in temporary pools and streams. A few even inhabit damp locations that are more terrestrial than aquatic.

Movement — Climbers, crawlers, sprawlers, clingers. Larvae of northern case makers live in a wide variety of portable cases. As a general rule, those that live in cool, flowing waters make their cases out of mineral pieces, while those that inhabit still or warm waters build with plant materials.

Feeding — Shredder-detritivores, collector-gatherers, scrapers, shredder-herbivores.

Other Biology — Although the populations of individual kinds of northern case makers are never particularly high, the diversity of kinds and the range of their habitats and habits are extraordinary. Some kinds diapause as eggs or larvae to survive in habitats that dry up or become too warm. Most northern case makers produce one generation per year, and adults of different kinds emerge throughout the warm months.

Stress Tolerance — Mainly facultative, others very sensitive to somewhat tolerant. This is a very diverse family, so a wide range of tolerance to pollution is to be expected among the different kinds of northern case makers.

Saddlecase Maker Caddisflies

Family Glossosomatidae; Plate 67

Different Kinds in North America — Moderate; 6 genera, more than 80 species.

Distribution in North America — Throughout.

Habitat — Lotic-erosional, occasionally lentic-erosional. Larvae of saddlecase makers occur primarily in cool, swift, flowing waters, ranging in size from springs to rivers. They reside primarily in medium to fast current on stones that are large enough to be stable.

Movement — Clingers. Saddlecase makers build a dome-shaped case made of small stones, which is not readily swept away in current. Larvae also hold on to the substrate with the claws on their legs and prolegs.

Feeding — Scrapers. Larvae of saddlecase makers feed under the protection of their dome-shaped cases. They mostly consume the thin layer of algae that grows on stones, but they also consume any fine detritus that settles onto the surfaces of stones.

Other Biology — Saddlecase maker larvae readily abandon their cases under stress and rebuild them later. They also remake their cases each time they shed their skin. When it is time to pupate, mature larvae usually move to the sides or bottom of a stone where there is less current. Larvae of saddlecase makers use their mouthparts to cut away the strap on the bottom of the case, then they seal the edge of the case to the rock. They use their silk to spin a semi-transparent, capsule-like, brown cocoon inside of the attached stone case, and transform into the pupa inside of the silk cocoon. Saddlecase makers usually produce two generations per year, and adults emerge from late spring to early fall.

Stress Tolerance — Very sensitive to somewhat sensitive.

Snailcase Maker Caddisflies

Family Helicopsychidae; Plate 62

Different Kinds in North America — Few; 1 genus, 4 species.

Distribution in North America — Throughout.

Habitat — Lotic-erosional, lentic-erosional. Larvae of snailcase makers occur in a wide variety of streams, including ones that range from clear to turbid and with current ranging from slow to swift. They can also be found along the wave-swept, rocky shores of lakes in water up to 3 m deep. Larvae are most commonly found on sand and gravel substrates, as well as on rocks and water-soaked logs. Snailcase maker larvae have a wide range of temperature tolerance and can live successfully in cold springs as well as geothermal springs up to 34 °C.

Movement — Clingers. Larvae live in a portable case.

Feeding — Scrapers. Snailcase maker larvae ingest mostly diatoms but also fine organic particles.

Other Biology — The helical case resembles a snail so much that it was originally described as a species of snail. The helical shape is more resistant to crushing, which may be an adaptation for snailcase makers spending some of their life down in the streambed among loose stones and gravel. Sometimes there is an egg diapause that lasts 5–6 months. Snailcase makers probably produce only one generation per year, but the emergence of adults is spread out from spring through early autumn.

Stress Tolerance — Facultative.

Strongcase Maker Caddisflies

Family Odontoceridae; Plate 71

Different Kinds in North America — Few; 6 genera, 13 species.

Distribution in North America — The only common kinds are restricted to the East, from New Hampshire through parts of

Quebec and Ontario to Wisconsin and south to Tennessee and North Carolina. Various other kinds are found throughout most of the continent, but they are always in isolated populations and are seldom collected.

Habitat — Lotic-erosional, lotic-depositional. Larvae of strongcase makers live in swift, cool streams of small to medium size. They are most often in locations with gravel and sand substrate. Some kinds occur within clumps of submerged moss.

Movement — Burrowers, crawlers, clingers. Larvae live in a portable case. For most of their life, strongcase maker larvae crawl on the gravel and sand substrate or burrow down into it. Their portable case is uniquely constructed to withstand the stress of burrowing. There is more silk on the outside holding the particles of rocks together, which gives the case a smooth appearance. In addition, the case is stronger because of silk braces on the inside between adjacent rock particles.

Feeding — Scrapers, collector-gatherers.

Other Biology — When it is time to pupate, the mature larvae of strongcase makers move to large stones (boulders, cobbles) in riffles, where they aggregate in places that are shielded from the full force of the current. They glue their case tightly to the rock surface and seal the end with one round, flat piece of gravel that is the same width as the diameter of the case. Cases with pupae are often cemented to each other in conspicuous clumps. Strongcase makers produce one generation per year, and adults emerge in late spring.

Stress Tolerance — Very sensitive.

Trumpetnet and Tubemaking Caddisflies

Family Polycentropodidae; Plate 76

Different Kinds in North America — Moderate; 7 genera, 76 species.

Distribution in North America — Throughout.

Habitat — Lotic-erosional, lotic-depositional, lentic-littoral. Larvae of trumpetnet and tubemaking caddisflies live in a wide range of freshwater habitats, including small streams, large rivers, ponds, lakes, and reservoirs. They occur on solid substrates, such as rocks, water soaked logs and branches, and submerged aquatic plants.

Movement — Clingers. Trumpetnet and tubemaking caddisflies live in a wide variety of silk retreats. All kinds remain in place by virtue of their retreat. They attach their retreat to various solid objects in flowing or standing waters.

Feeding — Engulfer-predators, collector-filterers. Most kinds of trumpetnet and tubemaking caddisflies feed primarily on small invertebrates, but the mechanisms for obtaining their food are varied. Those that build retreats resembling pouches or trumpets in slow currents of streams and rivers are collector-filterers. They reside in their retreats and feed primarily on the small invertebrates that are filtered from the water. Other kinds of tubemaking caddisflies use their retreats as a means of ambushing prey. These kinds live in shapeless, flat coverings or loose tubes on solid objects in habitats with little or no current. The larvae remain concealed in the retreat until an animal comes close to one of the openings, then they dart out and pounce on it. Sometimes there are silk threads extending out from the openings of the retreat. When an approaching prey touches the threads, the caddisfly larva feels the vibrations and attacks.

Other Biology — Trumpetnet and tubemaking caddisflies produce one or two generations per year, and adults of different kinds emerge throughout the warm months.

Stress Tolerance — Mostly facultative, others somewhat sensitive to somewhat tolerant.

Uenoid Case Maker Caddisflies

Family Uenoidae; Plate 68

Different Kinds in North America — Few; 5 genera, 46 species.

Distribution in North America — Most kinds only in the mountains of the West; a few kinds common in the mountains of the East in the region from Newfoundland to Minnesota and south to Missouri and Georgia.

Habitat — Lotic-erosional. The western species of uenoid case makers are confined to the upper reaches of cold, fast mountain streams. The common kinds in the East have a broader range of habitats from cold, mountain streams at high elevations to medium-sized and slightly warmer rivers at lower elevations. All kinds occur primarily on rocks.

Movement — Clingers. Larvae live in a portable case. For some of the common kinds of uenoid case makers, the larger pebbles along each side of the case act as ballast to stabilize the case and hold it down in current.

Feeding — Scrapers. Diatoms are the primary type of algae eaten by uenoid case makers.

Other Biology — When it is time to pupate, mature larvae of uenoid case makers glue the case to the rock surface and seal the end with a piece of gravel of appropriate size. They usually pupate on the undersides of rocks, often in large aggregations. Some common kinds of uenoid case makers in the East construct the protective pupal case in the spring and early summer, then diapause for several months as a larva while the water is warm and flow is low. Toward the end of summer and during fall, they wake up from diapause as larvae inside of the sealed case, molt to the pupal stage, and soon emerge as adults.

Stress Tolerance — Somewhat sensitive.

True Flies

Order Diptera; Plates 84 –100

True flies is a large order, in terms of diversity, and among the insects it is second only to beetles (Coleoptera). In the entire order of true flies there are 108 families and 17,000 species in North

America. Like the other large orders of insects, most members are terrestrial. However, there are 29 families with about 3,500 species in which the larvae are aquatic. Thus, the order of true flies has more diversity among its aquatic members than any other insect order, even the completely aquatic orders. The exceptionally high number of different kinds of true flies also brings about a high degree of ecological diversity, making it difficult to summarize information about the entire order. Much of the ecological information about this remarkable order is given in the part on biology of common families. This guide includes 17 families that are common in freshwater habitats. True flies have complete metamorphosis. All members of the order are aquatic as larvae and pupae and terrestrial as adults.

Distinguishing Features of Larvae — Bodies of true fly larvae are elongate, soft, and fleshy. Most are cylindrical and resemble maggots, which is the common name used for true fly larvae in some of the terrestrial families. Some aquatic larvae are flattened. The body length of mature larvae in different species usually ranges from 2 to 25 mm, excluding breathing tubes. Some families have large species that are 100 mm long as mature larvae. True fly larvae can be distinguished from other immature aquatic insects by the following combination of characteristics (Figure 33A, B):

- Head may be a typical capsule-like, separate structure with thick hard skin.

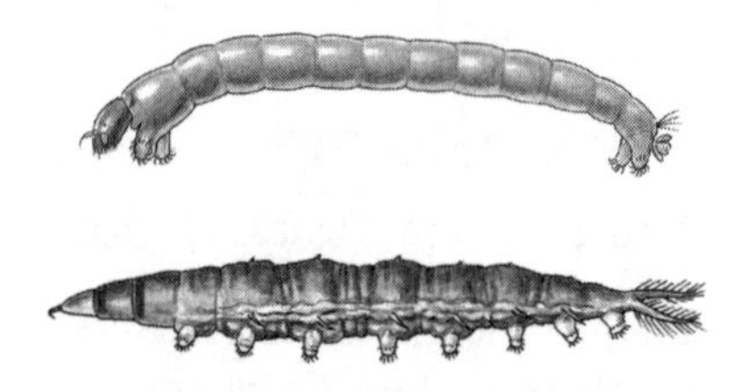

Figure 33. Representative true fly larvae.

- Head may be partially reduced on the rear margin so that it appears to be continuous with the thorax, or the head may be greatly reduced to just mouthparts that protrude from the thorax.
- No wing pads occur on the thorax.
- No segmented legs are present on the thorax.
- Prolegs may extend from various segments of the thorax and abdomen in some kinds.
- Thorax and abdomen are composed entirely of soft, thin skin in most kinds, but some kinds have separate hardened plates scattered on various body segments.

Distinguishing Features of Adults — Adult true flies have bodies that are elongate, cylindrical, and soft. Most are small, but some are medium in size. The body length of most aquatic species ranges from 1 to 12 mm. A few aquatic families have larger species with body lengths from 25 to 60 mm. True fly adults can be distinguished from other adult aquatic insects by the following combination of characteristics (Figure 34):

- Mouthparts are highly modified for acquiring fluids rather than chewing solid food, and may be sharp tubes for piercing or blunt pads for sponging.
- Only the front pair of wings is present on the thorax.

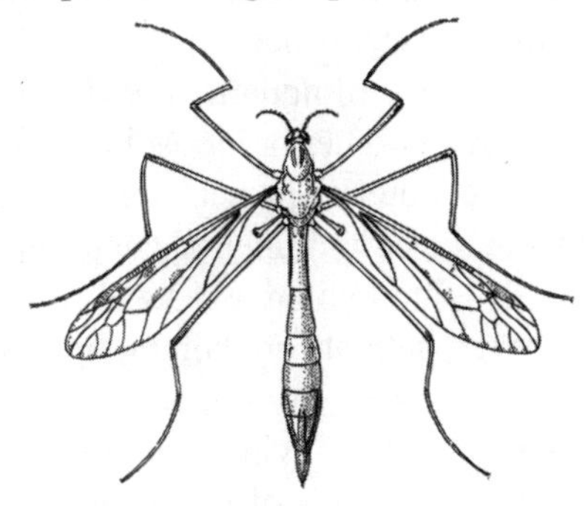

Figure 34. Representative true fly adult.

- Front wings are membranous with only a few veins.
- Hind wings are greatly reduced to small, thin stalks with knobs on the end (halteres), which have no resemblance to wings.

Explanation of Names — The only common name that is widely recognized for the order is true flies, or perhaps just flies. The better common name is true fly, because "fly" is used as part of the common name for many other orders and families (for example, mayfly, caddisfly, fishfly). The common name dates to before the twelfth century and refers to the verb for moving through the air. In Old English the verb was *fleogan*, then in Middle English it became *flie*. It is appropriate as a common name for this order because many species have astounding aerial abilities. There is an accepted custom for writing the common names of insects that have "fly" included. If the name is for an order or family other than Diptera, the name is written as one word (for example, dragonfly, stonefly, dobsonfly). If the organism is a member of the order Diptera, then the name is written as two words (for example, true fly, black fly, horse fly). All of the families have individual common names that are widely recognized, including the aquatic families. These are given in the part on biology of common families. The scientific name of the order is a combination of two Greek words: *di*, meaning twice or double, and *ptera*, meaning wings. The explanation for the scientific name is that adults in this order have only two wings, instead of the usual four for other insect orders.

Habitat — The larvae of aquatic true flies occupy a greater variety of habitats than any other order. At least some kinds can be found any place where water is present for at least a few weeks, no matter how hostile the environmental conditions might be. The families can be divided into two broad categories of habitat preferences, those that live in habitats containing open water and those that live in damp mud in semi-aquatic habitats without any open water. The many kinds of mud-loving organisms are a unique feature of this order. Different kinds of fly larvae live in all sizes and types of natural lotic and lentic habitats, including seeps, springs, streams, rivers, puddles, pools, marshes, swamps, ponds, and lakes. In addition, some kinds are adapted to develop in very small, temporary habitats, such as rainwater and dew trapped in plants (tree holes, leaf axils). These habitats are called phytotelmata and sometimes contain as little as 10 milliliters of water.

Some aquatic true flies are restricted to waters that are fresh, clear, and clean. However, other aquatic true flies are capable of living in some very harsh environments where hardly any other invertebrates can survive. In some cases the adverse conditions are the result of natural phenomena, but in other instances, pollution and environmental stress from human activities are responsible for the inhospitable conditions. Examples of naturally severe habitats where some types of true fly larvae live are wave-swept rocks in the intertidal zone along seacoasts, salt marshes, geothermal waters from hot springs and geysers (up to 49° C or 120° F), natural seeps of crude petroleum, alkaline ponds, saline lakes such as the Great Salt Lake, arctic bogs, deep lakes (up to 1,300 m or 4,265 ft), and torrential mountain streams. Examples of oppressive habitats that are the result of pollution or environmental stress from human activities are streams receiving mine wastes (pH as low as 2 and high concentrations of toxic heavy metals), severe organic pollution (sewage outfalls, settling ponds, active latrines), bodies of water choked with sediment from erosion, and streams and rivers that receive a mixture of toxic industrial wastes.

Movement — In conjunction with the muddy habitats that many true fly larvae live in, many of the larvae are burrowers, and some are sprawlers. A few that inhabit lotic-erosional habitats are clingers. Some of the best known true flies, the mosquitoes (Culicidae), are swimmers as larvae, but this is not a common habit in this order. Likewise, the family of phantom midges (Chaoboridae) is very successful with a planktonic habit in lakes and ponds, but that is unique in the order.

Feeding — True fly larvae also acquire their food in many different ways. Because many kinds live in soft, muddy habitats with an abundance of fine detritus lying on the bottom, the most common method of acquiring food is collector-gatherer. Some families are very efficient collector-filterers. There are a few scrapers, shredder-detritivores, and engulfer-predators as well.

Breathing — Both types of breathing systems, closed and open, are common among the true flies. Most of those with closed breathing systems acquire dissolved oxygen through all of the soft tissue covering their body. Some of the kinds with closed breathing systems also have gills. Larvae of some species in the family of non-biting midges (Chironomidae) are exceptional in that their blood can carry oxygen. They are able to do this because they have hemoglobin in their blood. Hemoglobin molecules bind oxygen molecules so that the blood can transport oxygen in a dissolved state, just like in vertebrate blood. Hence, the blood of some midge larvae is bright red, just like our blood. The bright red blood of these kinds of midges is readily apparent when they are alive. Transporting oxygen in the blood with hemoglobin as a carrier is exceptionally rare in insects, most of whom transport oxygen only as a gas in their internal network of finely branched air tubes. Under normal circumstances, midge larvae rely on oxygen transported in the normal way in their internal air tubes. They use oxygen transported in the blood with hemoglobin as an accessory mechanism to maximize the efficiency of their respiration. Having hemoglobin allows them to inhabit waters that are practically devoid of dissolved oxygen.

There are also numerous families of true flies in which the larvae have open breathing systems. Functional spiracles are located on various segments of the body. The most common location for functional spiracles is on the posterior segment of the abdomen. This position allows the larvae to keep their bodies submerged in the water or mud while their spiracles are in contact with the atmosphere. Some fly larvae have their posterior spiracles on extended breathing tubes. In this order, short breathing tubes, such as those on mosquito larvae, are often called siphons. Most of the fly larvae that are capable of getting oxygen from the atmosphere by means of open breathing systems, probably do not breathe exclusively by this method. These air breathing larvae are also likely to use some dissolved oxygen that diffuses from the water through the soft skin on their bodies.

Life History — The number of generations per year varies quite a bit among the different kinds of true flies. Most kinds complete their life cycle one time a year, but there are many kinds that pass through multiple life cycles in a year, as well as many kinds that require more than a year to complete one life cycle. The number of generations per year is often influenced by climate. Some kinds of mosquitoes that breed in the warm salt marshes of Florida have about 20 generations per year. Conversely, a kind of non-biting midge (Chironomidae) that inhabits Arctic tundra ponds that remain frozen for most of the year requires 7 years to complete its life cycle one time. Adult true flies emerge throughout the warm seasons. Some will emerge during winter if there is an unusual warm spell.

Female true flies lay their eggs in a variety of configurations and places. Some kinds of true flies deposit their eggs in a gelatinous mass that is attached to solid objects in the water. Others deposit their eggs individually or in loose clusters. Sometimes the single eggs or loose clusters are placed on the water surface. In other kinds of flies, they are put on solid objects, either just above or just below the water. The eggs of true flies usually hatch in several days or a few weeks, but there are some kinds of true flies that undergo a long diapause in the egg stage. When egg diapause occurs, it usually lasts several months in the true flies, but it can last for a year or more in the case of eggs deposited in temporary habitats that have dried up.

The duration of the larval stage in various kinds of aquatic true flies usually ranges from several weeks to 2 years, but there are a few kinds that spend a longer period as larvae. Most kinds of true flies molt either three or four times as larvae, but this ranges up to nine times in some kinds. Winter is usually spent as active larvae. Larval diapause is not common. In most kinds of true flies, the larvae transform to pupae in the same location where the larvae develop. Some migrate to the margin of their habitat. Most make no special preparation for the pupa, such as a cocoon. However, larvae in some families can spin silk from their lower lip, and those that do usually construct a cocoon for the pupa. Cocoons are usually situated in the same location where the larvae develop. A few of the aquatic true flies pupate inside of the last larval skin, which becomes dark and thick. This type of protective cover for the pupa is called a puparium.

In true flies, the duration of the pupal stage is usually fairly short, lasting several days to several weeks. The aquatic pupae breathe by either a closed or open system, according to the kind of true fly. Those with closed systems usually have gills. Those with open systems have only one pair of functional spiracles, and these are usually located at the end of some form of extension. Some pupae of true flies have long, thin breathing tubes, which are also called siphons, while others have short, stout, conical structures that are usually called horns or trumpets. Gills and breathing tubes are located on the thorax of pupae. Curiously, many families of true flies have similar structures while they are larvae, but they are located on the end of the abdomen. When it is time for the adult to emerge, most pupae float to the water surface by means of a gas bubble trapped under the skin. The adult emerges by using the skin of the pupa as a temporary raft.

Most kinds of true fly adults have short life spans, lasting several days to 2 weeks. Some kinds live for months as adults, but that is not common in many families. Most adult flies stay near their larval habitat, but some kinds that feed on blood often migrate many kilometers from their larval habitat. They usually spend most of their time perched on vegetation, except to find food or a mate. Some kinds fly only during the day and spend the night hiding in vegetation, while others have the opposite schedule for their activity. Almost all adult true flies feed exclusively on fluids, and their mouthparts are highly modified for fluid feeding, either by piercing, cutting, or sponging. In many kinds, the fluid is nectar or soft, decaying matter, but some kinds feed on the blood of vertebrates. Only female true flies feed on blood and usually only for part of their adult life. Blood meals are necessary for egg development. Females of blood-feeding species also consume nectar for a source of energy. Males feed only on nectar.

As in many of the aquatic orders, the only behavior that has been studied to any extent in the true flies is how the adult males and females find and recognize each other for mating. Some families of true flies engage in aerial swarming, similar to mayflies (Ephemeroptera) and caddisflies (Trichoptera). Non-biting midges (Chironomidae) and mosquitoes (Culicidae) are good examples of true flies that are frequently observed swarming. The males form the swarm, which is not as large and spectacular as in the mayflies. There is a characteristic motion of the swarm that is recognizable by females of the same species. Males often orient their swarms above an upright marker, such as a bush, stump, or fence post. A person standing in a field also makes a handy marker for a swarm, which explains why people often see hundreds of adult flies dancing in the air above their head. Female true flies recognize the swarming pattern of their own species and are stimulated to fly toward the swarm of males. In the families of flies that swarm, the males have an interesting way of recognizing females of their own species when they get close to the swarm. These males have antennae with many fine hairs sticking out in all directions from the main shaft, giving the antennae a feather-like appearance (plumose). There are very sensitive receptors at the bases of these hairs that can perceive the sound waves produced by the movements of the females' wings. Females of each species produce a unique sound with their wings, in terms of the frequency of their wing beats. Male true flies recognize the females of their own species, fly toward them, and grab them for mating.

Significance —Some of the adult true flies that feed on blood are serious pests, whereas the larvae of aquatic true flies are usually beneficial organisms. There is a wide range in the degree of harm caused by the blood feeding of different kinds of adult true flies. Some examples are: the mere nuisance of having many insects flying around people or animals, the aggravation of bites that itch and sometimes become infected, losses of income because livestock do not gain weight or tourists do not visit infested areas, and, most importantly, the transmission of life threatening diseases. Although the problem is more severe in tropical areas, disease transmission by blood-feeding true flies poses a serious threat to humans, livestock, pets, and wildlife throughout North America.

On the other side of the coin, larvae of true flies often have very important roles in ecosystems. Fly larvae often occur naturally in very high numbers. Therefore, some are important as shredder-detritivores in the breakdown of leaves and other large pieces of terrestrial plants that enter the water. Others are important as collector-filterers to remove fine particles of detritus suspended in the water and to make the nutrients and energy that this material contains available to other organisms. The extreme abundance of some aquatic fly larvae and adults make them very important as food for many species of fish and water birds. Some migratory waterfowl feed heavily on them when they are on the breeding grounds, and the protein contained in these insects is thought to be important for egg production by the female birds.

A few of the true fly families are very tolerant of pollution in their larval stages and, thus, can provide useful information in biomonitoring studies. A community composed mostly of pollution-tolerant fly larvae is a reliable indicator that some type of pollution or environmental stress has occurred. However, it should be kept in mind that a few pollution-tolerant organisms may be found in even the healthiest environments. There are also families of aquatic true flies that are very sensitive to pollution, and these provide similar information as the EPT families (Ephemeroptera, Plecoptera, Trichoptera).

Biology of Common Families

Aquatic Snipe Flies

Family Athericidae; Plate 97

Different Kinds in North America — Few; 2 genera, 4 species.

Distribution in North America — Throughout, except Florida and Texas, although less common in Central.

Habitat — Lotic-erosional. Larvae of aquatic snipe flies occur primarily in swift, well-oxygenated streams and rivers. They are usually found under stones in shallow riffles. Sometimes they occur among aquatic vegetation, especially moss, or on pieces of decomposing wood.

Movement — Primarily crawlers, also burrowers.

Feeding — Piercer-predators. Aquatic snipe fly larvae feed heavily on larvae of midges (Diptera: Chironomidae) and mayflies (Ephemeroptera), but they will prey on any soft-bodied aquatic invertebrates.

Other Biology — Adult females of aquatic snipe flies have a curious way of laying their eggs. They deposit their eggs in a mass on twigs over streams. However, the females remain with the egg mass and eventually die there. Many females of aquatic snipe flies are attracted to the same spot to lay their eggs, forming a clump of dead bodies and egg masses that is slightly elongate and about the size of a fist. When the larvae hatch they must crawl out of the cluster of corpses and drop into the water. Pupation takes place in the soil along the margin of streams. Females of all the common and widespread species of aquatic snipe flies are not blood feeders. There is one species that only occurs in southwestern Texas that does feed on blood. Aquatic snipe flies have one generation per year.

Stress Tolerance — Somewhat sensitive.

Biting Midges, No-See-Ums, Punkies

Family Ceratopogonidae; Plate 92

Different Kinds in North America — Many; 20 genera, 579 species.

Distribution in North America — Throughout.

Habitat — Lentic-littoral, lotic-depositional. There are aquatic and semi-aquatic species of biting midges. Aquatic species occur in the soft, find sediment on the bottoms of ponds, lakes, still areas of streams and rivers, swamps, salt marshes, and tree holes. Some of the aquatic species are especially common in thick growths of algae. Semi-aquatic species of biting midges inhabit the moist sand of the intertidal zone on ocean beaches, muddy areas that are rich in organic matter (especially around barns and feed lots), and wet vegetation at the margins of the water.

Movement — Primarily burrowers, some sprawlers. Many of the common species of biting midges are very long and thin, and they move through soft sediment by wiggling their bodies like snakes. Sometimes they swim through the water by the same motion.

Feeding — Engulfer-predators, collector-gatherers. Most biting midge species with mouthparts that project forward are engulfer-predators, feeding on the larvae of non-biting midges (Diptera: Chironomidae) and other small invertebrates. Many of these also feed extensively on the eggs of other aquatic insects. Most species of biting midges with mouthparts that project down are collector-gatherers.

Other Biology — Biting midge pupae float and swim in the water. They remain in contact with the air at the water surface by means of breathing horns on the thorax. Most adult females are blood feeders on mammals and birds, but some feed on reptiles and amphibians. Some species of biting midges take the blood of large insects, such as moths or dragonflies, often by piercing a wing vein. In some species of biting midges, adult females kill small, swarming adult aquatic insects, such as non-biting midges, mosquitoes, mayflies, and even their own species. They fly into the swarms, capture a prey, then consume its body fluids. Females of some species of biting midges feed on the male that they are mating with. Many species fly at night, especially at twilight, but some are active during the day. There can be many generations per year. Biting midges are considered pests as adults because some species attack humans. Their bite is much more aggravating than their small size would indicate, and they are difficult to keep away because they can pass through most screens and netting. Species that breed in the sand at the beach are often called sand flies, and these can be particularly troubling to the tourist industry. Although biting midges do not transmit diseases to humans in North America, some species do transmit serious diseases to livestock.

Stress Tolerance — Chiefly facultative, others somewhat tolerant.

Black Flies

Family Simuliidae; Plate 90

Different Kinds in North America — Moderate; 11 genera, 165 species.

Distribution in North America — Throughout.

Habitat — Lotic-erosional. Black fly larvae occur only in flowing waters, but different species can be found in all sizes of flowing waters from spring seeps to large rivers. Current preferences vary among species, from slow velocities along the margins of low-gradient streams to the maximum current speed in mid-channel areas of torrential streams. They attach to rocks, woody debris, vegetation, and any other stable, solid objects in the water. Larvae of black flies are also found on litter that people discard in streams, such as plastic sheeting, appliances, and parts of automobiles. Sometimes black fly larvae are so abundant that they almost completely cover suitable substrates.

Movement — Clingers. Larvae of black flies produce a silk thread from their mouth and put a pad of this sticky material on the substrate. They then attach themselves to the silk pad by a ring of minute hooks on the posterior of their abdomen (Plate 90B). This mechanism of attachment does not work if the substrate becomes covered with a slime from excessive growths of algae, fungi, or bacteria. Black fly larvae can move slowly by alternately grabbing the substrate with the posterior ring of hooks and the soft, fleshy, unsegmented leg (proleg) on the front of the thorax. Larvae immediately let out a silk thread if they lose contact with the substrate. This serves as a lifeline to return to their original position if they are swept away.

Feeding — Collector-filterers. In most species of black flies, larvae extend fan-like mouthbrushes into the current to filter fine particles suspended in the water. Periodically they "flick" these brushes into their mouths to remove the material that has been filtered from the water. Food consists of very fine detritus, algae, and bacteria. A few kinds that lack mouthbrushes feed on the film of organic matter that accumulates on the substrate that they are attached to.

Other Biology — Black fly larvae obtain dissolved oxygen through the surface of their body. Some kinds are territorial and separate themselves by at least one body length. They use their jaws to nip at any other black fly larva that comes within reach. Others exist in dense mats of larvae with their bodies packed tightly together. There is some evidence that they use their filter-feeding mechanism to capture the feces of other nearby black fly larvae. Pupation occurs in the same habitat where the larvae develop, and pupae and larvae of the same species can often be found together. Larvae prepare a rigid, sack-like cocoon, with the open end facing downstream (Plate 90C). Black fly pupae breathe with two tufts of branched, filamentous gills on the thorax. The heads of the pupae stick out of the cocoon, and the gills trail in the current. The adult emerges from the pupa in the cocoon. As the adult black fly comes out of the pupal skin, a bubble of air forms around it and carries it to the surface, where it flies away. Adults are known to migrate at least 80 km from where they emerge. Females of most kinds are blood feeders, and they are terrible pests of humans, livestock, and wildlife. Black flies emerge in enormous numbers, particularly in the northern United States and Canada. They fly only during daytime, and aggressively pursue their hosts in swarms. Any exposed skin is quickly subjected to many painful bites that continue to itch and ooze for a long time. In some areas, black flies practically eliminate tourism while the adults are present. They also significantly reduce profits from raising livestock because the loss of blood and constant aggravation from swarming and biting causes the animals to not gain weight, or in severe instances, to die. In the tropics, black flies transmit several debilitating diseases to humans, but in North America the diseases that they vector only affect wild and domestic animals. Most of the black fly species that bite humans most severely have ranges limited to the North. Most of the species that inhabit the South feed primarily on wild birds and mammals. Fortunately, northern species usually have only one generation per year, and the adults only emerge for 4–6 weeks in late-spring and early-summer. Many of the southern black fly species have several generations per year, and the adults emerge throughout the warm seasons.

Stress Tolerance — Primarily facultative, others somewhat sensitive. The abundance of facultative kinds of black fly larvae usually increases when moderate amounts of organic matter or nutrients are allowed to enter flowing waters. These types of pollution make more particles of food available to the filter-feeding larvae. When black fly larvae represent a majority of the community, that is a reliable indication of moderate organic or nutrient pollution. Larvae of common netspinner caddisflies (Trichoptera: Hydropsychidae) are often the other organisms that dominate under these circumstances of pollution. However, if these pollutants are so great that dissolved oxygen becomes excessively low or the substrate becomes covered with a thick growth of algae, bacteria, and fungi, then black flies will not survive. It is ironic that when grossly polluted streams and rivers are restored to a moderately clean state, black fly larvae repopulate those waters and the biting adults begin to be noticeable pests.

Crane Flies

Family Tipulidae; Plate 94

Different Kinds in North America — Many; 34 genera, 577 species (aquatic or semi-aquatic).

Distribution in North America — Throughout.

Habitat — Lentic-littoral, lotic-erosional, lotic-depositional. This is the largest family of true flies, in terms of number of species. Crane flies are also very diverse ecologically, including aquatic, semi-aquatic, and terrestrial species. Habitats of larvae include stones, leaf packs, and woody debris in swift riffles, as well as sand, gravel, and mats of algae along the margins of streams. A few larvae of crane flies live in accumulations of algal scum on rock outcrops, where only a trickle of water keeps the site moist. They also occur in rich mud and damp decaying vegetation at the edges of many standing-water habitats, including swamps and marshes. A few kinds of crane fly larvae inhabit intertidal zones and brackish waters.

Movement — Burrowers, sprawlers.

Feeding — Primarily shredder-detritivores and collector-gatherers, some engulfer-predators.

Other Biology — Larvae of most kinds of crane flies have an open breathing system with two spiracles on a flat, slightly recessed area, at the end of the abdomen (Plate 94B). Some have a closed breathing system and obtain dissolved oxygen all over the surface of their body. Some kinds of crane fly larvae are very common in leaf packs in streams, especially the large (up to 100 mm), gray ones that are sometimes called "leather jackets" (genus *Tipula*). Their shredder-detritivore feeding habits are significant in ecosystem dynamics. They break down the leaves that fall from trees on land and make the energy and nutrients contained in the leaves available to other aquatic organisms.

Some crane fly (genus *Hexatoma*) larvae are engulfer-predators that have an interesting behavior for feeding. None of the true flies have segmented legs that they can use to hold on to the substrate while they catch and subdue live prey. These particular crane fly larvae can form a large knot with the muscles at the end of their abdomen. When they catch an organism with their mouthparts, they enlarge the end of their abdomen and wedge the knot between stones in the riffle where they reside. Crane fly larvae do not leave the water to pupate (Plate 94C), but they do move to wet sand or gravel along the edge of their habitat. Adults probably feed on nectar. They do not feed on blood. Most kinds of crane flies produce one generation per year, but a few require 2 years to complete a generation.

Stress Tolerance — Mainly facultative, others somewhat sensitive to somewhat tolerant. The crane fly family is so ecologically diverse that different tolerances to stress are expected.

Dance Flies

Family Empididae; Plate 98

Different Kinds in North America — Moderate; 16 genera, 294 species (aquatic or semi-aquatic).

Distribution in North America — Throughout.

Habitat — Lotic-erosional, lotic-depositional, lentic-littoral. Most species of dance flies are terrestrial, and the aquatic species have not been studied very much. Most aquatic species live on the bottom of swift streams down to a depth of 1 m. Larvae of dance flies are found on rocks as well as in moss. Some live in wet marginal areas of streams and ponds, even in stagnant water. A few kinds of dance flies inhabit the surfaces of firm substrates that are covered with just a thin film of flowing water.

Movement — Primarily crawlers; also sprawlers, burrowers, and clingers.

Feeding — Primarily engulfer-predators, perhaps some collector-gatherers. Larvae of dance flies probably feed on any invertebrates that occur in their habitat, but they have been observed to prey on black fly larvae and caddisfly pupae.

Other Biology — Some dance fly larvae pupate in empty black fly (Simuliidae) cocoons, after they have eaten the black fly pupae. Some kinds have very distinctive pupae, with long filaments on the sides (Plate 98B). Much more is known about the adults of dance flies than the larvae. The common name comes from the swarming habits of some kinds of adults, which twist and turn in flight just above the surface of the water. Adult dance flies feed on live insects, dead insects on the water surface, nectar, and pollen, but they do not bite vertebrates. They often prey on adult mosquitoes (Culicidae) as they emerge from the pupal skin at the surface of the water. Adult dance flies also feed on pupae and adults of black flies. Because both the larvae and adults of dance flies act as natural control agents for some pest species of true flies, dance flies are considered to be beneficial organisms. In some species of dance flies, there is an elaborate courtship ritual in which the male captures a prey and presents it to a female to entice her to mate.

Stress Tolerance — Facultative.

Dixid Midges

Family Dixidae; Plate 89

Different Kinds in North America — Few; 3 genera, 45 species.

Distribution in North America — Throughout.

Habitat — Lentic-littoral, lotic-depositional, lotic-erosional. Larvae of dixid midges are most commonly found in shallow water with lush growths of aquatic plants. They live in marshes, along marshy borders of lakes, in shallow ponds, bogs, and other relatively stagnant or slow-flowing bodies of water. In all of these habitats, dixid midge larvae occur in calm, protected areas near the margins.

Movement — Swimmers, climbers. Dixid midge larvae are

most frequently observed resting on the surface film. They bend their body into an inverted "U" shape, with the head in the water, the posterior of the abdomen touching the water surface, and the middle segments of the body protruding above the water. They swim on the surface by alternately straightening and bending the body. If disturbed, dixid midge larvae submerge and swim by the same movements. They sometimes climb on damp surfaces at the margin of their habitat, such as accumulations of detritus or rocks.

Feeding — Collector-filterers, scrapers. Larvae of dixid midges usually feed in their normal resting position at the water surface. Brushes on their upper lip establish a current of water toward their mouth, and food is strained out of the water by other brushes on their jaws. The food obtained in this manner consists of algae, various other microorganisms, and fine particles of detritus. Older larvae also swim down and scrape microorganisms from submerged vegetation and rocks.

Other Biology — Dixid midges larvae breathe air with the spiracles on the end of their abdomen. Mature larvae crawl 1–5 cm from the water for the pupal stage. They pupate in a vertical position, either on a bank or emergent vegetation. Larvae of dixid midges attach themselves by their sides with a glue-like secretion before they transform to pupae. They choose humid, shady areas where they are splashed occasionally in order to prevent drying out during the pupal stage. Adult dixid midges do not feed on blood.

Stress Tolerance — Somewhat sensitive.

Horse Flies, Deer Flies

Family Tabanidae; Plate 96

Different Kinds in North America — Moderate; 14 genera, 317 species.

Distribution in North America — Throughout.

Habitat — Lentic-littoral, lotic-depositional, a few lotic-erosional. There are both aquatic and semi-aquatic species in this family. Larvae of the aquatic species of horse flies and deer flies occur in the sediment on the bottom of ponds, marshes, and pools of streams. Semi-aquatic species are found in muddy situations near all types of aquatic habitats, as well as wetlands. Several species of deer flies are very abundant in salt marshes. A few species live in sand and gravel on the bottom of swift streams.

Movement — Primarily burrowers, some sprawlers. Larvae of horse flies and deer flies creep through the mud by telescoping their soft, flexible body. The welts that encircle their body help to push against the mud.

Feeding — Primarily piercer-predators, some collector-gatherers. The mouthparts of horse fly and deer fly larvae consist primarily of two hard, sharp vertical hooks. They use these to slash holes in the bodies of their prey, then they insert their head and consume the fluids and soft tissues. Aquatic worms are frequently consumed, but horse fly and deer fly larvae will probably eat any invertebrate that they come upon in their habitat. There is a published record of young toads being eaten, so small vertebrates may occasionally be a part of their diet.

Other Biology — Mature larvae of horse flies and deer flies crawl out of the water to pupate in the soil. Adult females are vicious biters, with mouthparts that first cut the skin then sponge up the blood. The bites of large species often leave wounds with blood trickling out. Adults of horse flies and deer flies are strong fliers and range far from the larval habitat. They are only active during the day and prefer hot, humid conditions. When not in flight, they rest in direct sunlight. Although adult horse flies and deer flies do not transmit diseases in North America, they are significant pests in some regions because they have a very nasty bite and emerge in fairly high numbers throughout the warm seasons. They are particularly troublesome on beaches and golf courses in the Southeast. Most kinds of horse flies and deer flies produce one generation per year, but some species in the South have two per year, while some in the North require 2 or more years to complete a generation.

Stress Tolerance — Somewhat tolerant to very tolerant.

Mosquitoes

Family Culicidae; Plate 87

Different Kinds in North America — Moderate; 12 genera, 166 species.

Distribution in North America — Throughout.

Habitat — Lentic-littoral, lotic-depositional. Mosquito larvae can live in almost every type of still-water habitat. Common habitats are woodland pools, marshes, swamps, ponds, lakes, and backwaters and pools of streams and rivers. Some species prefer small temporary habitats such as ditches, puddles, pools left by melting snow or streams rising above their banks, and water trapped in cavities in trees. Unfortunately, mosquitoes also breed in temporary habitats created by humans when they carelessly discard items such as automobile tires and food and beverage containers. Some species are very abundant in the brackish waters of salt marshes.

Movement — Swimmers, planktonic. Larvae and pupae (Plate 87B) of mosquitoes are both very active, hence they are called wrigglers and tumblers, respectively. Most species stay near the water surface for feeding and breathing, but if mosquito larvae or pupae are disturbed they quickly swim down by alternately bending and unbending their body. Mosquito larvae also spend time floating in the water column, probably when it is necessary to look for a more suitable place in the habitat.

Feeding — Collector-filterers, collector-gatherers. Most mosquito larvae use brushes on the mouthparts to set up currents that direct algae, bacteria, fungi, protozoa, and very fine particles of detritus suspended in the water into their mouth. Some species swim to the bottom and gather detritus from the sediments. A few kinds of mosquitoes are predators as larvae, often feeding on larvae of other mosquito species.

Other Biology — Most species of mosquito larvae breathe air from just below the water surface by means of a short siphon on the rear of the abdomen. A few species breathe air with a pair of simple spiracles on the rear rather than a siphon. A few mosquito species stick their breathing siphon into underwater parts of live plants to obtain their oxygen. Pupae of mosquitoes breathe air from the water surface with horns on the thorax, sometimes called trumpets (Plate 87B). Eggs are laid singly or in groups, called rafts, either directly on the water or in places out of the water that will be flooded later. Eggs of some kinds of mosquitoes can remain dormant for months, or even years, in the dry sediment of temporary habitats, then hatch successfully when they become wet from rain, snowmelt, or flooding. Mosquitoes require only a few days to 2 weeks (usually 7–10 days) to complete their life cycle, hence some species can produce many generations per year. Species in temporary habitats usually produce no more than one generation per year, and sometimes none if the habitat does not become flooded. Adult females of most mosquito species require a blood meal in order for the eggs to develop. Females attack a wide range of vertebrates, and some are capable of transmitting serious diseases to humans, livestock, pets, and wildlife. The nuisance of biting mosquitoes can take away the pleasure of outdoor recreation and cause devastating economic effects on regions that depend on tourism.

Stress Tolerance — Somewhat tolerant.

Moth Flies

Family Psychodidae; Plate 93

Different Kinds in North America — Moderate; 6 genera, 112 species.

Distribution in North America — Throughout.

Habitat — Lentic-littoral, lotic-depositional and erosional. Moth fly larvae usually live in semi-aquatic places near the margins of their habitats. Some preferred locations include decaying vegetation, mud, sand, floating mats of algae, or algae growing in splash zones. They are sometimes found in tree holes. A few kinds of moth flies are truly aquatic and live on the bottom of streams, while a few kinds are nearly terrestrial and develop in barely moist sand. Moth fly larvae are most often encountered in habitats associated with the treatment of human wastes, such as wastewater lagoons, trickling filter beds, or in discharges of inadequately treated wastes. Larvae of some common species of moth flies develop in household drainpipes because they are capable of withstanding hot water, soap, and disinfectant.

Movement — Primarily burrowers, a few clingers.

Feeding — Primarily collector-gatherers, a few scrapers.

Other Biology — Moth fly pupae are aquatic and lie on the bottom without a cocoon. The common name for the family comes from the adult, which is very hairy. Adult moth flies are sometimes seen in bathrooms of homes or public facilities because the larvae can breed in drainpipes. Adults of most species do not feed on blood, but those that develop in moist sand do bite and sometimes become pests on beaches of southern states. For this reason, they have their own common name, sand flies.

Stress Tolerance — Very tolerant.

Mountain Midges

Family Deuterophlebiidae; Plate 85

Different Kinds in North America — Few; 1 genus, 6 species.

Distribution in North America — Western Canada and United States.

Habitat — Lotic-erosional. Larvae of mountain midges inhabit fast streams with high concentrations of dissolved oxygen. They occur in small cracks and depressions on the tops of rocks that are otherwise smooth. The rocks that they inhabit are usually uniformly light in color. Mountain midge larvae are usually restricted to sections of streams where there is a large volume of flow and the current velocity is at least 75 cm per second. The greatest numbers of larvae are found just below the surface, where the water is splashing on the rocks. Pupae of mountain midges occur in the same place as the larvae. Although mountain midges are most common in the mountains, they also occur at lower elevations in western streams if there are sections of fast current.

Movement — Clingers. Larvae of mountain midges are able to hold on by means of several rows of small hooks on the bottom of each of the prolegs that extend from the sides of their body (Plate 85B). Their flat body shape also helps them cling to rocks in very fast current without being swept away.

Feeding — Scrapers.

Other Biology — Mountain midges spend much of the year in the egg stage. The eggs do not hatch until summer, then larvae develop to the adult stage in 3 or 4 weeks. The unusual flat pupae (Plate 85B) are attached to rocks underwater. The pupae occur in the same places as the larvae, except the pupae are sometimes in slight depressions in the stones. This family rivals mayflies (Ephemeroptera) in the short duration of the adult stage. Adult mountain midges emerge in early morning and live only 1 or 2 hours. They do not bite. Species that live in streams at low elevations produce several generations per year, whereas species in high elevation streams only have one generation per year.

Stress Tolerance — Very sensitive.

Net-Winged Midges

Family Blephariceridae; Plate 84

Different Kinds in North America — Few; 5 genera, 27 species.

Distribution in North America — Throughout the mountainous areas of East and West, and to a lesser extent in northern Central. These do not occur in Florida or Texas.

Habitat — Lotic-erosional. Larvae of net-winged midges are only found in cool, rapidly flowing streams, primarily in the mountains. Larvae are found on the surfaces of large stones such as cobble, boulder, and outcrops, in the fastest water, including torrential waterfalls. Larvae tend to aggregate in groups that may number as high as 100 individuals.

Movement — Clingers. Net-winged midge larvae attach themselves tightly to stones by means of six suckers on the bottom of their body (Plate 84B). The suckers are arranged in a row running the length of the body, with one sucker on each division of the body. The flat shape of the larvae also helps them to be effective clingers. Larvae of net-winged midges move slowly by alternately releasing and reattaching the suckers from front to rear. Their movement is something like an "inchworm," which is the caterpillar of a terrestrial moth.

Feeding — Scrapers. Larvae of net-winged midges mainly consume diatoms, but they also eat other types of algae and some fine detritus.

Other Biology — When net-winged midge larvae are disturbed, they move sideways, only slightly faster than their normal forward movement. They accomplish this by releasing most of their suckers except for those on one end, swinging their body to the side, reattaching those suckers, and repeating the process from the other end. The blackish, flattened pupae of net-winged midges (Plate 84C) also occur on large rocks in fast current, usually in cracks and depressions. The pupae are glued tightly to the rock surface by silk adhesive pads. Pupae of net-winged midges have two conspicuous groups of plate-like gills, arranged in layers on the thorax. Adult net-winged midges feed on the blood of other small true flies or nectar, if they feed at all. It is interesting how much of their life history is associated with changing water levels. Females lay the eggs on rock surfaces that have been exposed by falling water levels associated with the dry season. The eggs hatch when they are submerged by the rising waters of the wet season. Receding water seems to be a stimulus for the adult net-winged midges to emerge from the pupae. Adults emerge very rapidly because the wings expand to full size while the adults are still inside the pupal skin. As the adult net-winged midges suddenly burst from the pupal skin, they wrap their wings around their body and grab a bubble of air, which they ride in the current until they can take flight.

Stress Tolerance — Very sensitive.

Non-Biting Midges, Midges

Family Chironomidae; Plate 91

Different Kinds in North America — Many; more than 200 genera, more than 1,000 species.

Distribution in North America — Throughout.

Habitat — All categories. Larvae of non-biting midges occur on live aquatic vegetation, rocks and logs, accumulations of coarse detritus, and on soft, fine sediment. Different kinds live in the complete range of aquatic habitats, including flowing and standing, permanent and temporary, fresh and saline, and pristine and severely degraded.

Movement — Primarily burrowers, some clingers. Some of these worm-like larvae roam freely, but most non-biting midge larvae construct fragile tubes composed of very fine particles of silt, sand, detritus, or algae that are held together with silk. The tubes are usually in the upper portion of fine sediment and can be oriented either horizontally or vertically. Some kinds of non-biting midge larvae attach their tube to solid objects, such as plants, rocks, logs, or large pieces of detritus. A few kinds attach their tube to the body of other organisms, including mayflies, stoneflies, hellgrammites, and snails. This simply provides a good attachment site for the midge larva and does no harm to the other organism, a condition known as commensalism. The tube of a non-biting midge larva is usually open at both ends, and the larva inside maintains a current of water through the tube by wiggling its body.

Feeding — Primarily collector-gatherers; some collector-filterers, scrapers, engulfer-predators. Larvae of most kinds of non-biting midges merely consume the organic component of the fine sediment in which they live or which surrounds them on firm substrates. Some sediment-inhabiting species build vertical tubes, from which they extend their body and gather up the detritus in a circle around the tube (Figure 2T). Larvae of some tube-building species that live in standing water are adapted to be collector-filterers. These kinds of non-biting midges spin a fine-mesh capture net across the inside of their dwelling. They draw water through the net by wiggling their body, then they consume the entire capture net and its contents of fine detritus and algae. Some species that live in flowing water are also collector-filterers, but they let the current do the work of bringing their food to them. These kinds of non-biting midges attach their tube to stable objects, such as rocks or large woody debris. They construct arms that stand up from the substrate at the front of their tube, then they stretch sticky silk threads between the arms. The sticky silk threads catch very fine particles of food that are suspended in the water. Periodically the midge larva consumes the silk threads and their contents of fine detritus and algae. Some species of non-biting midges that live on rock surfaces are scrapers, while others are free-ranging engulfer-predators that consume other midge larvae and small, soft invertebrates such as worms (Oligochaeta).

Other Biology — The family of non-biting midges, without a doubt, is the most diverse and abundant individual family of aquatic insects, although they are often overlooked because of their small size. There are over 100 genera and 1,000 species in North America, making this family larger than some of the aquatic orders. Non-biting midges often account for as much as 50% of the species present in a community of invertebrates. It is not uncommon for the abundance of midges to reach 50,000 larvae per square meter of bottom, particularly in the deep water of lakes.

Larvae and pupae of non-biting midges have closed respiratory systems. Larvae acquire dissolved oxygen all over the surface of their body. Pupae have horns or filamentous gills on the thorax to obtain dissolved oxygen (Plate 91B). Some kinds of larvae have hemoglobin in their blood that stores oxygen and allows them to exist, at least temporarily, in environments with little or no dissolved oxygen, such as the deep water of lakes. These bright red species are often called bloodworms. Tube-building species pupate in their tubes. Species of non-biting midges that roam freely as larvae have swimming pupae, like mosquitoes (Culicidae). Because of their abundance and diversity, midge larvae are very important in aquatic food webs. They play important roles in the break down of organic matter and the cycling of nutrients in the sediment. They are eaten by many kinds of invertebrates and vertebrates. Larvae of non-biting midges are the major prey consumed by many species of fish. Even some large species of fish depend upon them for food when the fish are young and in their nursery habitat. Migratory waterfowl consume a lot of midge larvae in the shallow water of ponds and wetlands. Adult midges are short-lived and probably do not feed. Even though they do not bite, they sometimes cause a significant nuisance by virtue of the large numbers that emerge and form mating swarms near their larval habitat. Pest management is sometimes implemented in lakeside developments. Some kinds produce many generations per year, and some will even emerge during the winter if there is an unusual warm spell.

Stress Tolerance — Mostly facultative, others very sensitive to very tolerant. The great diversity in the family of non-biting midges accounts for the range in tolerance to stress. A categorization for the entire family may not be meaningful because there are so many differences in the ecology of the various kinds. In general, if midge larvae are very numerous and account for the majority of the community, that is an indication of poor environmental health caused by some type of pollution. The kinds of midge larvae that are bright red often thrive where organic wastes or nutrients pollute the water and reduce dissolved oxygen concentrations. Some kinds of midge larvae are very tolerant of toxic substances, such as heavy metals and petroleum products. They often develop deformities in their mouthparts when they live in an environment polluted with toxic substances, and scientists sometimes examine these deformities as a biomonitoring technique.

Phantom Crane Flies

Family Ptychopteridae; Plate 86

Different Kinds in North America — Few; 3 genera, 17 species.

Distribution in North America — Throughout.

Habitat — Lentic-littoral, lotic-depositional. Larvae of phantom crane flies are usually found in stagnant, muddy areas at the edges of shallow ponds and marshes. Sometimes they live in depositional areas of seeps and slow-flowing springs. They reside in the upper 1–3 cm of very fine, mucky sediment that is rich in detritus. Often there is a thick accumulation of decaying leaves and needles lying over the sediment.

Movement — Burrowers.

Feeding — Collector-gatherers.

Other Biology — Phantom crane fly larvae have an open breathing system and acquire air by means of a long, telescoping tube on the posterior of their abdomen. They can also acquire dissolved oxygen through their body surface, but this is only adequate during periods of inactivity. For example, larvae of phantom crane flies usually burrow deep into the sediment during periods of the winter when the water freezes over. Pupation occurs in the same habitat where the larvae develop, without any special preparation. Pupae have a long breathing tube on the thorax. The common name "phantom" comes from the flight habits of the adults. They extend their long legs away from the body and float through the air with little movement of their wings. Adults of phantom crane flies occur in damp areas among shrubby vegetation where the larvae breed. They are not blood feeders.

Stress Tolerance — Somewhat tolerant.

Phantom Midges

Family Chaoboridae; Plate 88

Different Kinds in North America — Few; 3 genera, 15 species.

Distribution in North America — Throughout.

Habitat — Lentic-littoral, lentic-limnetic, lentic-profundal; sometimes lotic-depositional. Some of the most common kinds of phantom midges live in the deepest sections of ponds and lakes. Other kinds live in cold springs, small temporary pools, and water from melting snow.

Movement — Planktonic, sprawlers. Common species of phantom midges have two dark air sacs in their body, one in the thorax and one near the posterior of the abdomen, which can be used to regulate their buoyancy. Mature larvae have a daily vertical migration pattern. Phantom midge larvae alternate between sprawling on the sediment during the day, then floating up in the water to become planktonic at night. Young larvae are planktonic all of the time.

Feeding — Engulfer-predators, piercer-predators. Phantom midge larvae have a special type of antenna, called prehensile, that can grab small animals and bring them to their mouth. Much of their prey is microscopic animals, especially crustaceans such as water fleas (Cladocera), in the zooplankton. Phantom midge larvae also prey on a variety of other small insect larvae in the water, especially mosquitoes.

Other Biology — The common name phantom midge comes from the transparent color of the larvae and their habit of slowly rising and sinking in the water. The daily vertical migration pattern of phantom midge larvae, up at night and down during the day, is partly associated with their zooplankton food supply, which follows the same pattern of movement. Breathing is the other factor that accounts for the daily vertical migration. The deep water of ponds and lakes has practically no dissolved oxygen. The common species of phantom midges that inhabit deep water have a closed breathing system and they depend on dissolved oxygen, which they obtain all over the surface of their body. The two air sacs in their body hold a reserve of oxygen that allows them to extend their stay in deep water, but eventually they must move to the upper waters that are saturated with dissolved oxygen to replenish their reserves. It is safer for phantom midge larvae to make this trip up at night, when hungry fish are not as likely to see them. Other kinds of phantom midge larvae that live in shallow water breathe air with a siphon on the end of their abdomen, much like mosquitoes (Culicidae). If phantom midge larvae are abundant in a pond or lake, their feeding determines the structure of the zooplankton community. They selectively eat the largest kinds of zooplankton, such as water fleas (Cladocera). Water fleas eat smaller zooplankton, especially rotifers. When phantom midge larvae are present, they reduce the number of water fleas and rotifers become more abundant. If phantom midge larvae are not present, water fleas become more abundant and the number of rotifers is reduced.

The pupal stage of phantom midges is also spent floating in the water. Pupae breathe with two horns on the thorax. In the kinds that live in deep water the breathing horns are closed and they obtain oxygen dissolved in the water, whereas in the kinds that live in shallow water the horns are open for breathing air. There can be enormous populations of phantom midge larvae in the depths of large lakes. They are very important food for some species of fish. Fish feed voraciously on them when the pupae float to the surface for the adults to emerge. Adult phantom midges emerge synchronously in vast numbers in spring or early summer. However, the adults are short lived and do not feed on blood, so they have little significance, other than providing food for birds.

Stress Tolerance — Somewhat tolerant.

Rat-Tailed Maggots, Flower Flies

Family Syrphidae; Plate 99

Different Kinds in North America — Moderate; 15 genera, 165 species (aquatic or semi-aquatic).

Distribution in North America — Throughout.

Habitat — Lentic-littoral, lotic-depositional. Most members of the flower fly family are terrestrial as larvae. Larvae of the aquatic species, which are known as rat-tailed maggots, inhabit the shallow marginal areas of ponds, marshes, and pools of streams and rivers. Rat-tailed maggots are usually found in waters where there is a lot of decaying organic matter. Some are found in tree holes. Rat-tailed maggots can inhabit water completely devoid of oxygen, thus, they are often found immediately below sewage discharges and in oxidation lagoons.

Movement — Burrowers.

Feeding — Collector-gatherers.

Other Biology — Rat-tailed maggots have open respiratory systems and breathe by means of a single, long, telescoping tube at the end of the abdomen. They pupate in the same habitat. Adults are most commonly found on or around flowers, where they hover motionless in the air. Adult flower flies are not blood feeders.

Stress Tolerance — Very tolerant. Rat-tailed maggots are one of the few invertebrates that will live in ponds and lagoons that are engineered for the safe disposal of organic wastes. Thus, if rat-tailed maggots are abundant in natural surface waters, that is a reliable indication of intense pollution from organic wastes.

Shore Flies, Brine Flies

Family Ephydridae; Plate 100

Different Kinds in North America — Many; 69 genera, 445 species.

Distribution in North America — Throughout.

Habitat — Primarily lentic-littoral; also lotic-erosional and depositional at margins. The family of shore flies and brine flies includes both aquatic and semi-aquatic species. Larvae of different species can be found in just about all aquatic habitats. Salt marshes are a common habitat. Most kinds are found in muddy marginal areas. Many shore flies and brine flies live in floating mats of algae or in various types of plant debris that have accumulated at the edge of the water. Some reside within the tissues of live aquatic plants, either by mining in the leaves or boring in the stems. This family is noteworthy for the harshness of the environments that are inhabited by the larvae of certain species. No other form of insect life can tolerate some of the places where shore flies thrive. Some examples of these extremely inhospitable environments are The Great Salt Lake, the Rancho la Brea tar pits, pools of crude petroleum in oil fields, alkaline springs in the West, and pools of geothermal water around geysers in Yellowstone National Park.

Movement — Burrowers.

Feeding — Primarily collector-gatherers; also scrapers, piercer-predators, and shredder-herbivores. The food of shore fly and brine fly larvae includes semi-liquid decaying organic matter, microscopic bacteria and fungi, and algae.

Other Biology — Pupae of shore flies and brine flies are protected inside a puparium. Pupae float in the water with a pair of spiracles always projecting above the surface for breathing air. In habitats where shore flies are exceptionally abundant, such as The Great Salt Lake, the wind often pushes vast numbers of pupae together and they wash on shore in rows. Adults are common along shores, hence the common name for the family. Adult shore flies and brine flies can be seen skating on the water surface, walking on the ground near the water, or flying in thick swarms. Adults of some species congregate on floating mats of algae, where they lay eggs. Shore fly adults feed on algae and dead insects floating on the water, and they do not bite. Native Americans in the West used to collect the pupae of shore flies for food when they washed ashore around salt and alkaline lakes. Larvae, pupae, and adults of shore flies are probably important in the diet of many animals that frequent their habitat, especially migratory waterfowl.

Stress Tolerance — Chiefly facultative, others very tolerant. Some of the kinds of shore flies and brine flies that live in the naturally adverse habitats described above may also live successfully in habitats that are environmentally stressed from human activities.

Soldier Flies

Family Stratiomyidae; Plate 95

Different Kinds in North America — Moderate; 11 genera, 178 species (aquatic or semi-aquatic).

Distribution in North America — Throughout.

Habitat — Lentic-littoral. Only about one-half of the soldier fly species are aquatic or semi-aquatic. Most of the aquatic kinds are found at the edges of shallow ponds and marshes. Most of the common kinds live on organic muck, decaying vegetation, or on floating mats of algae, particularly in shady spots among emergent plants, such as cattails. Other kinds of soldier fly larvae can be found in marginal areas among damp moss, loose bark, and decaying wood. Some kinds are able to develop in highly saline water, hot springs, or sewage discharges.

Movement — Primarily sprawlers, also burrowers.

Feeding — Collector-gatherers. Soldier fly larvae eat fine detritus and the algae that often grows profusely in their shallow, nutrient-enriched habitats.

Other Biology — Larvae of soldier flies have an open breathing system. They acquire air at the surface by means of two spiracles at the end of their abdomen. The spiracles are recessed in a chamber that is often surrounded by non-wettable hairs. Adult soldier flies spend most of their time on vegetation and flowers in the vicinity of the larval habitat. They do not feed on blood.

Stress Tolerance — Somewhat tolerant to very tolerant.

References on Information about Different Kinds of Freshwater Invertebrates

Borrer, D. J., C. A. Triplehorn, and N. F. Johnson. 1989. *An Introduction to the Study of Insects.* 6th edition. Saunders College Publishing, Philadelphia, Pennsylvania. 875 pages.

Brigham, A. R., W. U. Brigham, and A. Gnilka, editors. 1982. *Aquatic Insects and Oligochaetes of North and South Carolina.* Midwest Aquatic Enterprises, Mahomet, Illinois. 722 pages.

Corbet, P. S. 1999. *Dragonflies: Behavior and Ecology of Odonata.* Comstock Publishing Associates, Ithaca, New York. 829 pages.

Dillon, R. T., Jr. 2000. *The Ecology of Freshwater Molluscs.* Cambridge University Press, Cambridge, United Kingdom. 509 pages.

Edmunds, G. F., Jr., S. L. Jensen, and L. Berner. 1976. *The Mayflies of North and Central America.* University of Minnesota Press, Minneapolis, Minnesota. 330 pages.

Hickin, N. E. 1967. *Caddis Larvae. Larvae of the British Trichoptera.* Fairleigh Dickinson University Press, United Kingdom. 480 pages.

McAlpine, J. F., B. V. Peterson, G. E. Shewell, H. J. Teskey, J. R. Vockeroth, and D. M. Wood, Coordinators. 1987. *Manual of Nearctic Diptera.* Volume 2. Research Branch Agriculture Canada Monograph 28. Canadian Government Publishing Centre, Supply and Services Canada, Hull, Quebec, Canada. Pages 1–674.

McAlpine, J. F. (editor), B. V. Peterson, G. E. Shewell, H. J. Teskey, J. R. Vockeroth, and D. M. Wood, Coordinators. 1981. *Manual of Nearctic Diptera.* Volume 1. Research Branch Agriculture Canada Monograph 27. Canadian Government Publishing Centre, Supply and Services Canada, Hull, Quebec, Canada. Pages 675–1332.

McCafferty, W. P. 1981. *Aquatic Entomology. The Fisherman's and Ecologists' Illustrated Guide to Insects and Their Relatives.* Science Books International, Boston, Massachusetts. 448 pages.

Merritt, R. W., and K. W. Cummins, editors. 1996. *An Introduction to the Aquatic Insects of North America.* 3rd edition. Kendall/Hunt Publishing Company, Dubuque, Iowa. 862 pages.

Needham, J. G., M. J. Westfall, Jr., and M. L. May. 2000. *Dragonflies of North America.* Scientific Publishers, Gainesville, Florida. 939 pages.

Peckarsky, B. L., P. R. Fraissinet, M. A. Penton, and D. J. Conklin, Jr. 1990. *Freshwater Macroinvertebrates of Northeastern North America.* Cornell University Press, Ithaca, New York. 442 pages.

Resh, V. H., and D. M. Rosenberg, editors. 1984. *The Ecology of Aquatic Insects.* Praeger Publishers, New York. 625 pages.

Smith, D. G. 2001. *Pennak's Freshwater Invertebrates of the United States: Porifera to Crustacea.* 4th edition. John Wiley & Sons, Inc., New York. 648 pages.

Stehr, F. W., editor. 1987. *Immature Insects.* Kendall/Hunt Publishing Company, Dubuque, Iowa. 754 pages.

__________. 1991. *Immature Insects.* Volume 2. Kendall/Hunt Publishing Company, Dubuque, Iowa. 974 pages.

Stewart, K. W., and B. P. Stark. 1988. *Nymphs of North American Stonefly Genera (Plecoptera).* Entomological Society of America, College Park, Maryland. 460 pages.

Thorp, J. H., and A. P. Covich, editors. 2001. *Ecology and Classification of North American Freshwater Invertebrates.* 2nd edition. Academic Press, San Diego, California. 1056 pages.

Usinger, R. L., editor. 1956. *Aquatic Insects of California, With Keys to North American Genera and California Species.* University of California Press, Berkeley, California. 508 pages.

Ward, J. V. 1992. *Aquatic Insect Ecology. 1. Biology and Habitat.* J. Wiley & Sons, New York. 439 pages.

Ward, J. V., B. C. Kondratieff, and R. Zuellig. 2002. *An Illustrated Guide to the Mountain Stream Insects of Colorado.* 2nd edition. University Press of Colorado, Niwot, Colorado. 248 pages.

Westfall, M. J., Jr. and M. L. May. 1996. *Damselflies of North America.* Scientific Publishers, Gainesville, Florida. 649 pages.

Wiggins, G. B. 1996. *Larvae of the North American Caddisfly Genera (Trichoptera).* 2nd edition. University of Toronto Press, Toronto, Ontario, Canada. 457 pages.

Williams, D. D., and B. W. Feltmate. 1992. *Aquatic Insects.* C·A·B International, Wallingford, United Kingdom. 358 pages.

Subject Index

Page numbers in boldface identify where a term is explained in detail. Page numbers in italics identify where the explanation of a term is illustrated in a figure.

Subject Index

Common Names Index

Page numbers in italics identify where the distinguishing features or the biology of the organism are shown in a figure. Page numbers in boldface identify where detailed information about the biology of the organism is provided.

Common Names Index

Common Names Index

Scientific Names Index

Page numbers in italics indicate where the distinguishing features or the biology of the organism are shown in a figure. Page numbers in boldface indicate the most important reference, where the name is introduced or explained.

Scientific Names Index